DJ Dodds
October 1993

THE GARDENER'S DICTIONARY OF HORTICULTURAL TERMS

The Gardener's Dictionary of Horticultural Terms

COMPILED BY

Harold Bagust

CASSELL

CASSELL PUBLISHERS LIMITED
Villiers House, 41/47 Strand, London WC2N 5JE

Illustrations by Julie Williams

First published 1992

Distributed in the United States by
STERLING PUBLISHING CO. INC.
387 Park Avenue South, New York, NY 10016-8810

Distributed in Australia by
CAPRICORN LINK (AUSTRALIA) PTY LTD
PO Box 665, Lane Cove, NSW 2066

British Library Cataloguing-in-Publication Data
BAGUST, Harold
The Gardener's dictionary of horticultural terms
I. Title
635.0321

ISBN 0–304–34106–1

Typeset by Pure Tech Corporation, Pondicherry, India
Printed and Bound in Great Britain by
Hartnolls Limited, Bodmin, Cornwall.

CONTENTS

PREFACE

The purpose of this book is to provide the enthusiastic gardener or student of horticulture with a basic guide through the labyrinth of technical terms in use today. A few historical details have been included and reference is sometimes made to words or expressions now obsolete or seldom used. No attempt has been made to include a list of plant names and their origins, a subject covered elsewhere (see in particular Stearn & Smith *A Dictionary of Plant Names for Gardeners*, 1972, revised edition 1992, Cassell); those that are mentioned are intended to illustrate a particular feature. The origin of words has not been dealt with in any detail except in a few cases where it was felt to be of particular interest or importance.

The headwords in this dictionary have been culled from a great body of gardening, horticultural and botanical literature, much of it going back fifty years and more. The oldest book consulted was James Lee's *Introduction to Botany*, published in 1765 and dedicated to The Very Eminent DR. LINNAEUS, Knight of the Polar Star, first Physician to the King of Sweden, Professor of Botany at Upsal, Fellow of the Academies of Upsal, Stockholm, Petersburgh, &c. &c.

The dictionary has taken more than four years of intensive research to collate. I have been fortunate, over many more years than that, in having the advice and cooperation of many eminent and knowledgeable people in the horticultural and botanical fields, including Gervas Huxley, Percy Thrower, Fred Loads, Dr D. P. Holdgate (formerly of Neville Orchids and the Guinness Organisation), Dr R. K. Robinson of Oxford University, and Anthony Huxley, during the period when I was a regular contributor to *Amateur Gardening* under his brilliant editorship. I am particularly grateful to Dr Richard White of Southampton University for checking and correcting the manuscript, a time-consuming task, and to Julie Williams, the illustrator, whose patience with a pernickety and very demanding author was at times almost angelic.

No work of this kind can ever claim to be complete. Inevitably, odd words will be found to have been omitted, either deliberately or in error, but I am confident that the dictionary covers most of the terms likely to be encountered by those to whom it is addressed. It is my hope the the dictionary will fulfill a need in horticultural circles and prove a valuable source of reference to amateurs and professionals alike.

Harold Bagust
Holcombe Rogus, 1992.

A

Abatis, Abattis

Barricade or fence constructed of felled trees or dense hedging with branches pointing outward.

Abaxial

Referring to or attached to the side facing away from the stem or axis, e.g. the lower side of a leaf: *cf.* Adaxial.

Abrupt

Truncated, looking as if the tip or end had been cut off, as in some types of foliage.

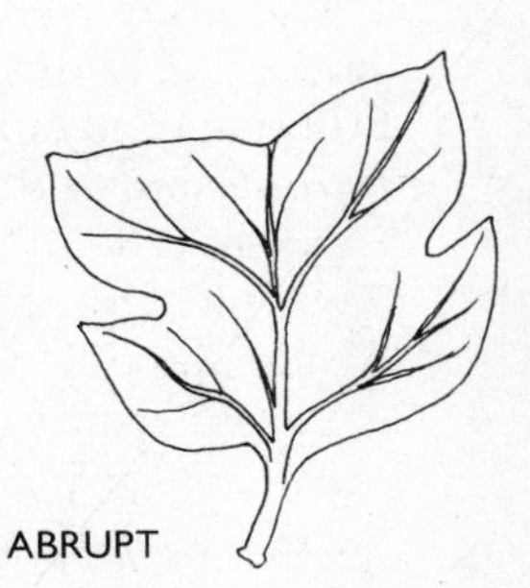

ABRUPT

Abscission Layer

Corky layer, formed at the base of the leaf stalk on deciduous trees, sealing off the sap flow and producing a change in leaf coloration and eventual fall in autumn.

Acaulescent, Acauline

Stemless or having a very short, sometimes subterranean stem: *cf.* Caulescent.

Accelerator

Any substance used to assist the decomposition of organic matter in the preparation of garden compost.

Acclimation

The natural acclimatization process of plants without human aid: *cf.* Acclimatization.

Acclimatization

The process whereby plants are accustomed to a new environment under human management: *see also* Hardening off: *cf.* Acclimation.

Acephalous

With head cut off or aborted.

Achene

Small dry single-seeded fruit which does not split to distribute seed as, for example, in clematis and sycamore.

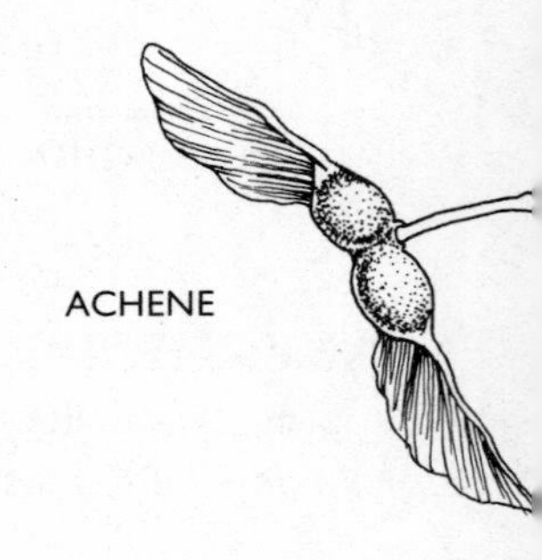
ACHENE

Achlamydeous

Of flowers that lack petals and sepals, such as willow (*Salix*).

Acicular

Needle-shaped, as applied to some kinds of foliage, including certain grasses. *Same as* Aciform.

ACICULAR

Acid

Soils with little or no lime content are described as 'acid', the degree of acidity or alkalinity being measured on the pH scale; pH7 is neutral, figures below 7 are increasingly acid, those above similarly alkaline. Acid-loving plants include rhododendron and camellia, acid-hating alpines and most Cruciferae.

Aciform

See Acicular.

Acinaciform

Shaped like a scimitar or short curved sword, broadest at the pointed end, as of some leaves.

Acotyledon

Plant with no obvious cotyledons or seed leaves, such as moss.

Acrogen

Cryptogam – fern or moss – with growing point at the tip.

Acropetal

Growth or development of organs in the direction of the apex, as in hyacinths, where the flowers at the base of the inflorescence open first: *cf.* Basipetal.

Acrospire

The first leaf to issue from a seed upon germination.

Actinomorphic

See Radial.

Actinomyces

A genus of minute fungi or bacteria with radiating mycelium.

Aculeate

Pointed or prickly.

Acuminate

Tapering to a point with slightly concave sides. Usually applied to leaves: *cf.* Acute.

ACUMINATE

Acute

Tapering to a point with slightly convex sides, as of leaves: *cf.* Acuminate.

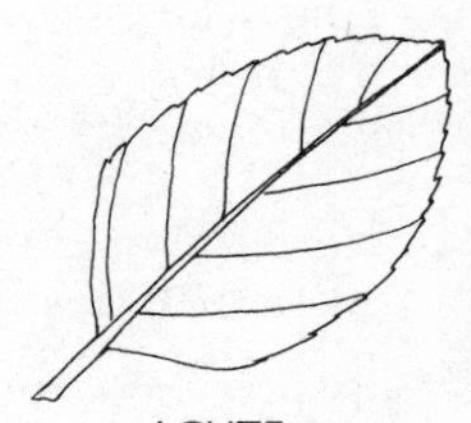

ACUTE

Adaptation

The process by which species adjust to environment in the course of evolution.

Adaxial

Referring to or attached to the side facing the stem or axis, e.g. the upper side of a leaf: *cf.* Abaxial.

Adnate

Attached to another organ, especially down the entire length.

Adpressed

Closely pressed together but not united.

Adpressed-pilose

Having soft hairs lying flat in contact with the leaf surface.

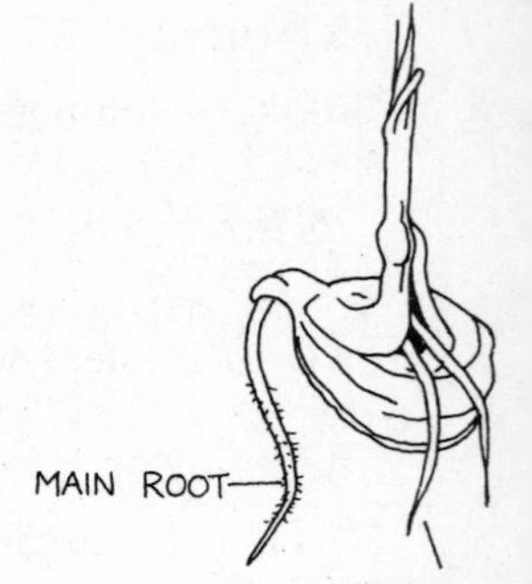

ADVENTITIOUS

Adventitious

Occurring in unusual or unexpected locations such as roots on aerial stems, buds on leaves, etc., e.g. the plantlets which develop on the leaves of kalanchoe (*Bryophyllum*).

Aecidial Cups

The visible 'blisters' on the underside of rusted foliage.

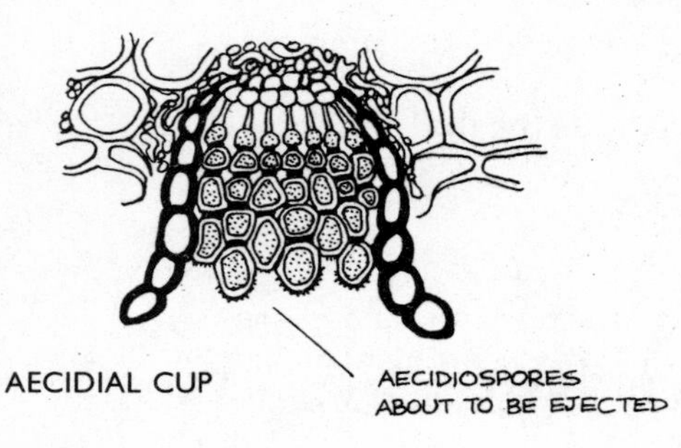

Aecidiospore

A type of spore produced by a rust fungus prior to the production of uredospores.

Aeration

Permeation of soil by air; exposure of soil to the action of the elements by rough digging.

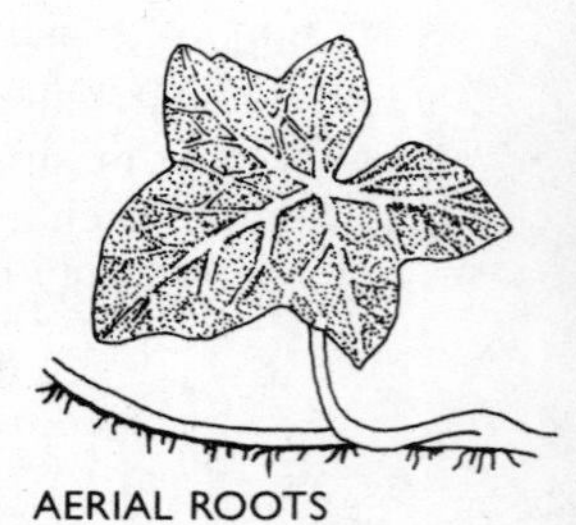
AERIAL ROOTS

Aerial Roots

Roots produced above ground level, as with those that develop on the stems of ivy.

AESTIVATION

PEA

ROSE

Aerobe

Micro-organism which needs free oxygen to exist: *cf.* Anaerobe.

Aestivation

The arrangement of the parts of a flower or leaves within the bud.

Afforest

To plant with trees, to create a forest.

After-grass

Grass that grows among the stubble after the crop has been harvested. *Also called* Fog.

Agar-agar

Gelatinous substance derived from seaweed and used as a solidifying agent in culture media. Sometimes written simply agar, or nutrient agar when nutrients have been added.

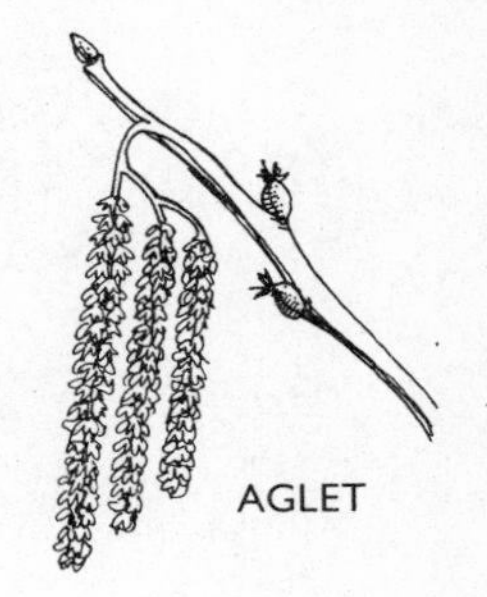

AGLET

Aglet, Aiglet

Dangling or pendulous catkin, of birch, hazel, etc: *see also* Ament, Amentum.

Agronomy

Applied agricultural science dealing with rural economy and husbandry. In recent years it has concentrated mainly in the theory and practice of crop production and soil management.

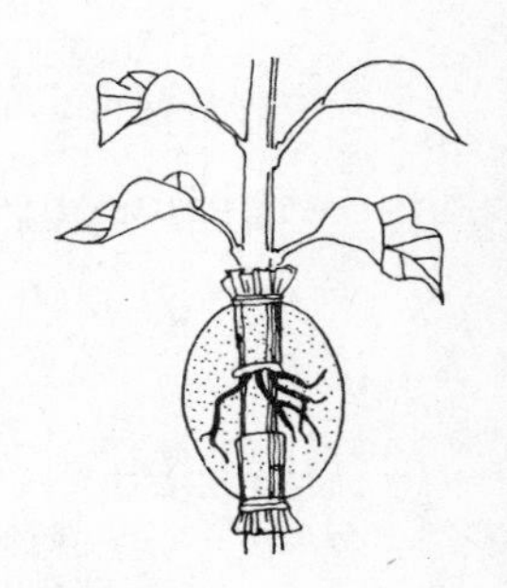

AIR-LAYERING

Air-layering

A method of propagating certain types of plant such as holly, lilac, azalea, etc. A portion of stem is wrapped in damp moss or similar material and sealed in a waterproof membrane such as plastic. The moss must be kept moist, and rooting may take up to two years or more with some subjects.

Alar

Wing-shaped, wing-like.

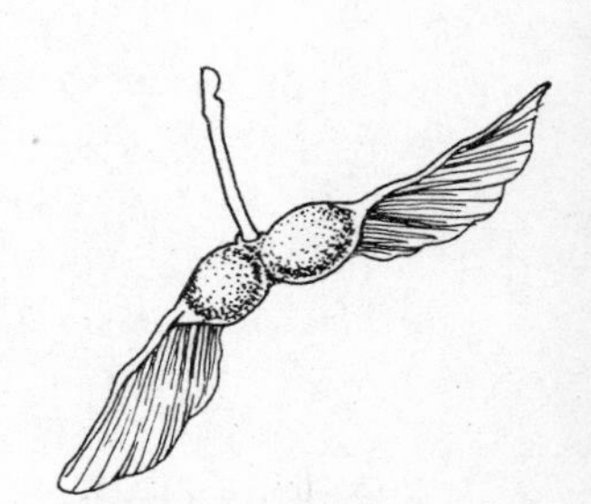

ALATE (SYCAMORE)

Alate

Winged; having wings or wing-like appendages, as the seeds of sycamore.

Albino

A plant or offshoot lacking normal green colouring. Sometimes called a ghost or ghost shoot.

Albumen

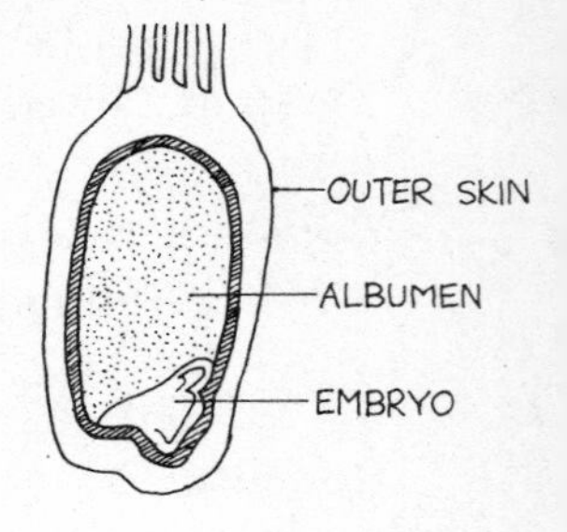

Starchy material in the seed, usually between the skin and embryo.

Alburnum

Sapwood, or the most recently formed wood in a tree.

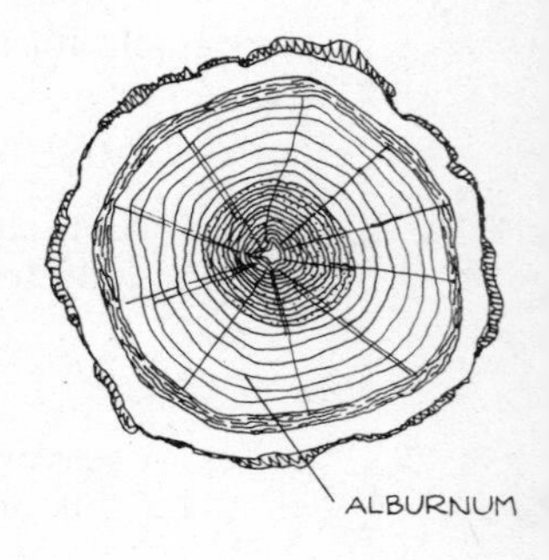

Alcove

Recess in a garden hedge or wall.

Aleurone Layer

The outermost layer of the endosperm.

Algae

A division of primitive cryptogamic plants including pond-scums and numerous microscopic water plants, seaweeds, etc.

Alkaline

Soils containing lime are described as 'alkaline', the degree of alkalinity being measured on the pH scale – pH7 being neutral, the higher figures denoting the scale of the increase in alkalinity. Some plants (such as azaleas) dislike alkaline soils and will not thrive where lime is present.

Alkaloid

A group of nitrogen-containing compounds present in some plants e.g. poppies, often with poisonous or medicinal properties (e.g. cocaine, nicotine, quinine).

Allele

A specialized form of gene which is responsible for a particular inherited characteristic.

Allopolyploid

An organism in which two different species have each contributed one or more sets of chromosomes, e.g. cultivated wheat.

Allotment

Small portion of rented land where the tenant may produce fruit, flowers and vegetables for home consumption.

Alluvial

A type of soil formed from the silt left behind by rivers or tides.

Alpine

Originally a mountain plant but now loosely applied to plants suitable for rock gardens and similar conditions.

Alpine House

A greenhouse, usually unheated, specially ventilated and equipped mainly for the cultivation of alpine subjects.

Alternate

Branches or leaves attached to the main stem singly at different heights and on different sides of the stem: *cf.* Opposite.

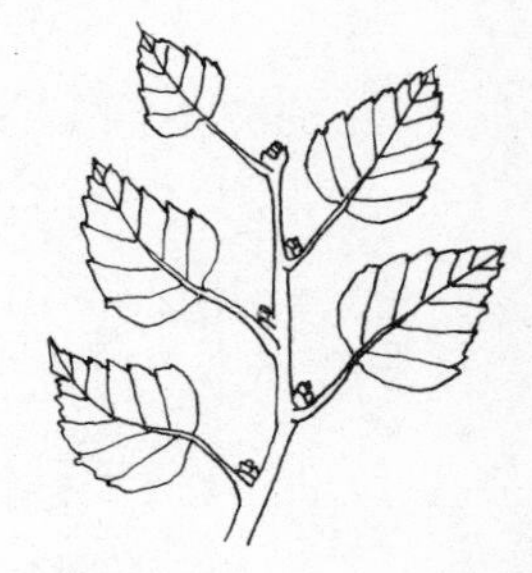

ALTERNATE

Alternate Host

Either of two different host plants of a pest or pathogen that requires both to complete its life cycle.

Alternation of Generations

The occurrence in a life cycle of two or more different forms in successive generations, the offspring being unlike the parents and usually reproducing by a different method. It is a feature of ferns where the large spore-bearing generation alternates with the smaller sexually reproducing generation.

Alternative Hosts

Host plants other than the main host on which an organism can develop.

Ament, Amentum, *pl.* Amenta

Catkin: *see also* Aglet.

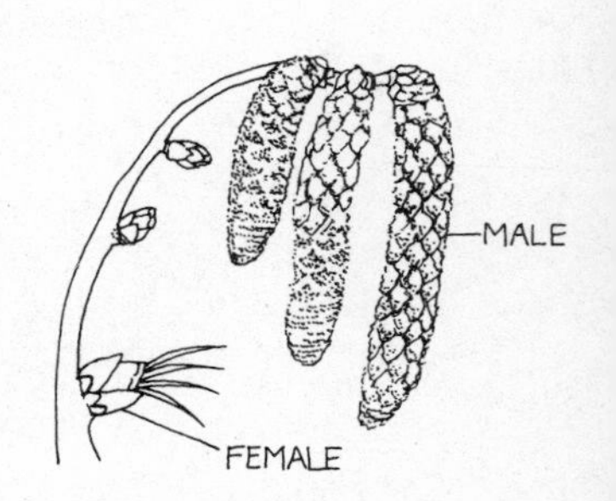

AMENT, AMENTUM

Amino Acids

Organic acids containing nitrogen in a certain configuration. There are about 20 ubiquitous amino acids which are constituents of proteins, but many more so-called 'non-protein' amino acids which occur in particular plant species.

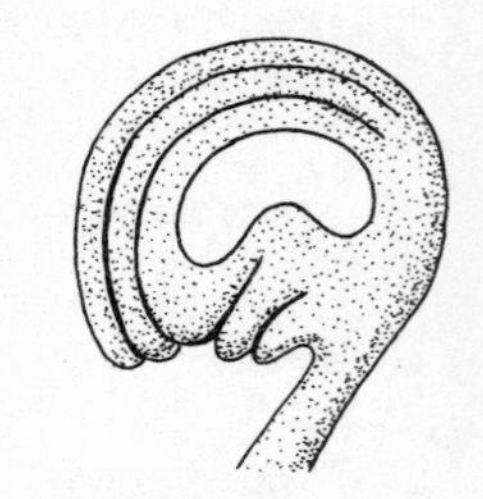
AMPHITROPOUS

Amoeba

Primitive microscopic single-celled organism which reproduces by division.

Amphibious

Capable of growing in either water or soil. Watercress is such a plant.

Amphitropous

Of an ovule inverted and attached near its middle.

AMPLEXICAUL

Amplexicaul

Of a leaf, sessile with the base clasped around the stem, as in tradescantia.

Anaerobe

A plant which can exist without oxygen, e.g. yeast: *cf.* Aerobe.

Anatomy

As applicable to plants: the study of the internal structure of an organ: *cf.* Morphology.

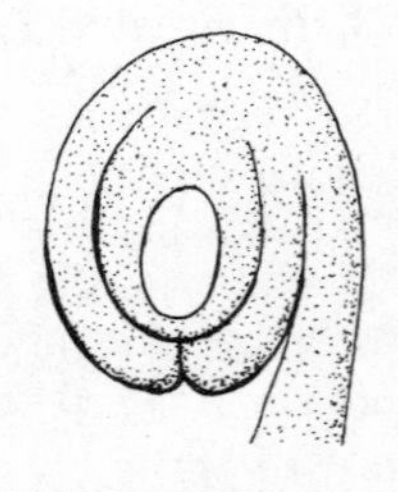
ANATROPOUS

Anatropous

Having an ovule turned over to lie alongside its funicle (stalk).

Anbury

A disease of some root crops such as turnips.

Androdioecious

Having male (and hermaphrodite) flowers on separate plants.

Androecium

The stamens of a flower collectively; the male reproductive organs.

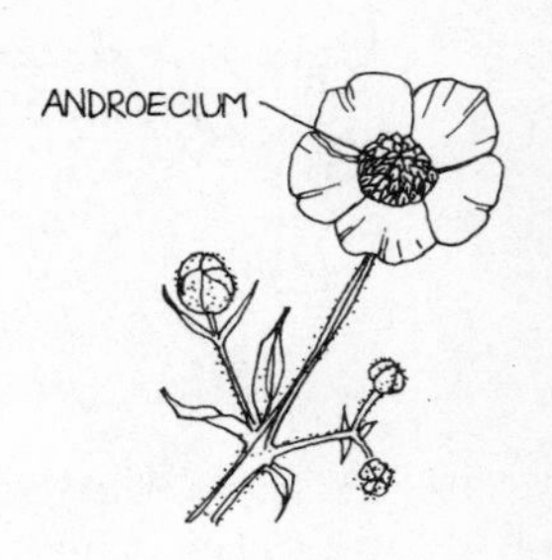

Androphore

An elongation of the receptacle carrying the stamens and carpels.

Androspore

Flagellated spore produced by some algae.

Anemone-centred

A term applied to flowers, e.g. some dahlias and chrysanthemums, where the central disc is prominent, rather in the form of a cushion.

ANEMONE-CENTRED

Anemophilous

Wind-pollinated.

Angiocarpous

Having the fruit enclosed in a special casing, e.g. acorn.

Angiosperm

A member of the Angiospermae class in which the seeds are enclosed in an ovary. One of the main divisions of flowering plants.

Annelid

A member of the Annelida, the major group of segmented worms which includes the earthworm.

Annual

Plant which survives for one season only from germination to death. *cf.* Perennial, Biennial.

Annual Ring

The ring of wood developed during each growing season in trees and shrubs from which the age of the plant can be calculated.

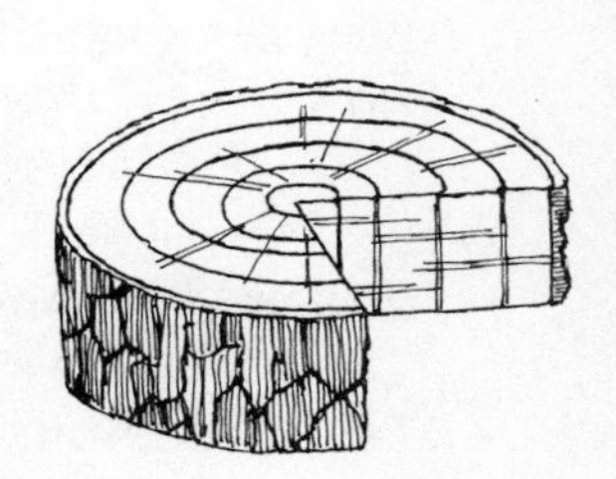
ANNUAL RING

Annular

Relating to or forming a ring.

Annulate

Consisting of rings.

Annulus, *pl.* Annuli

Ring of cells around sporangia in ferns.

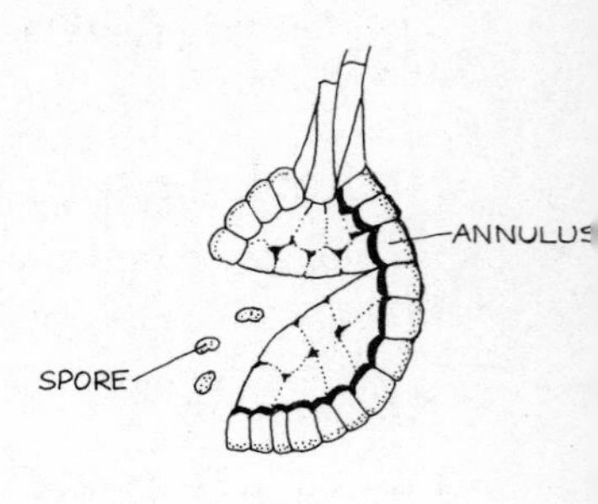

Antemarginal

Contained within the margin, usually toward the edge of a leaf, e.g. the leaf-zoning of zonal pelargonium 'Distinction'.

Anther

That part of the stamen carried at the top of the filament and bearing the pollen.

Anther Sac

The part of the anther containing the pollen.

ANTEMARGINAL

Antheridium

The male cell in cryptogams that produces sperm; corresponds to the anther in flowering plants.

Antheriferous

Anther-bearing; having anthers.

Anthesis

The period from the opening of the flower bud to the setting of the seed.

Anthocyanin

The pigment in flowers responsible for the pink, red, purple and blue shades.

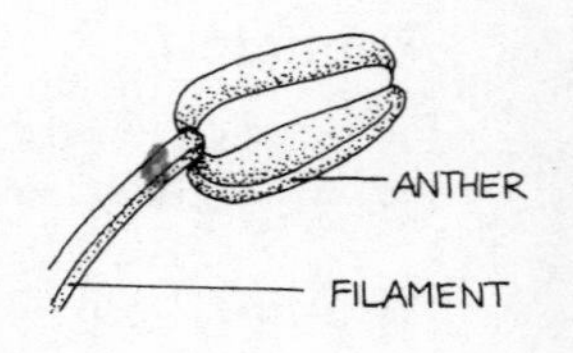

Antipetalous

Opposite a petal.

Antisepalous

Opposite the sepals.

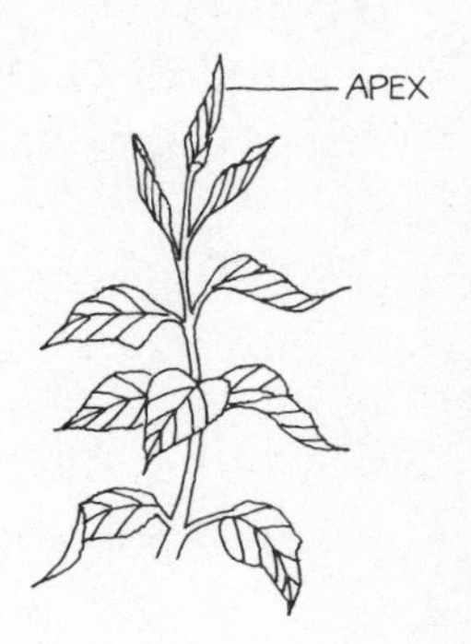

Antrorse

Pointing forward or upward.

Aperturate

Having one or more openings, usually applied to pollen.

Apetalous

Lacking petals.

APHID

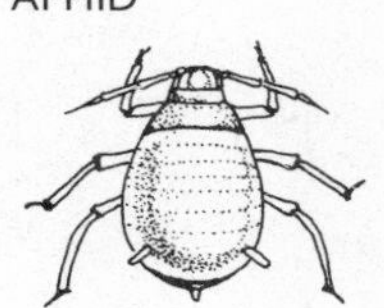

Apex

The tip or growing point of an organ such as a shoot.

Aphid, *pl.* Aphids

Small homopterous insects, including greenfly and blackfly, that congregate in large numbers on young plant growth. They excrete a sticky honeydew which can form a black sooty mould on the foliage, and are also one of the main carriers of virus diseases; they are often tended by ants who feed on the honeydew. Aphids can be destroyed by spraying or dusting with a suitable insecticide, the application of a systemic insecticide, or by a variety of smokes and fumigants. They can also be washed off the plant by syringing with warm soapy water but avoid modern detergents, some of which can be harmful to plants.

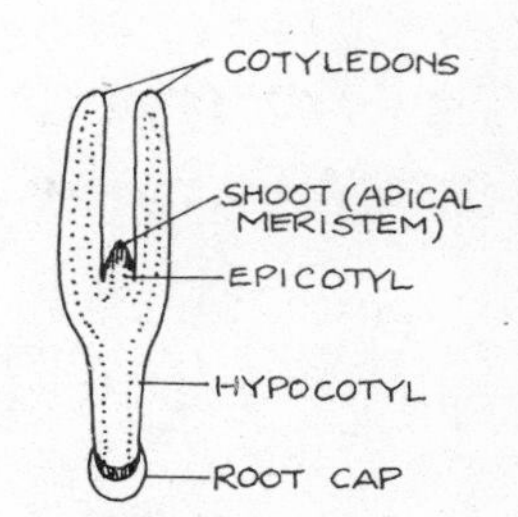

MAIN PARTS OF A GERMINATING SEED SHOWING POSITION OF THE APICAL MERISTEM

Aphyllous

Lacking leaves.

Apical

Pertaining to the apex; situated at the tip.

Apiculate

Terminated by an apicula.

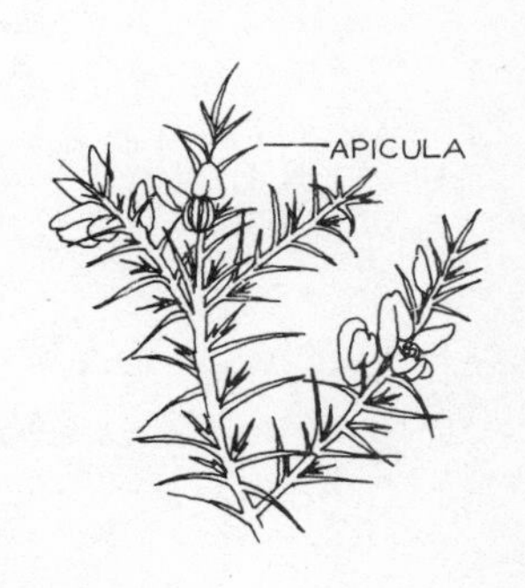

Apiculus, Apicula

Sharp flexible point, usually on leaves.

Apocarpous

Having the carpels separate.

Apomictic

Asexual reproduction, without fertilization.

Apothecium, *pl.* Apothecia

Cup-shaped fruit body, often brightly coloured as in some fungi; a particular characteristic of many lichens.

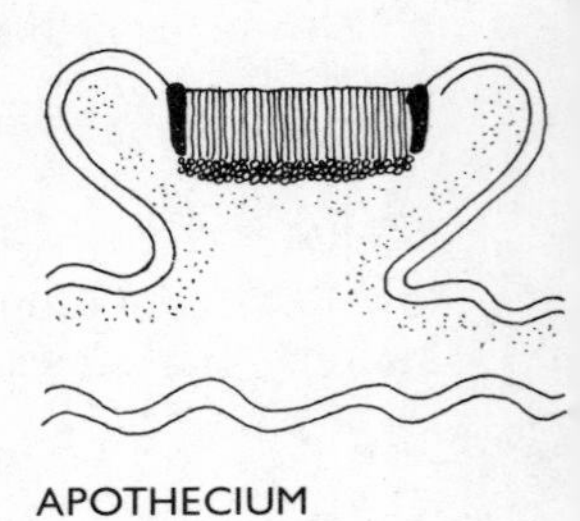

APOTHECIUM

Arachnoid

Having soft cobwebby hairs.

Arboraceous

Tree-like.

Arboreal

Tree-like; living in or on trees.

Arboreous

Wooded, abounding in trees; woodland.

Arborescent

Branching like a tree.

Arboretum

Tree-garden or man-made plantation of (correctly) broad-leaved trees; a tree museum or collection: *cf.* Pinetum.

Arboriculture

The cultivation of trees and shrubs.

Arbour

Shady retreat among trees, shrubs and climbing plants.

Archegonium

The flask-shaped female reproductive organ of mosses and ferns.

Arcuate

Curved or bent like a bow in archery, usually applied to leaves.

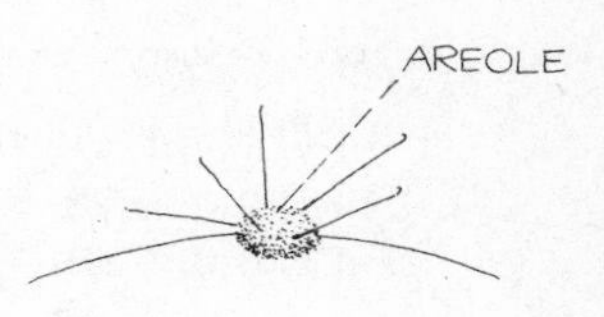

Arcure

An elaborate European system, seldom seen in Britain, of training fruit trees, said to increase the crop: *see also* Bending.

ARCURE

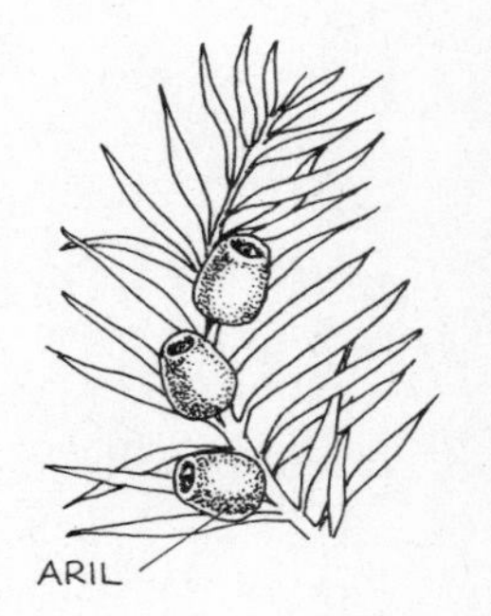

Arenaceous

Sandy, as applied to soil structure, etc.

Areole

A small pit or cavity on cacti from which a flower develops and to which the spines are attached.

Aril

Extra seed covering, often brightly coloured as in the yew.

Arista

See Awn.

Aristate

Tapering to a narrow elongated apex or spike (arista), usually applicable to leaves: *see also* Awn.

Articulate

A natural joint or node where separation is easiest.

Arundinaceous

Relating to or resembling a reed.

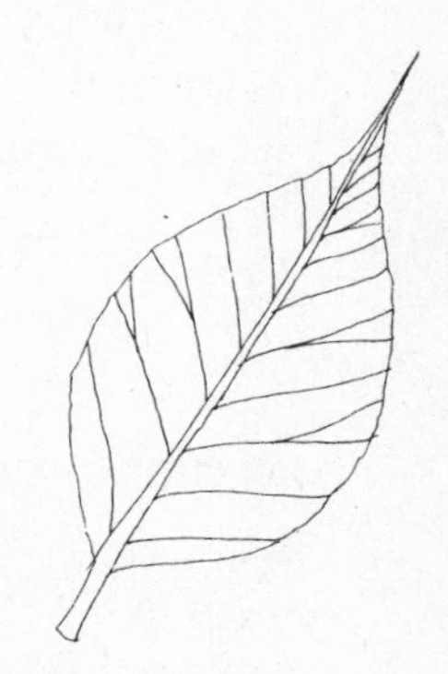
ARISTATE

Ascocarp

The vessel containing the asci and ascospores.

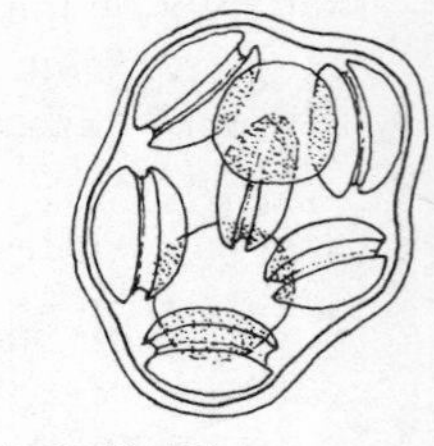

ASCOCARP CONTAINING ASCI

Ascomycete

Fungus producing ascospores, e.g. cup fungus often found growing on decaying trees.

Ascospore

A spore developed in an ascus.

Ascus, *pl.* Asci

Round or elongated sac-like organ in some fungi and lichens, containing spores.

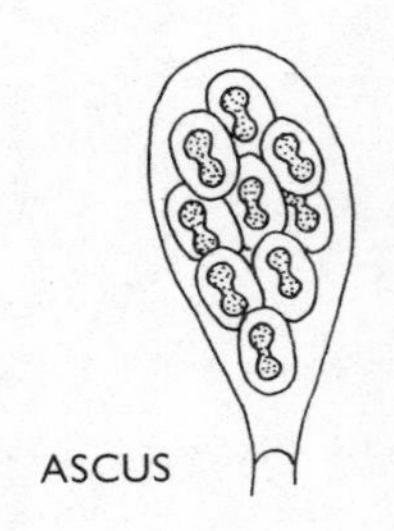

ASCUS

Asepalous

Lacking petals.

Asymbiotic

A plant which is hostile to certain other plants, such as horse-radish (*Cochlearia armoracia*).

Atavistic

Reverting to an earlier (and usually remote) form.

Attenuate

Gradually long-tapering, as of a leaf.

ATTENUATE

Auricled

Having ear-like outgrowths, e.g. the lobes on some leaves.

Auriculate

Having an ear-shaped lobe or lobes, as in some leaves and petals.

Autosome

A chromosome other than a sex-chromosome.

Autotrophic

A plant capable of obtaining its food supply from inorganic matter. The nitrogen-fixing bacteria which live sym-

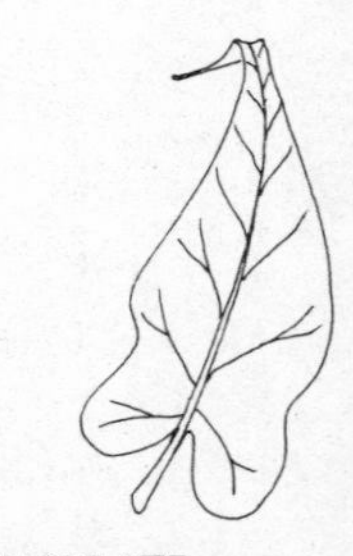

AURICULATE

biotically on the roots of leguminous plants, for example, are autotrophic regarding nitrogen, and heterotrophic respecting carbon. Green plants are autotrophs.

Auxanometer

A device for measuring plant growth under laboratory conditions.

Auxin

Any of several growth-controlling hormones present in plants.

Awl-shaped

See Subulate.

Awn

The beard or bristles on the seeds of some grasses and cereals, such as barley; probably a dispersal mechanism enabling the seed to be carried by the wind or on the fur of animals.

Axil

The angle between the stem of a plant and the leaf stalk growing out of it. A bud growing in this angle is called an axillary bud.

Axis

The whole main stem of a plant or the receptacle of a flower.

Azotobacter

A genus of nitrogen-fixing bacteria.

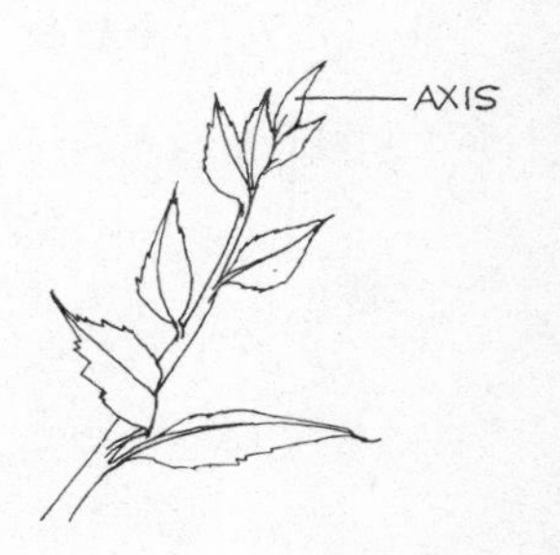

B

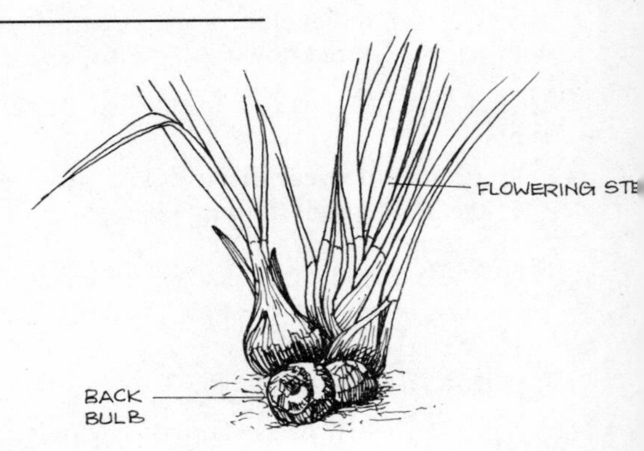

Baccate

Like a berry, having berries.

Bacciferous

Producing berries.

Back Bulb

A pseudobulb produced at the back of an orchid's flowering stem; one which has borne a flower in a previous season.

Back-cross

Progeny derived from a cross between a hybrid and one of its parents.

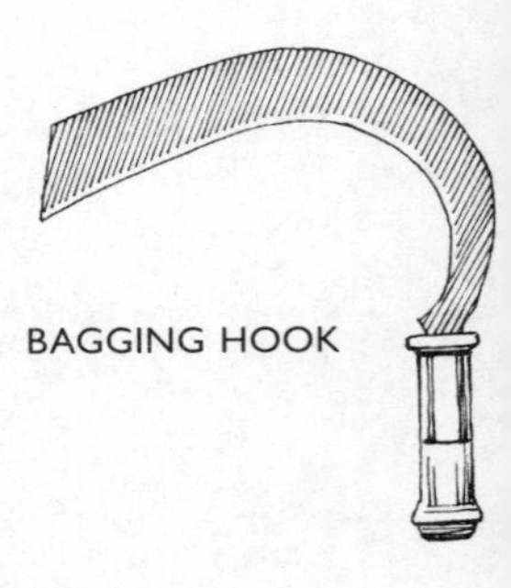

Bacteria

Microscopic single-celled organisms occurring on or in the tissues of plants (and animals), many of which bring about chemical changes such as decay, whereas others are beneficial. There are also many free-living bacteria not associated with plant or animal tissue.

Bagging Hook

A type of sickle or billhook.

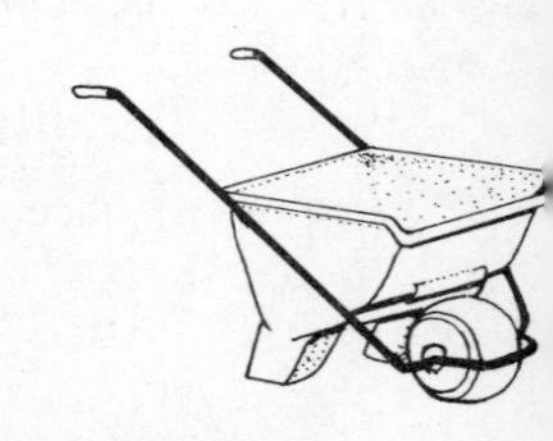

Ball-barrow

A type of wheelbarrow in which the wheel is replaced by an inflated plastic ball. Said to be easier to handle than a conventional barrow, especially across soft or rough ground.

Ballast

Gravel, rubble, small stones, broken rock, etc. used as a base for a path, road, wall, etc.

Ballerina

Trade name for a type of fruit tree developed by East Malling Research Station, in Kent; usually apple, but others are being developed, which grows as a single spike with fruit clustered down the entire length of the stem.

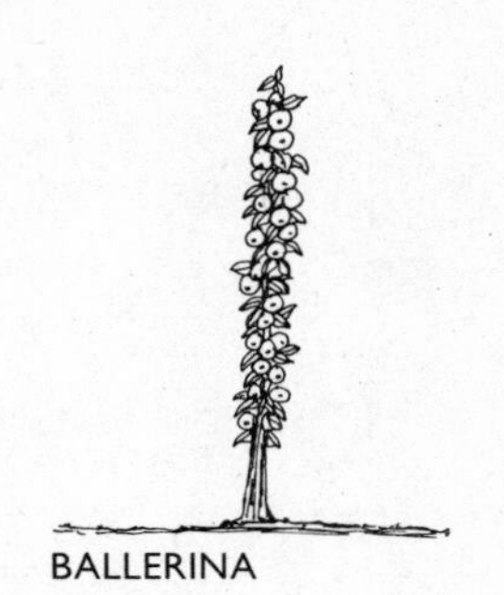

BALLERINA

Balling

(1) A physiological disorder of roses which results in the flowerbuds failing to open, turning brown and rotting.
(2) The method of enclosing the roots and soil-ball of trees in hessian or strong netting for transport.

Barbed

Having short stiff hairs slanted downward or backward.

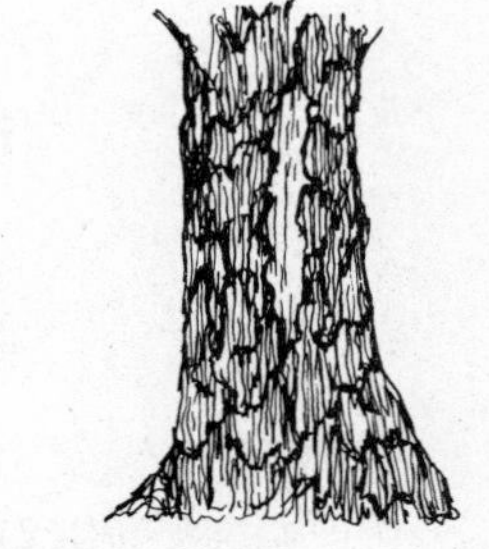

BARK-BOUND

Bark

The dead corky cells on the exterior of stems and branches to protect trees, etc. against predators.

Bark-bound

A condition in which the bark becomes too hard to expand in the normal way, thus strangling the tree. Lack of moisture in the soil is the most common cause. Occasionally the stress is relieved by the bark being split vertically, either naturally or by human action.

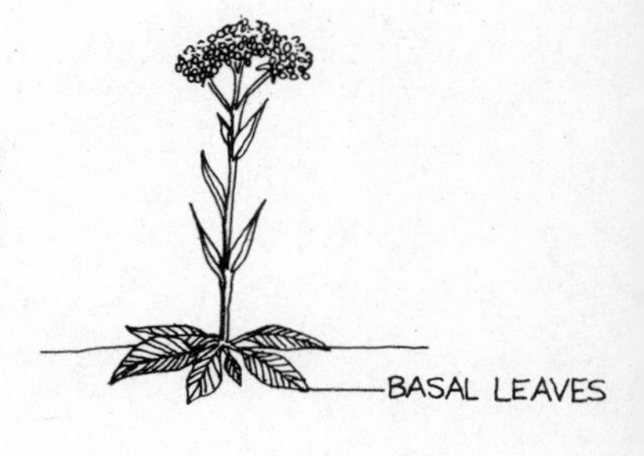

Basal

Botanically, almost synonymous with radical. Applies to the lower leaves of a plant which often differ from those higher up the stem.

Basal Placentation

Having the placenta at the base of the ovary.

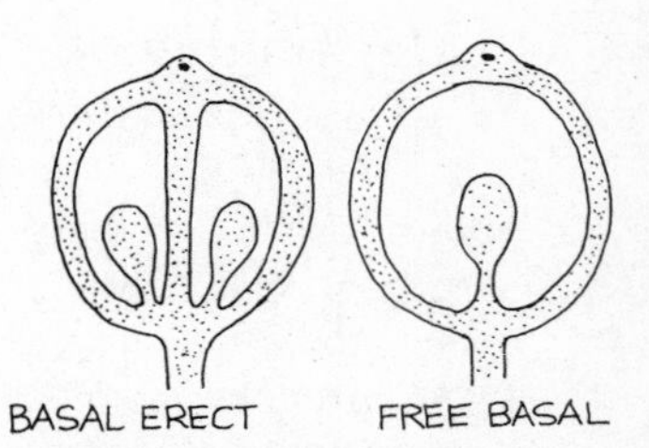

BASAL ERECT FREE BASAL

BASAL PLACENTATION

Basic Slag

A slow-acting fertilizer high in phosphorus and manganese content; a by-product of the iron and steel industry.

Basidiomycete

Fungus producing basidiospores, e.g. mushrooms.

BASIDIOMYCETE

Basidiospore

A spore borne at the end of a basidium.

Basidium

Spore-bearing part of some fungi.

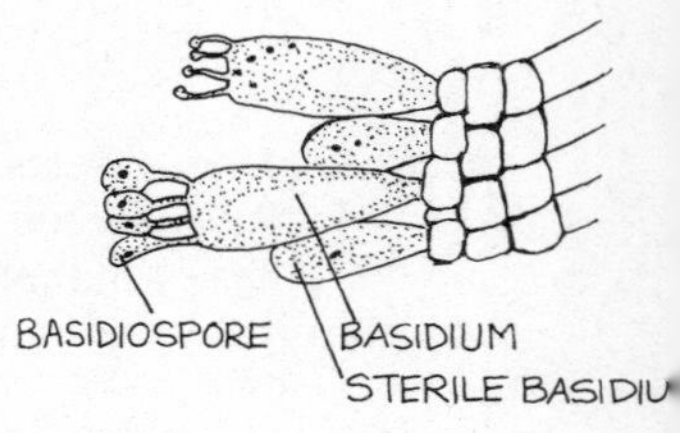

Basifixed

Anthers attached to the filament at the base and therefore incapable of independent movement.

BASIFIXED

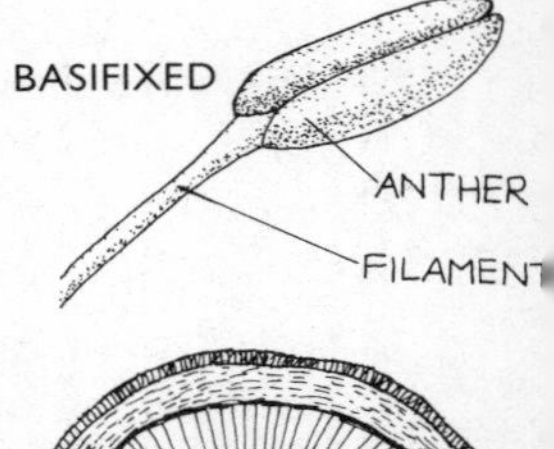

Basipetal

Development of repeated organs such as flowers proceeding toward the base, with the oldest at the apex: *cf.* Acropetal.

Bast, Bass

Phloem or inner bark, especially of lime trees, which when cut into strips, dried and plaited, can be used for coarse matting.

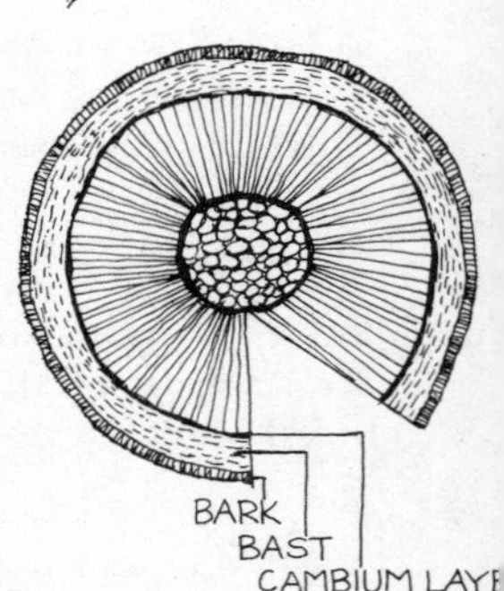

Bastard Trenching

Double digging: *see also* Digging.

Batter

The inward slope from base to top of a hedge created by controlled clipping to give extra stability.

BATTER

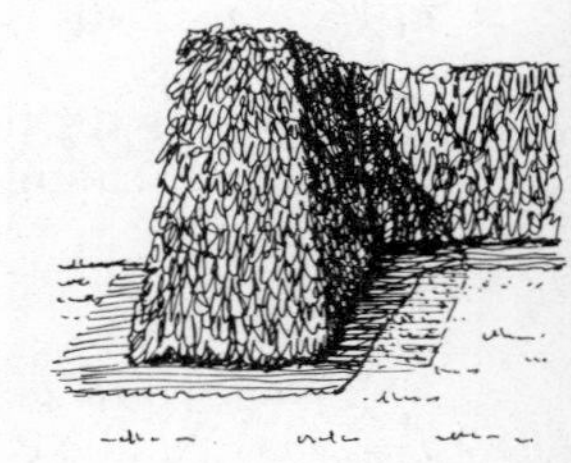

Beak

Long extended point. A term mainly used to describe the elongation of fruits and pistils.

Beard

The hairs which grow on the lower petals of some irises. The awn on some cereals and grasses, e.g. barley, is sometimes called the beard: *see also* Awn.

Bedding Plant

Plant used for temporary bedding displays, often raised under glass before being 'bedded out' in the open ground.

BEETLE

Beetle

(1) Group of insects, many beneficial, including the ladybird (ladybug in USA).
(2) Very heavy mallet used for driving-in fencing posts, etc.

BEETLE

Bell Glass

An early form of crop protection for tender subjects such as lettuce. Extensively used in France in earlier times and very popular with English Victorian gardeners; now largely superseded by various types of cloche.

Bending

A method of increasing the production of fruit and flowers by the bending or arching of young branches. Arcure is an elaborate system of bending used mainly in France.

BENDING

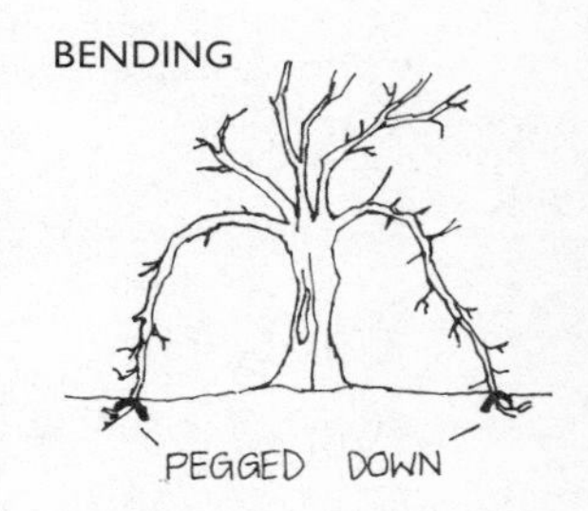

Benthos

Plants living on or attached to the sea bed; opposite to plankton which drifts on the surface.

Berry

A pulpy fruit containing one or more seeds but no true stone, e.g. tomato, grape, gooseberry and raspberry.

Besom

Bundle of twigs tied to a stick for sweeping up fallen leaves, distributing worm-casts on lawns, etc.

BESOM

Betalains

Red and yellow pigments present in the members of the *Dianthus* family, e.g. cloves and sweet williams.

Bi-

A prefix meaning two or twice.

Biauriculate

Having two auricules (ears), as of some leaves.

Bicarpellate

Derived from two carpels.

Bicephalous

Double-headed or having heads in pairs.

Bicoloured

A plant having two colours on the same petal or in the same flower.

Biennial

A plant sown one year to flower or fruit the next, then dying or being discarded. Many vegetables are biennial but are treated as annuals and harvested in their first year before they have flowered: *cf.* Annual, Perennial.

Bifid

Forked or split with two points.

Bifoliate

Having two leaves.

Bifoliolate

Leaves consisting of two leaflets.

Bifurcate

Forked or Y-shaped; divided into two branches: *cf.* Furcate.

BIFURCATE

Bigeneric

Derived from two different genera – a hybrid.

Bilabiate

Having two lips, an upper and lower part, as in antirrhinums.

Billhook

Type of small hatchet, usually with hooked end, used in hedging and heavy pruning operations.

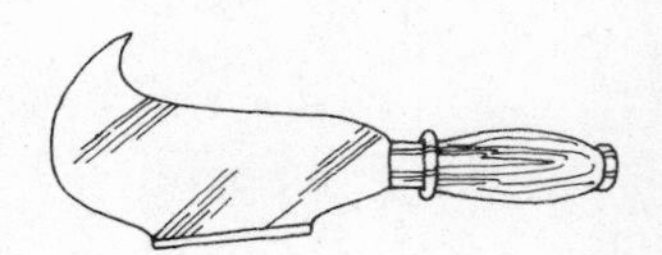

BILLHOOK

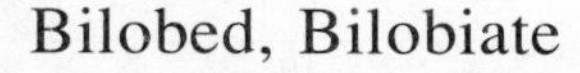

Bilobed, Bilobiate

Divided into two lobes or consisting of two lobes.

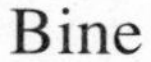

Bine

See Haulm.

Binominal, Binomial

The system of nomenclature devised by Carl von Linné (Linnaeus) in the eighteenth century in which each subject has two names, generic and specific. The generic name (which is written with an initial capital letter) is followed by the specific name which can be either an adjective complementing the generic name or another substantive in apposition to it.

Biological Control

Systems of controlling pests by the use of predators and parasites that feed upon them.

Biovulate

Having two ovules.

Bipartite

Divided into or consisting of two parts.

Bipinnate

Twice pinnate, as applied to leaflets that are divided into secondary leaflets: *cf*. Pinnate, Tripinnate.

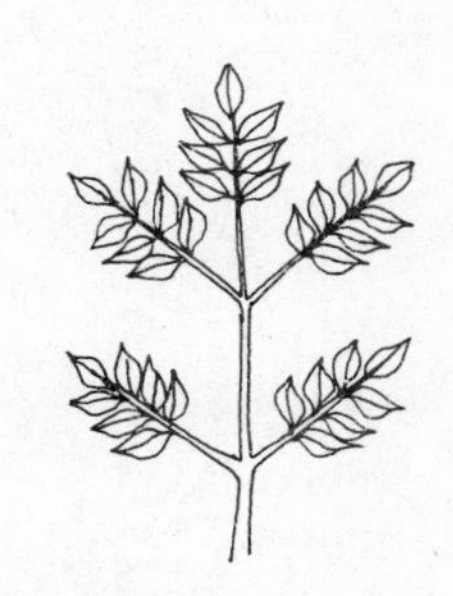

BIPINNATE

Biseriate

In two rows or two series, as a perianth consisting of a calyx and a corolla.

Biserrate

Serrate leaf with small teeth on larger teeth, as on silver birch: *cf.* Doubly serrate.

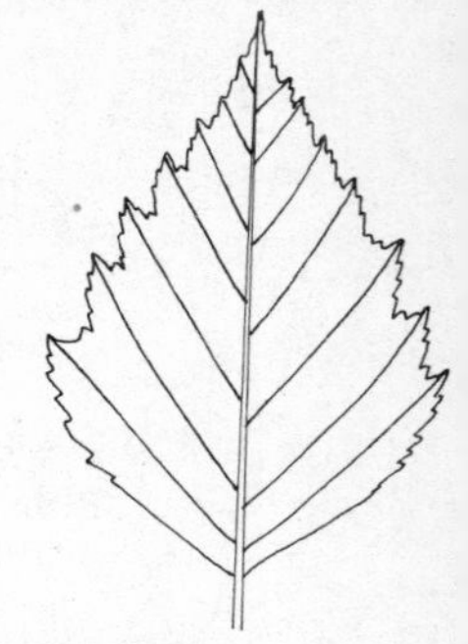
BISERRATE

Bisexual

Having both sexes in the same flower; hermaphrodite.

Biternate

Twice ternate: applied to leaves where the primary divisions are again divided into three parts: *cf.* Ternate, Triternate.

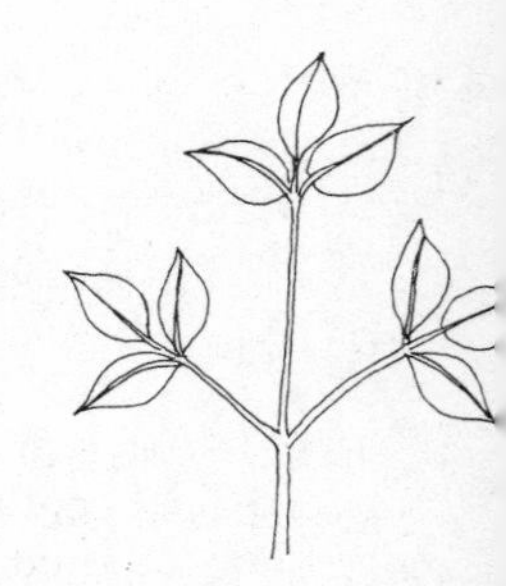
BITERNATE

Black Rot, BLACK ROOT ROT (USA)

A widespread disease caused by the soil-borne fungus, *Thielaviopsis basicola*. Roots of infected plants show lesions, which are black or dark brown because of tissue discoloration caused by prolific production of dark, thick-walled chlamydospore chains and pigmented hyphae: *cf.* Stem rot.

Black Spot

A disease affecting several plants such as roses, and showing as small black marks on the foliage. Caused by a fungus which survives mainly in clean air – gardens in city centres with polluted atmospheres are seldom troubled by black spot.

Blackfly

See Aphid.

Blackleg

See Collar rot.

Blade

(1) The main broad part of a leaf or petal.
(2) The leaves of certain plants are called blades, e.g. iris and gladioli.

BLADE (IRIS)

Blanching

Making plants whiter and less bitter by the partial or total exclusion of light. A process widely used in the production of chicory, celery, etc.

Bleeding

The loss of sap from plant tissues that have been damaged or cut. Rubber trees are bled to extract the latex.

Blight

(1) General but incorrect term often used in reference to aphid infestation.
(2) Type of fungus which attacks potatoes, tomatoes, etc.

Blind

(1) Any plant which fails to produce a flower.
(2) Seedlings where the growing point has been damaged or destroyed, usually by garden pests such as maggots or caterpillars. Once the growing point has gone, no further growth is possible and the plant is useless.

BLIND SEEDLING

Blood

Dried blood is a fairly fast-acting fertilizer containing about 10 per cent nitrogen. It can be applied to the soil in dry form or used as a liquid manure when mixed with water.

Bloom

(1) A flower or flowerhead.
(2) The very fine powdery or waxy deposit covering the surface of certain leaves, stems and fruits such as grapes and plums: a protective coating easily rubbed off if the fruit is badly handled.

Blossom

(1) To open into flower.
(2) A flower, or the mass of flowers on fruit trees, etc. in spring.

Bodge

Simple water-cart used widely in early Victorian times before the advent of the garden hosepipe.

BODGE

Bog Garden

A type of garden where the soil is kept permanently very damp but not waterlogged.

Bole

The trunk of a tree between soil level and the first main branch.

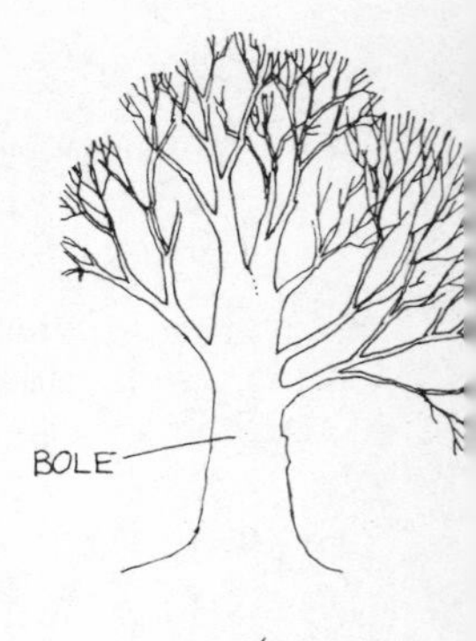

Boll

Spherical seed-vessel of some plants such as cotton.

Bolting

Premature flowering. Under normal conditions all vegetables will eventually flower and set seed but most are harvested and used in their immature state. Some, however, e.g. lettuce, will often bolt in hot weather, producing flower spikes but no hearts.

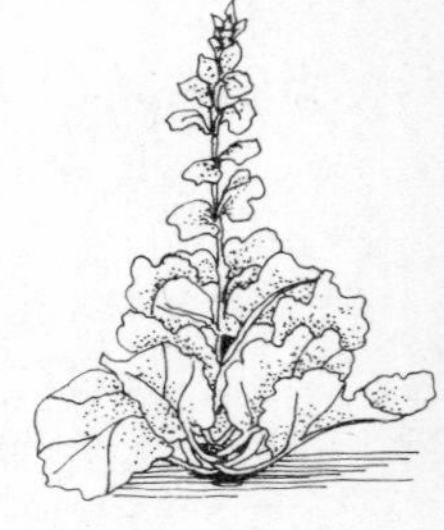
BOLTING LETTUCE

Bonemeal

A slow-acting fertilizer containing phosphorus, nitrogen and some calcium produced by grinding down animal bones. The finer grades are slightly faster-acting than the coarse.

Bonsai

A method of dwarfing certain trees and shrubs, developed and favoured by the Japanese. It involves special shoot and root pruning coupled with complicated training systems.

BONSAI
(*Pinus thunbergii*)

Bootlace Fungus

See Honey fungus.

Borax, Boron

Boron is one of the essential plant foods required in such minute quantities that it is usually regarded as one of the trace elements. Garden soils are seldom deficient in boron and an excess is poisonous to plants, especially pot plants.

Bordeaux Mixture

A copper sulphate preparation which originated in the Bordeaux region of France, much used in the past against potato blight and on flower and fruit crops. Still available, but largely superseded by more efficient modern chemicals.

Boscage

Wooded scenery; a dense area of trees or shrubs.

Bostryx

A cymose inflorescence in which the lateral branches develop on one side of the stem only, as found in some of the Hypericaceae family.

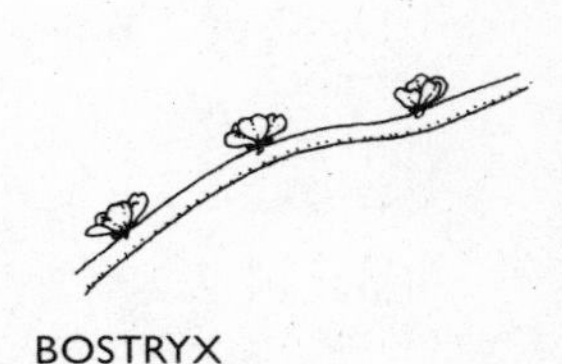

BOSTRYX

Botanize

To seek or collect plants for study.

Botany

The science of plants; the flora of a particular area.

Botrytis Cinerea

A fungus which attacks lettuce, strawberries, pelargoniums, etc., especially when grown under glass. It can be countermanded to some extent by fungicidal sprays, but improving the environment by better ventilation and careful watering, coupled with an increase in temperature, will usually overcome it if caught in the early stages. Often called grey mould.

Bottle

A bundle of hay or straw, sometimes reeds.

BOTTLE OF STRAW

Bottle Garden

A specialized form of indoor plant cultivation. Many plants thrive in totally sealed glass containers, the moisture transpired condensing on the inner surface of the glass and being returned to the soil. Large carboys are excellent for this purpose and, carefully planted, make ideal bottle gardens, or special glass containers are obtainable from some garden centres: *cf.* Terrarium, Wardian case.

BOTTLE GARDEN

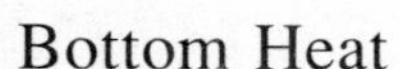

Bottom Heat

Gentle heat applied to the base of a cutting, seedling or plant to assist rooting or development. In the past animal manure was used to provide bottom heat but electric soil-heating cables are the modern equivalent.

Bough

Branch of a tree, especially a mature one larger than a twig.

Bracken

Type of large fern found mainly on heathland: the commonest British fern, *Pteris aquilina.*

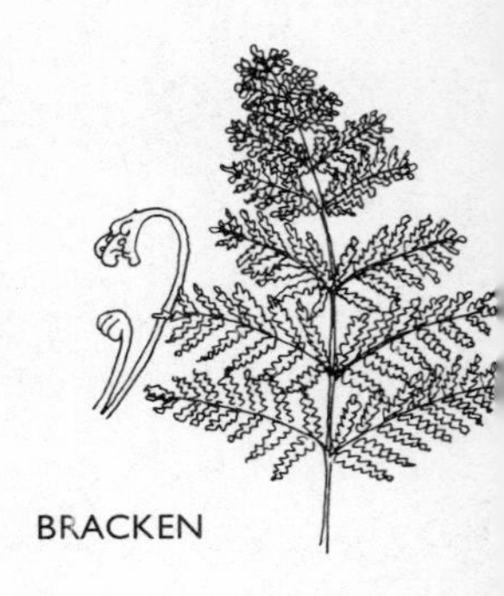

BRACKEN

Bract

A leaf-like structure arising from the stem of a flower and often apparently forming part of the flowerhead, usually in cases where the flower itself is insignificant. Sometimes it resembles ordinary leaves but is more brightly coloured, as in the poinsettia.

BRACT (POINSETTIA)

Bracteate

Having bracts.

Bracteole

A bractlet or very small bract on the axis of a flower. Also used to refer to bracts within inflorescences, where the term 'bract' would be reserved for the bract, if any, at the base.

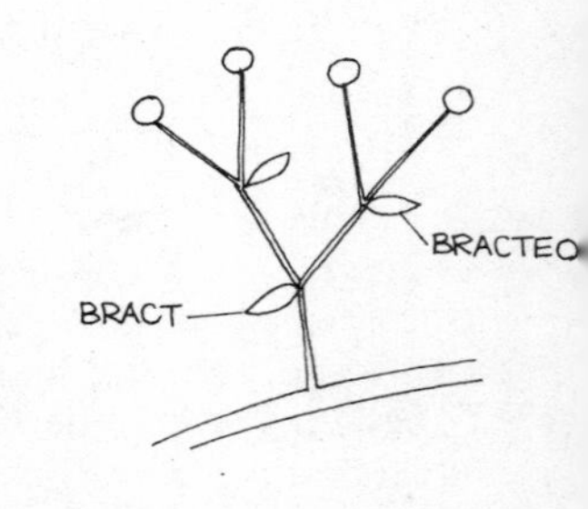

Brake

(1) Thicket; dense growth of short shrubs and undergrowth.
(2) An area of bracken usually covering heathland.

Bramble

Coarse prickly shrub, usually invasive, e.g. blackberry.

Bramble Scythe

Similar to a standard scythe but with a shorter, wider and less flexible blade specially developed for dealing with brambles.

Branch

Limb of a tree, shrub, etc., more substantial than a twig but less so than a bough.

Branchlet

Small branch, in size between a branch and a twig.

Brash

Hedge-clippings and similar refuse.

Brassica

A member of the cabbage family including cauliflower, turnip, swede, etc.

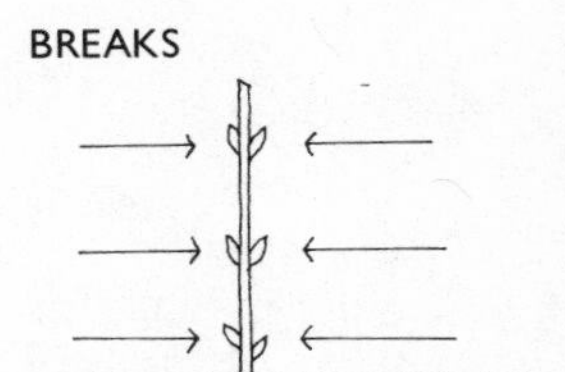
BREAKS

Break

A side-shoot obtained by removing the growing point of a developed plant to force it to produce more lateral shoots (breaks) and thereby increasing its bushiness.

Bridge Grafting

A method of grafting to overcome the total girdling of a tree either by disease, animals, accident or vandalism. *See* Grafting.

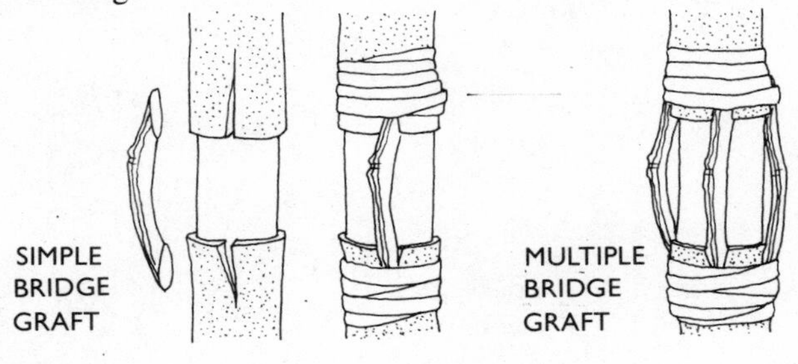

Bristly

Bearing strong stiff hairs or bristles.

Broadcasting

The method of random sowing of seed by casting it freely on to prepared ground without recourse to pre-drilling or lining.

BROADCASTING

Brushwood

Small trees and shrubs, underwood.

Brutting

The fracturing – but not severing – of the young new shoots of fruit trees to restrict late growth.

BRUTTING

Bryophyte

A member of one of the main groups of the vegetable kingdom – mosses, etc.

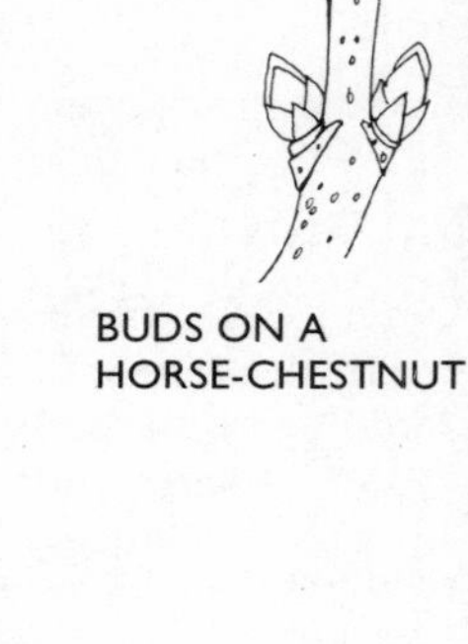

BUDS ON A HORSE-CHESTNUT

Bud

(1) Dormant rudimentary projection on a stem from which growth may develop.
(2) An embryonic flower.

Budding

A grafting method often used on roses and fruit trees. A well-developed bud is taken from one plant and inserted under the bark of another. In this way a flower produced by a delicate plant can be grown on to a more robust rootstock. *See* Grafting.

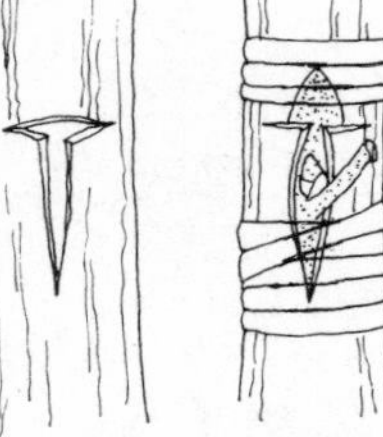

BUDDING

Bug

A term used loosely – especially in USA – to describe any small insect but should strictly be limited to members of the order Hemiptera which includes aphids, bed-bugs, pond-skaters, etc.

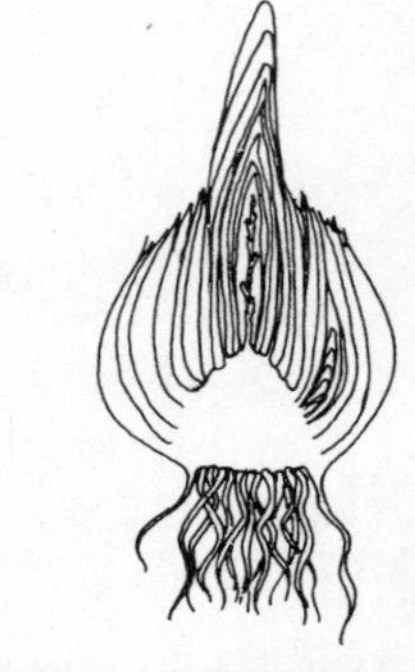

BULB (NOTE SCALES)

Bulb

A food storage organ consisting usually (but not always) of an underground type of bud with a short thick stem and tightly packed, fleshy scales. Corms, tubers and rhizomes are often erroneously described as bulbs.

CORM (NO SCALES)

Bulbil

(1) A small immature bulb, usually attached underground to a parent bulb, as in gladioli.
(2) A small bulb-like organ produced on a stem near a flowerhead, as in some types of onion.
(3) A small bulb produced in a leaf axil, e.g. lily.

Bulblet

Small immature bulb usually produced at the base of a mature one; alternative name for a bulbil.

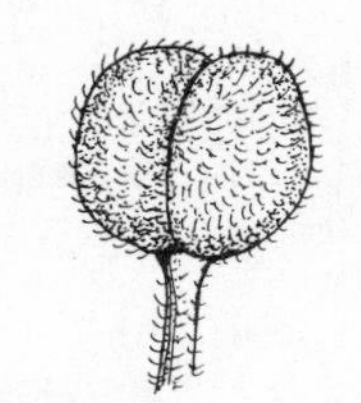
BURR OF GOOSEGRASS

Bullate

Crinkled or puckered, as, for example, the savoy cabbage.

Burgundy Mixture

A general purpose fungicide consisting of a combination of ordinary washing soda and copper sulphate: a favourite with Victorian gardeners but now largely superseded by proprietary products.

BUSH

Burr

(1) Hooked or barbed fruit or flowerhead which attaches itself to a passing animal as a means of dispersal.
(2) A woody projection or outgrowth found on the stems of some types of tree.

Bush

Short dense shrub with no definite trunk.

Bushel

A dry measure of 8 gallons for grain, fruit, sand, soil, etc. no longer official. Imperial dry measure tables were as follows

2 gallons	=	1 peck	=	9.0919 litres
4 pecks	=	1 bushel	=	3.637 decalitres
8 bushels	=	1 quarter	=	2.909 hektolitres
5 quarters	=	1 load	=	1.454 kilolitres

Butt

That part of a tree trunk immediately above ground level.

Buttoning

A condition of cauliflowers resulting in a 'loose head': usually caused by lack of moisture at some stage of development.

BUTTONHEAD CAULIFLOWER

C

Caber

Roughly trimmed trunk of a small pine tree, used in the Highland sport of 'tossing the caber'.

Cactus

A member of the family Cactaceae, fleshy succulents whose stems store water and do the work of leaves, often carrying spines. There are about 200 genera and over 2000 species in cultivation. A characteristic of cacti is that they bear areoles.

SOME TYPES OF CACTUS

Caducous

Perishable; falling when their work is finished, as of leaves in autumn. Falling early or prematurely.

Caespitose

See Cespitose.

Calcarate

Spurred.

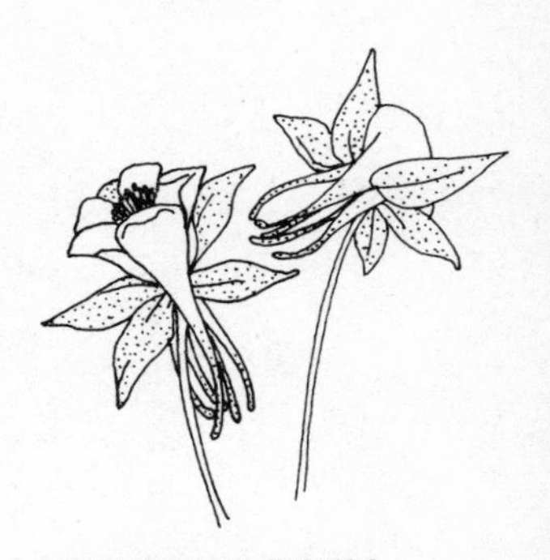

AQUILEGIA FLOWERS ARE CALCARATE

Calcareous

Chalky.

Calcicole

A lime-loving plant. Calcicolous plants grow best when lime is present in the soil. Limy soils are alkaline with a pH above 7: *cf.* Calcifuge.

Calciferous

Having a chalk content.

Calcifuge

A lime-hating plant. Calcifugous plants are those that dislike lime or chalk in any form and will seldom survive for long in a limy soil. They require an acid soil with a pH below 7: *cf.* Calcicole.

Caliciform

Resembling or having the form of a calyx or cup.

Callosity

A callus or thickening; a protuberance.

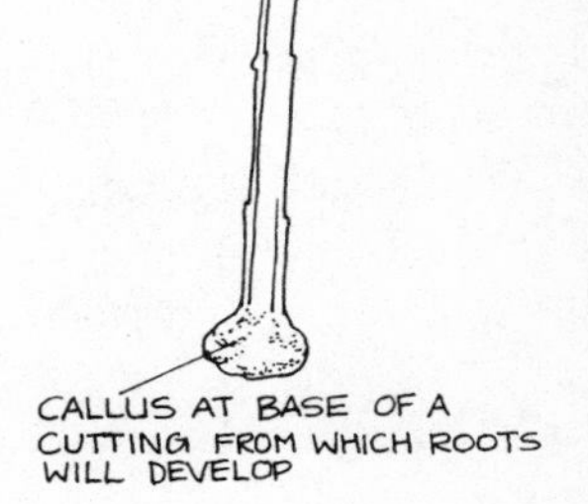

CALLUS AT BASE OF A CUTTING FROM WHICH ROOTS WILL DEVELOP

Callus

The soft tissue that forms over a cut surface to seal it, often becoming corky with age. Also applied to the undifferentiated tissue in artificial tissue culture, from which young plants can be induced to develop under laboratory conditions.

CALLUS AROUND A TREE WOUND

Calycanthemy

Having a calyx-like corolla.

Calyciform

Having the form of a calyx.

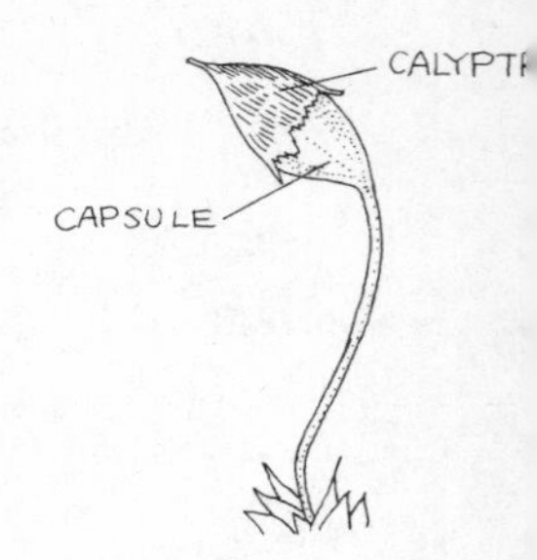

MOSS PLANT

Calycinal

Pertaining to a calyx.

Calyculate

(1) In the form of a calyx.
(2) Having a calycule.

Calycule

A whorl of small bracts resembling a calyx.

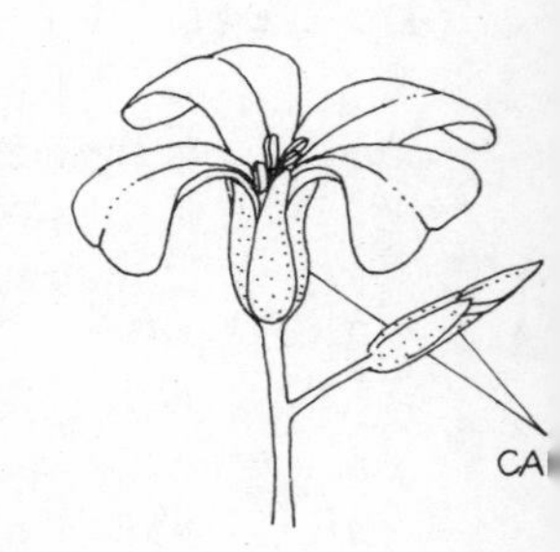

Calyptra

Lid-like or cap-like, as in the cap or hood of a moss capsule or of a root-tip.

Calyptrogen

The group of cells forming the root-cap.

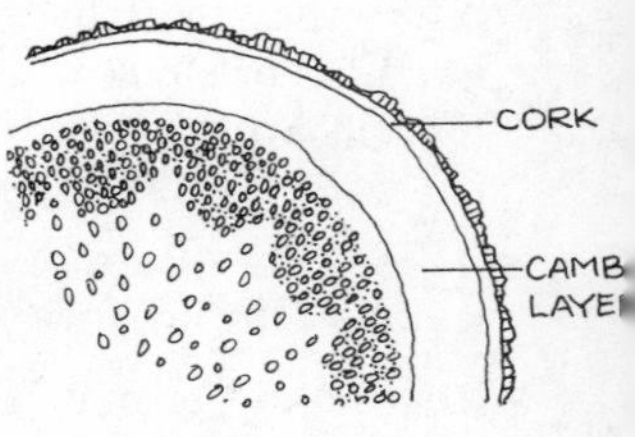

Calyx, *pl.* Calyces

The group or cluster of modified leaves enclosing a flower bud, each segment being a sepal. In some plants the sepals partially replace the flower petals (as in clematis) and are brightly coloured.

Cambium Layer

The living cells lying immediately beneath the bark of a tree, and at the growing tips of roots and shoots.

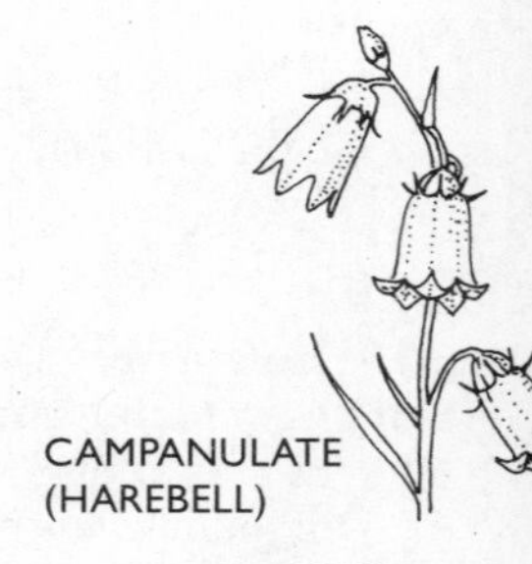
CAMPANULATE (HAREBELL)

Campanulate

Bell-shaped flowers, e.g. harebells and Canterbury bells.

Campylotropous

Ovules bent over at 90 degrees to the stalk which appears to be attached to the side of the ovule.

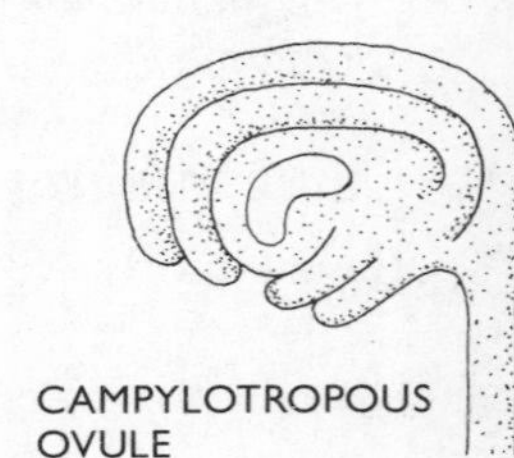
CAMPYLOTROPOUS OVULE

Canal

Duct in a plant through which passes food, air, liquids, etc.

Canaliculate

Having a lengthwise groove or channel.

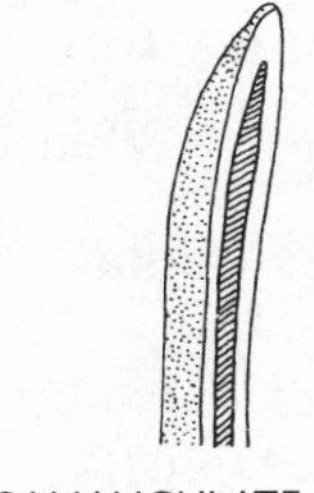
CANALICULATE

Cane

Hollow or solid straight stem of certain plants such as some reeds and grasses, raspberries, etc.

Cane Brake

American term for a thicket or overgrown area of land.

Canescent

Covered in short, fine hairs, usually grey or white; grey or hoary in appearance.

Canker

(1) A disease of some root crops, including parsnip. The roots are invaded and pitted by a brown corky rot.
(2) A fungus disease of trees, including certain fruit trees.

Canker Worm

The larva of a moth which destroys leaves and buds of fruit trees and attacks other plants.

Cap

A calyptra, the top of a fungus (toadstool, etc.) or the hood covering a moss capsule.

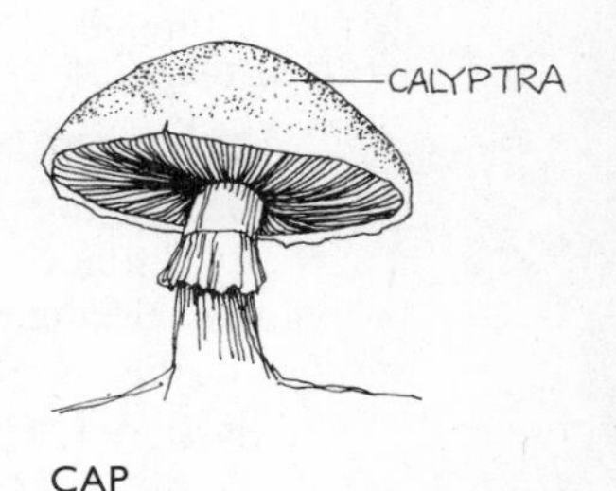

CAP

Capillary

Very thin hair-like tube, enabling fluids to be raised in soil and plants. Capillary action is the force that causes liquids to rise in wicks and blotting paper.

Capitate

Forming a very dense or compact cluster, usually of flowerheads; literally head-like or with a knob-like head.

MOST ONION FLOWERS ARE CAPITATE

Capitellate

Diminutive of capitate.

Capitulum, *pl.* Capitula

A dense cluster of small stalkless flowers, as in clover. The 'flowers' of Compositae, e.g. daisies, are often referred to as capitula.

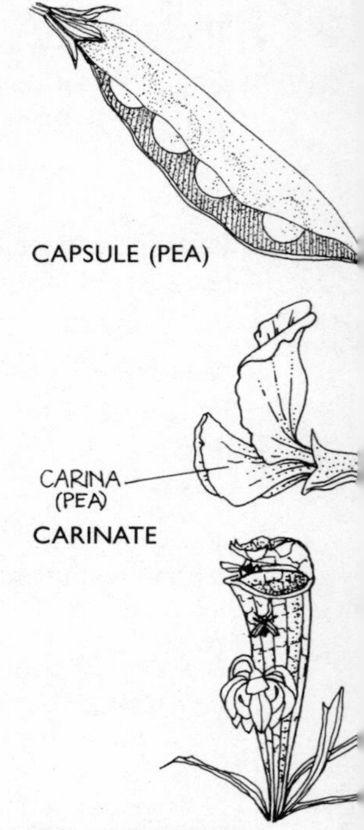

CAPSULE (PEA)

CARINATE

THE CARNIVOROUS PITCHER PLANT

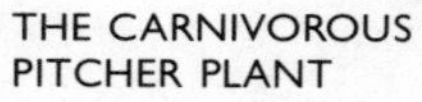

Capsule

The envelope containing seeds which dries and splits when ripe to discharge its contents. The capsule consists of carpels joined together, several in some cases, as in the poppy, or two only, as in peas.

Carbon Dioxide

A colourless, odourless gas expired by humans and absorbed, in daytime only, by plants during photosynthesis.

Carinate

Having a central ridge or keel, as in the boat-like formation of the two lower petals in flowers of the pea family.

Carnivorous

Meat-eating. In botany the term is usually applied to insect-eating plants such as the Venus's flytrap (*Dionaea muscipula*).

Carpel

A plant's female reproductive organs consisting of the ovary, stigma and style; the seed-bearing part of a flower. Part of a pistil or ovary; a pistil may have one (simple carpel) or several (compound carpel).

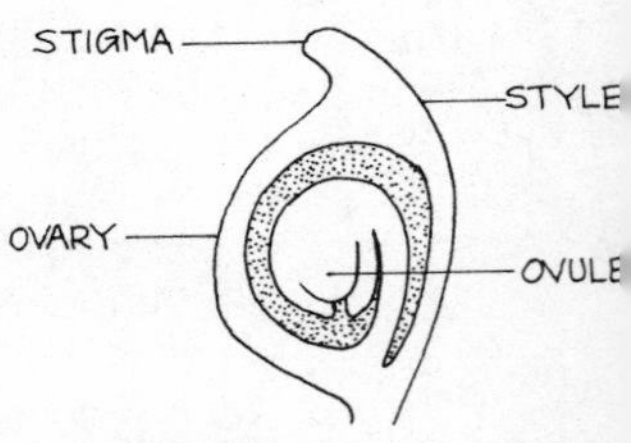

CARPEL OF A BUTTERCUP CUT IN HALF LONGITUDINALLY

Carpellate

Composed of carpels. Bi-carpellate, composed of two carpels; tri-carpellate, composed of three carpels, etc.

Cartilaginous

Cartilage-like; tough but flexible.

Caruncle

An outgrowth on the seeds of certain plants, e.g. castor-oil plant (*Ricinus communis*).

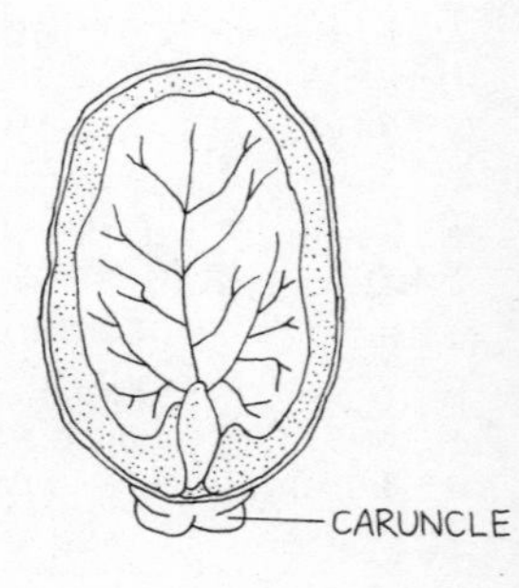

Caryopsis

A single-seeded indehiscent fruit; a term usually associated with grasses such as wheat and barley.

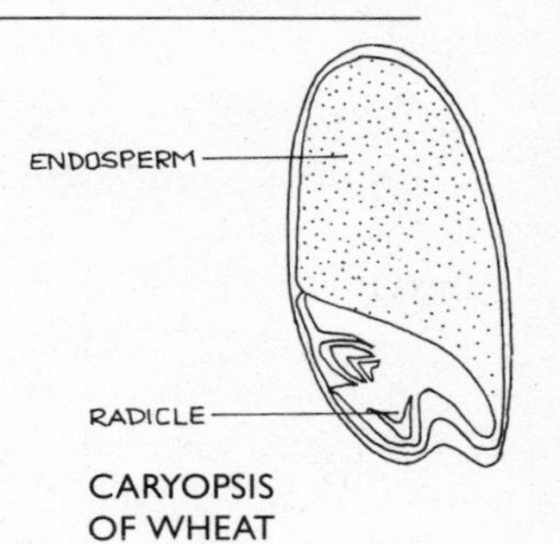

CARYOPSIS OF WHEAT

Cast

The earth secreted by earthworms – worm-cast.

Catch Crop

A fast-maturing crop raised on ground where a later and usually slower-growing crop is to be planted: *cf.* Intercropping.

Catkin

An ament. Mostly pendulous, stalkless flowers, often without petals and usually of individual sexes, the males being quite different to the females. The hazel is a pendulous catkin-bearing tree, while the pussy-willow (*Salix caprea*) has erect catkins: *see also* Ament.

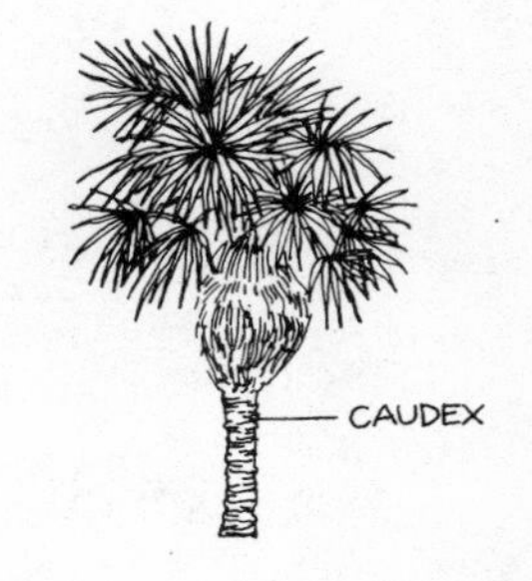

Caudate

Bearing a tail or tail-like appendage(s).

Caudex

(1) The bole or trunk of a palm.
(2) The underground stem of a perennial herb from which the annual shoots arise.

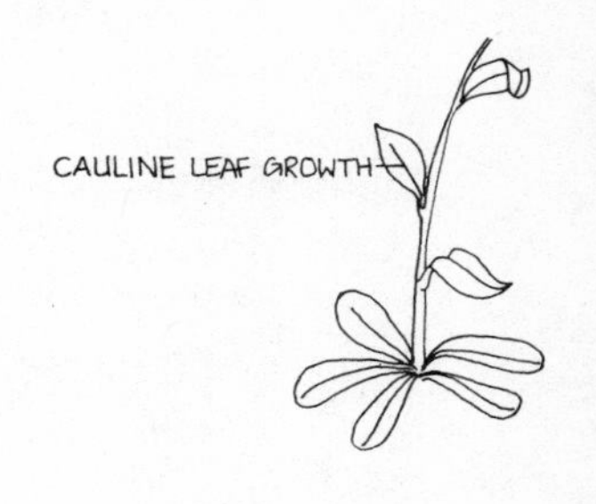

Caulescent

Having a well-developed stem above ground: *cf.* Acaulescent.

Cauline

Related to or attached to a stem or stalk; as of leaves that grow on an extended stem.

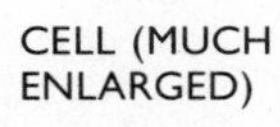

CELL (MUCH ENLARGED)

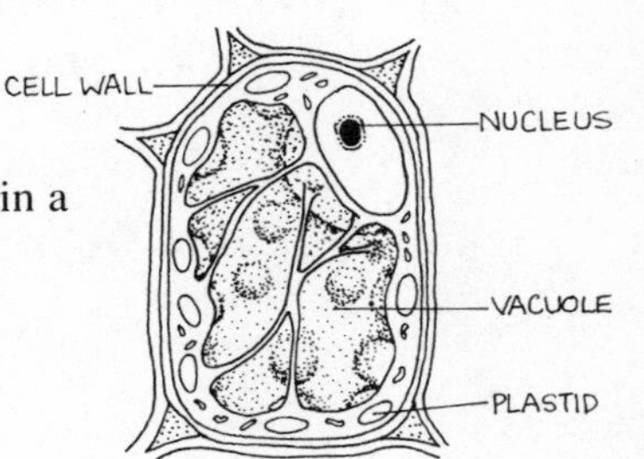

Cell

(1) Unit of structure of living matter; a nucleus within a mass of protoplasm bounded by a membrane.
(2) The cavity containing pollen in an anther lobe.
(3) One chamber of an ovary.

Cellulose

The main component of the cell-walls of plants.

Centipede

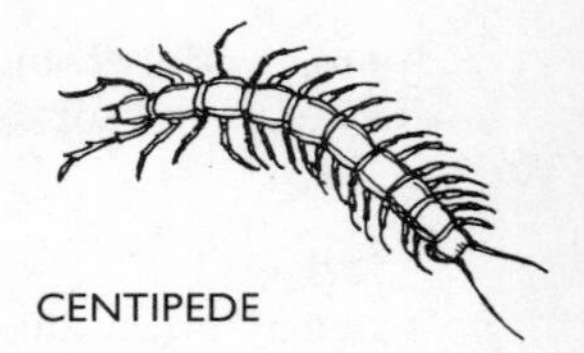
CENTIPEDE

Carnivorous flattened worm-like animal having many joints and a pair of feet to each body segment. Harmless and generally beneficial to plants because they feed on small garden pests. Light brown or golden in colour, they are not to be confused with millipedes which are black or dark grey, curl up when disturbed and may attack the roots of plants.

Central Placentation

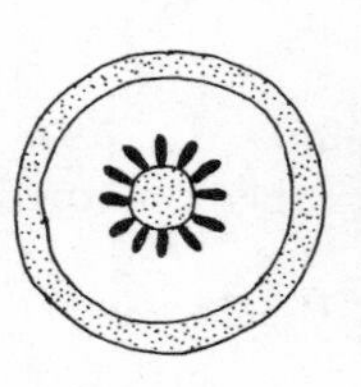
FREE CENTRAL PLACENTATION

Having the placenta centrally positioned in the ovary.

Centrifugal

Progressing from the centre outward toward the margin: *cf.* Centripetal.

Centripetal

Progressing from the margin inward toward the centre: *cf.* Centrifugal.

Cereal

Edible grain such as wheat, barley, etc.

Cernuous

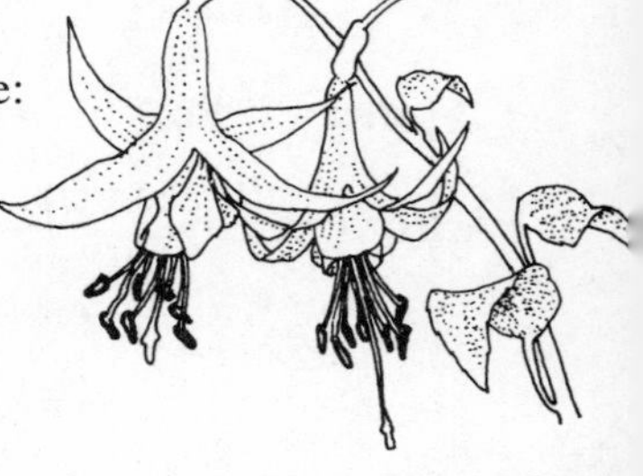
CERNUOUS FUCHSIA

Pendant or drooping, as in some varieties of fuchsia, etc.

Certified Stock

Plant material guaranteed to be free from virus infection and other diseases.

Cespitose, Caespitose

Forming dense mats or tufts.

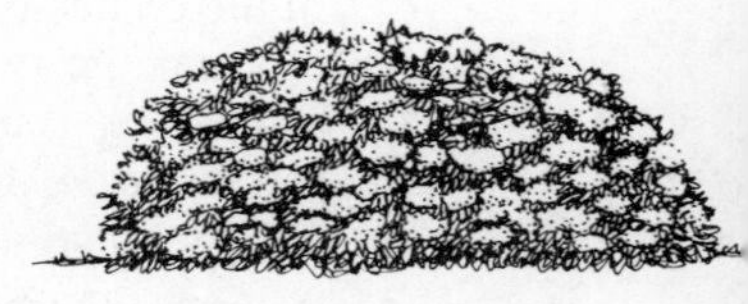
CESPITOSE (SPIREA)

Chaff

Thin dry scales or bracts.

Chaffy

In texture similar to chaff.

Chalk

White soft limestone consisting mainly of calcium carbonate. Powdered form often incorporated in soil mixtures.

Channelled

Organs hollowed in the form of guttering, as in long, narrow leaves with edges upturned, and stems deeply grooved on one side.

Chaparral

American term for dense, tangled brushwood or scrubland.

Chartaceous

Paper-like, usually crisp texture and brown or transparent.

Chasmogamous

Type of inflorescence in which pollination takes place while a flower is open or expanded: *cf.* Cleistogamous.

Chat

A small undersized potato.

Chemotactic Movements

The movement of plant cells in response to chemical stimulation.

Cheshunt Compound

A mixture of copper sulphate and ammonium carbonate used to protect seedlings from fungal attack.

Chimera, Chimaera

A plant derived from more than one genetic composition. The significant difference between a chimera and a hybrid is that in the former the two (or more) genetic constitutions occur in separate tissues or parts of the same plant; if propagated from the tissue of only one type, therefore, the chimera fails to develop.

CHIMERA
(*Sansevieria trifasciata laurentii*)

Chitting

(1) Promoting the germination of seed before sowing.
(2) Sprouting of seed potatoes prior to planting.

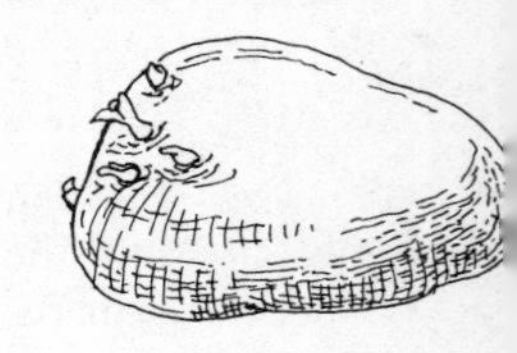
CHITTED POTATO

Chlamydospore

A type of fungal spore that is resistant to adverse environmental conditions, including desiccation.

Chlorophyll

The green colouring matter in vegetation is due to a mixture of four pigments – chlorophyll-a, chlorophyll-b, carotin and xanthophyll. Chlorophyll is not merely a pigment, it is the medium through which a plant captures energy from light to manufacture food.

Chloroplast

The green-pigment-bearing body, a chlorophyll-bearing plastid.

Chlorosis

The blanching of the green parts of plants, often due to iron deficiency.

Choripetalous

Having separated petals.

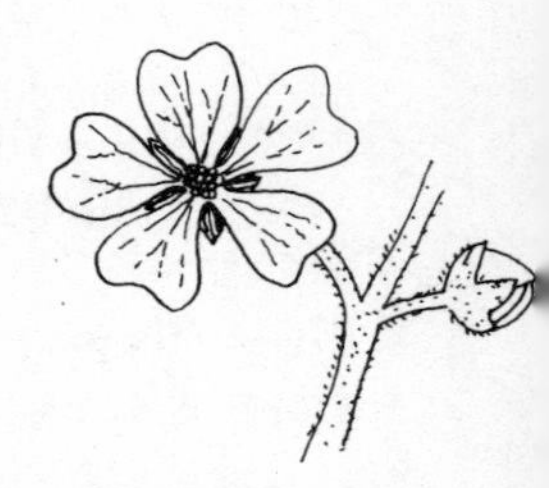
CHORIPETALOUS

Chromatin

An easily stained substance in the nucleus of a cell.

Chromatophore

A plastid within a pigment-cell in protoplasm or bacteria.

Chromoplast

A chromatophore or pigment-cell.

Chromosome

One of the microscopic bodies present in all living cells which control development and determine characteristics of the offspring both of plants and animals.

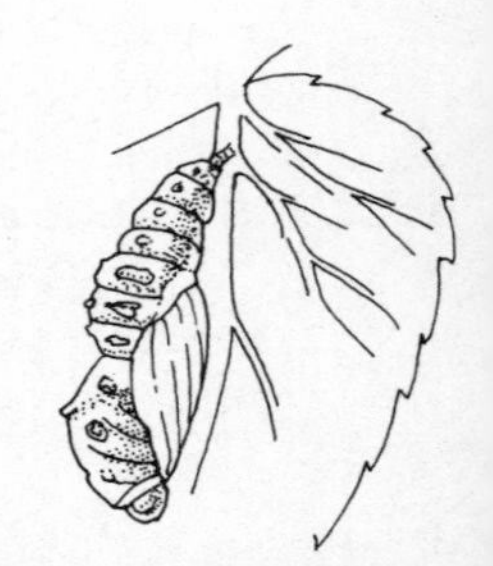
CHRYSALIS OF LARGE WHITE BUTTERFLY (*Pieris brassicae*)

Chrysalis

The inactive penultimate stage of certain insects between the larva and the adult, especially that of a butterfly. The insect within its cocoon, not the cocoon itself. *Same as* Pupa.

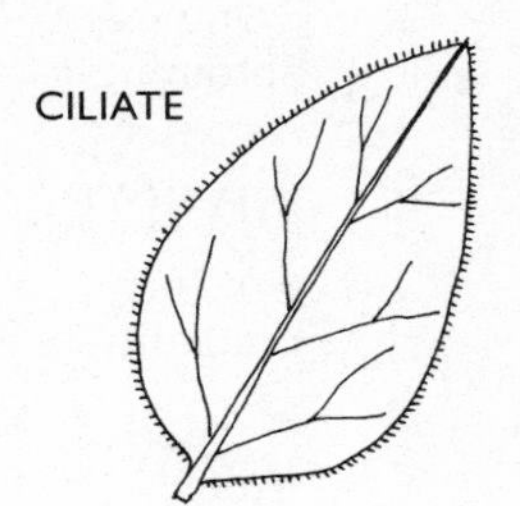
CILIATE

Cicatrix

Scar left where a leaf has fallen.

Ciliate

Fringed with cilia or hairs; usually applied, in botany, to leaves.

Ciliolate

Diminutive of ciliate – fringed with minute hairs.

Cilium, *pl.* Cilia

(1) A fine marginal leaf-hair.
(2) Whip-like filament attached to some spores and sperms to assist distribution.

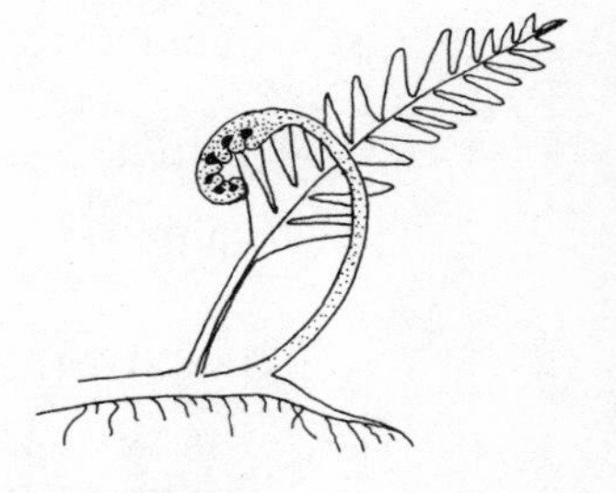
CIRCINATE

Cinereous

Powdery-grey, the colour of wood ash.

Circinate

Coiled from the top downward, as in fern fronds before they expand completely.

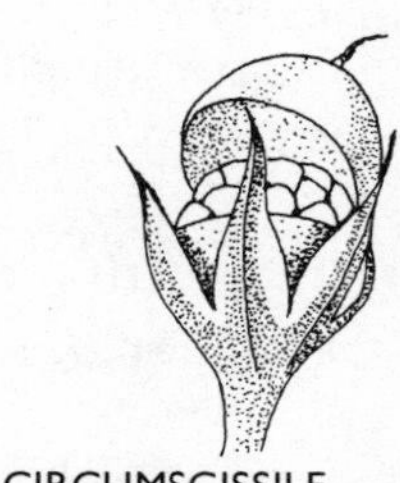
CIRCUMSCISSILE (PIMPERNEL)

Circumscissile

Opening by means of a split around the circumference so that the top comes off like a lid, as in some seed capsules. The process is also known as equatorial dehiscence.

Cirrhous

Tendril-like.

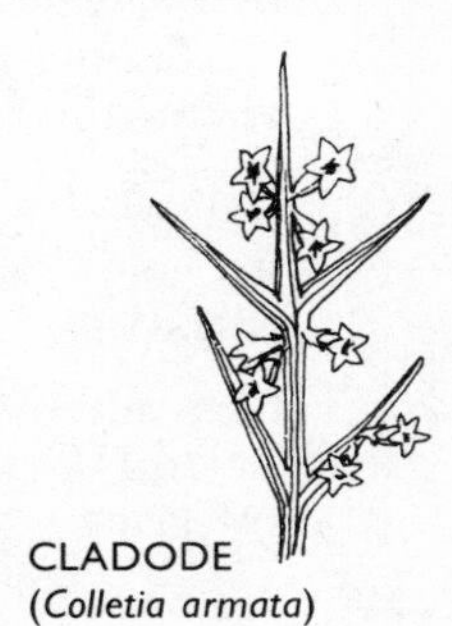
CLADODE (*Colletia armata*)

Cladode

A cladophyll or flattened stem which functions as a leaf, most often seen on plants in areas where hard, dry conditions prevail.

Cladophyll

A flattened branch or stem which looks and acts like a leaf.

Clambering

Climbing with or without the aid of tendrils or twining stems.

Clamp

A storage mound used for the preservation of harvested root vegetables such as carrots, potatoes, etc., made by piling straw and soil over the roots to keep them frost-free. Also known as a pie, pit or hog in some country districts.

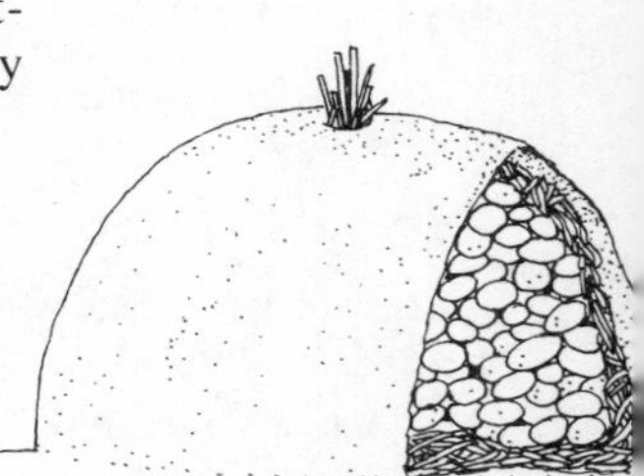

CLAMP

Clasping

See Amplexicaul.

Classification

The act of arranging all plants in groups with a name which is understood internationally; Latin is the universally accepted language for the precise specification and naming of these groups. Botanists divide and subdivide the classes until the system is quite bewildering to the ordinary gardener, but the following is a simplified explanation:–

(1) The Species is the basic unit and although it has several definitions, the one which is generally accepted in the plant world is that species can readily interbreed with one another but not with members of another species. In those rare cases when they do, the hybrid offspring will be sterile.
(2) A group of species (being closely related) forms a Genus.
(3) A group of genera forms a Family.
(4) Closely related families make an Order.
(5) Related orders form a Class, and above that category are Division (or Phylum) and Kingdom.

Vines & Rees (*Plant and Animal Biology*, Longman) give an excellent example of the classification breakdown of the common dog rose as follows:–

KINGDOM : Plantae
DIVISION : Spermatophyta
CLASS : Angiospermae

ORDER : Rosales
FAMILY : Rosaceae
GENUS : *Rosa*
SPECIES : *Rosa canina*

There are numerous subgroupings: a Tribe may be inserted between genus and family, and a Suborder between family and order. Calssification is subject to opinion, debate and consensus, with no overall authority.

Clavate

Club-shaped.

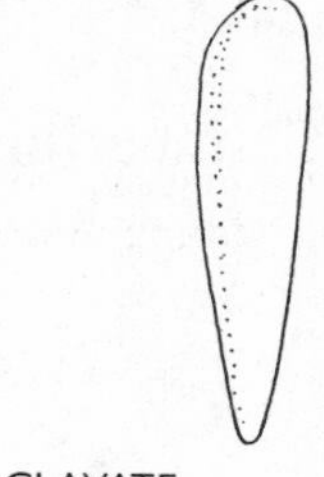
CLAVATE

Clavellate

Diminutive of clavate – shaped like a small club.

Claw

The narrow basal part of some petals and sepals, including those of the *Dianthus* family.

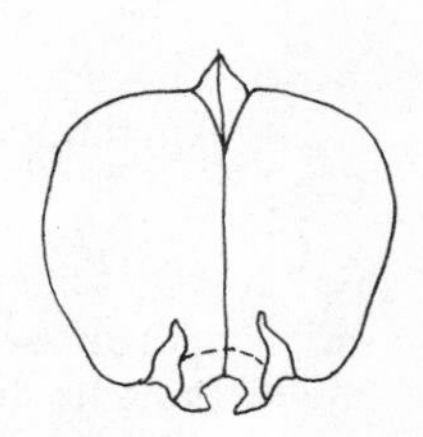
CLAW AT BASE OF PETAL

Clay

Tenacious and impervious soil substance, difficult to break down in the garden but excellent for growing certain types of plant, e.g. roses.

Clearing

An area of land cleared for cultivation, often surrounded by woodland or scrub.

Cleft

A leaf with lobes divided almost to the middle.

Cleistogamous

Type of inflorescence where self-pollination occurs in the flower while it is still closed or unexpanded: *cf.* Chasmogamous.

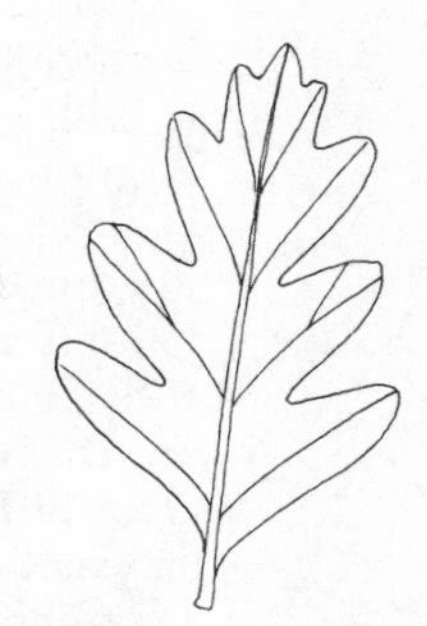
CLEFT

Climax

The final stable vegetation type which develops naturally in a given environment.

Climber

A plant that climbs by its own efforts, using other plants or objects, such as fencing or trellis, for support, twining by means of stems, suckering pads, tendrils, etc. A plant that has to be trained up a support and attached by artificial means is not a true climber but a trailer.

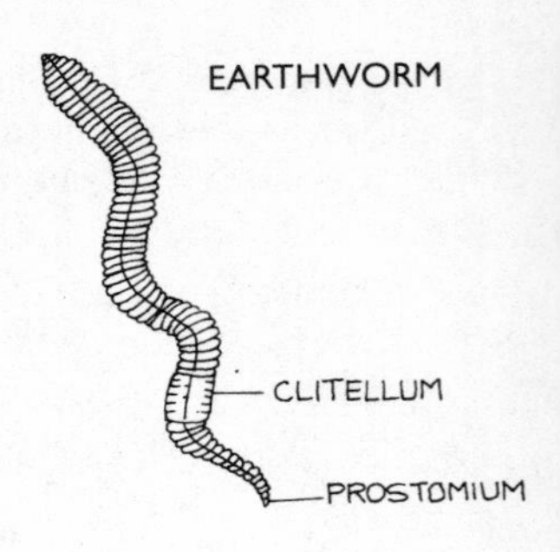

Clitellum

The glandular belt on an earthworm.

Cloche

The French word for 'bell' or 'dish-cover' was originally applied to the Victorian bell-glass but now usually refers to the 'continuous' cloche, a protective tunnel constructed of glass, polythene or similar transparent materials for raising early crops such as lettuce, radish, strawberries, etc.

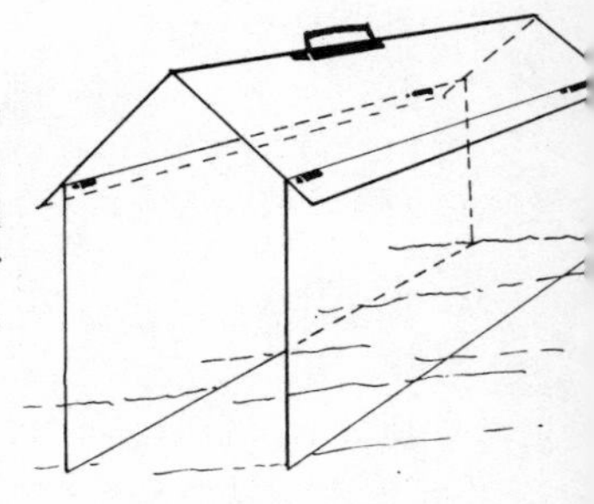

BARN CLOCHE

Clod

A lump or spadeful of soil or clay. A ploughed furrow consists of a row of clods when the ground is wet, hence the old countryman's expression 'clod-hopper', meaning a farm labourer.

Clone, Clon, Cloane

Botanical term to describe plants reproduced by vegetative means such as budding, layering, cuttings, etc. from a single parent and which therefore are identical genetically.

CLOVES (SHALLOTS)

Close

Atmospheric conditions in which high humidity is present, such as in a propagator or a sealed greenhouse on a warm day.

Clove

(1) One of the young bulbs of shallots or garlic developed as an offshoot of a larger bulb.
(2) A type of strongly scented carnation.

Club Root

A fungus disease which attacks cruciferous plants, including brassicas. The fungus lives in badly drained acid soils

CLUB ROOT IN BRASSICAS

and may persist for many years. Prevention is by dressing the soil with lime before planting, and allowing a lapse of at least three years before replanting brassicas on the same piece of ground.

Clump

(1) A group of dense shrubs or trees.
(2) A cluster of entangled roots.

Cluster Cups

Common name for burst rust-producing bodies of a fungus after the spores have been dispersed. The cups resemble tiny volcanic craters left when the spores rupture the epidermis.

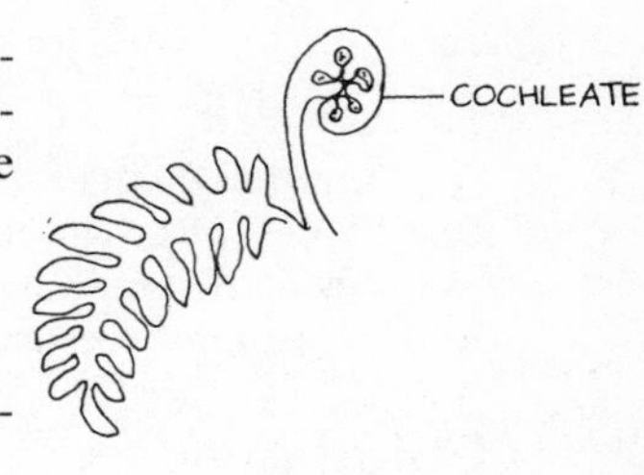

Cochleate

Coiled or twisted like a snail shell, as of a furled fern-frond.

Cock

Conical heap of hay in a field after cutting; a haycock.

Cockroach

Large beetle-like insect which can damage seedlings, young shoots, etc. in the greenhouse, and houseplants indoors. Numerous proprietary insecticides are available for the control of these pests, or they can be trapped in jars containing jam, honey or any similar sweet substance, sunk into the soil.

Coconut Fibre

Coarse material derived from the outer husk of the coconut, widely used overseas in the past for incorporating into the soil, and finely chopped, as a rooting medium. In recent years it has become popular in Britain as a substitute for peat and is available in several grades. It is similar to peat in most respects, including moisture-retention, and is just as difficult to redampen if allowed to dry out. Also known as coir.

Cocoon

A silken case or envelope produced by a caterpillar or similar insect larva to protect the pupal stage.

Codling

Immature apple (fruit).

Codling Moth

Moth whose larvae, commonly called maggots, feed on apples.

Coir

See Coconut fibre.

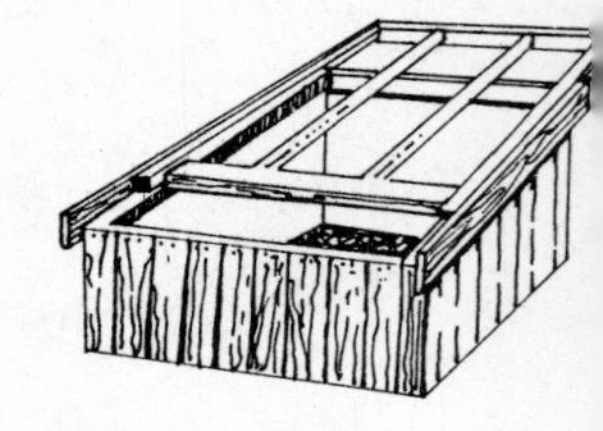
COLD FRAME

Colchicine

An extremely poisonous alkaloid (said to be more toxic than strychnine) which can be used in minute concentrations in plant-breeding experiments to produce offspring with more than the usual two sets of chromosomes.

Cold Frame

A glazed garden frame providing protection but not heated; often used for 'hardening-off' tender subjects before planting out.

Cold House

A greenhouse, unheated except by natural sunlight.

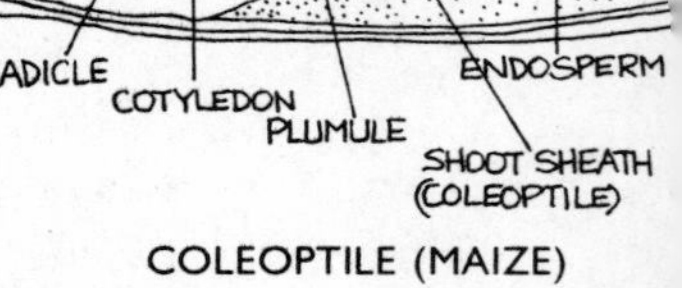

COLEOPTILE (MAIZE)

Cole

Cabbage, especially the smaller varieties.

Coleoptile

Hollow organ produced by cereal grains as they germinate; the sheath enclosing the plumule.

Coleorhiza, Coleorrhiza, *pl.* Coleorrhizae

Protective coat or sheath on the radicle of some plants.

Collar

(1) The junction of root and stem in a plant.
(2) A protective band placed around plants to protect against damage by slugs, etc, or around the stems of young trees to discourage rabbits, etc.

COLEOPTILE
BASE OF PLUM
COLEORHIZA
RADICLE

COLEORRHIZA (MAIZE)

Collar Rot

The rotting of plants and young seedlings at soil level, caused by soil-borne fungi. Stem rot, foot rot, blackleg and damping-off are a few of the names applied to this problem. There is no cure but some control can be obtained by the timely application of a fungicide.

Collenchyma

The strengthening tissue of cell walls developed mainly in young shoots and leaves.

Colour Break

A 'sport' or mutation caused by a spontaneous accidental change in the chromosomes or genes of a plant, usually affecting the colour or colour-pattern of the flower.

Colpate

Of pollen having one or more elliptic apertures in the pollen-wall.

Column

The structure formed by the combination of the style and stamen in orchids.

Coma

(1) A tuft of bracts or leaves at the apex of an inflorescence, such as in pineapple.
(2) A tuft of soft hairs or down on a seed.
(3) The crown or head of some palms.

COMA (PINEAPPLE)

Commensalism

See Symbiosis.

Comose

Any plant carrying a coma.

Companionate

Of plants which are grouped together and which benefit one another either by enhancing growth or more often by combating pests. French marigolds planted close to regal pelargoniums will keep the latter free from whitefly or at least reduce the infestation, and carrots sown between rows of onions are usually free from carrot fly.

Compatible

Capable of free cross-fertilization. Fruit trees that can pollinate each other are termed compatible and varieties are selected with this in mind. Grafts also have to be compatible or the scion will not take.

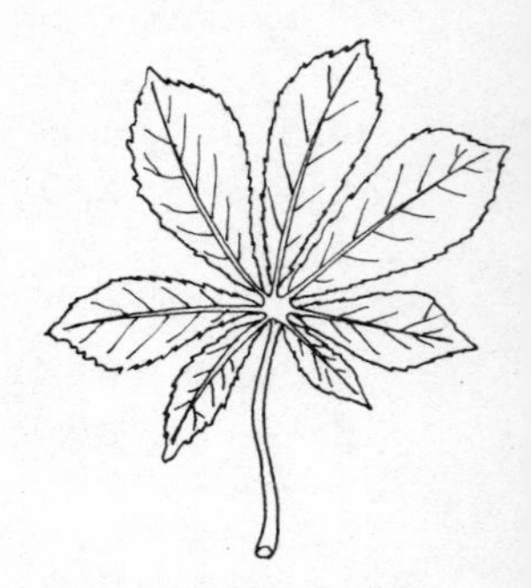

COMPOUND LEAF (HORSE-CHESTNUT)

Compost

(1) Decayed vegetable matter used as an organic manure.
(2) In Britain, a mixture of loam, peat and sand plus plant nutrients, used for the culture of pot plants (rarely used in this sense in USA or Australasia).
(3) Mushroom compost is a mixture of peat, chemicals, etc. prepared specially for raising mushrooms: *see also* John Innes composts and Soilless composts.

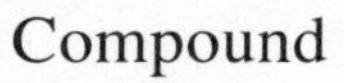

Compound

Of leaves, consisting or two or more similar parts; divided into two or more leaflets. The leaves of the horse-chestnut, for example, are palmately compound.

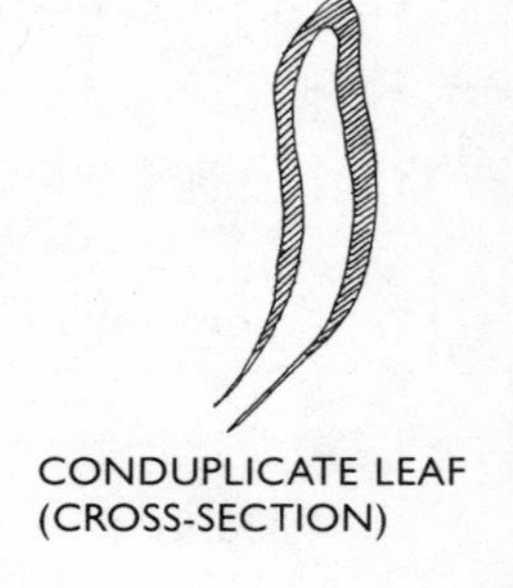

CONDUPLICATE LEAF (CROSS-SECTION)

Compressed

Flattened or narrowed laterally.

Conceit

See Folly

Concolorous

Of a uniform single colour.

Conduplicate

Folded once lengthwise, as in the leaves of many orchids, and certain seeds, including those of brassicas.

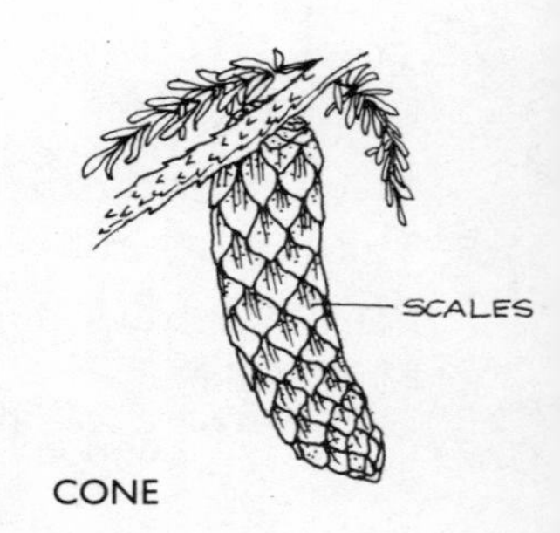

CONE

Cone

A cluster of scales, usually dense and elongated; the flower and fruit of some coniferous trees such as pines and spruces. The familiar hard cones are the females, the scales of which, in most cases, open to release the seeds; but in others the seeds are retained and rely upon the cone rotting or other methods of dispersal.

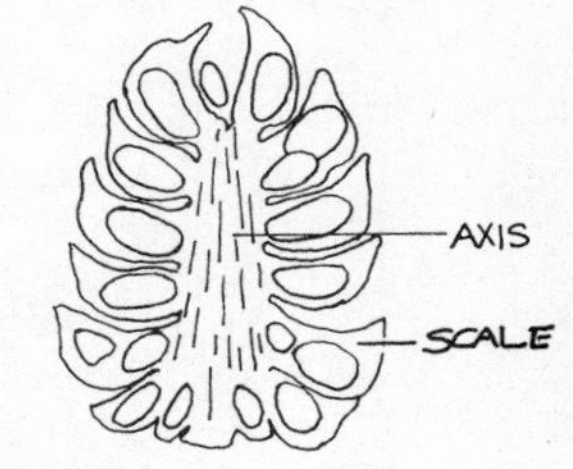

CONE

Cone Axis

Central core of a cone to which the scales and seeds are attached.

Cone Scale

The woody, flaky structures that radiate from the cone axis enclosing the seeds.

CONIFER (NORWAY SPRUCE)

Confluent

Merging together.

Congeneric

Of the same genus.

Congested

Overcrowded; packed closely together; dense growth, tightly formed: *cf.* Lax.

Conidium, *pl.* Conidia

An asexually produced fungal spore.

Conifer

Usually a cone-bearing tree such as spruce and pine, but sometimes descriptive of others of the Coniferae group which does not produce cones, e.g. the ginkgo.

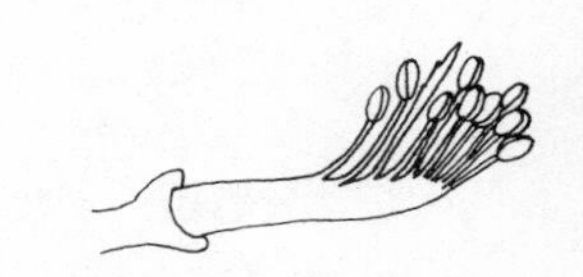
CONNATE STAMENS

Coniferous

Bearing cones, as applied to a conifer.

Connate

Joined or attached, e.g. stamens fused into a tube.

Connate Perfoliate

Joined at the base, as of pairs of leaves.

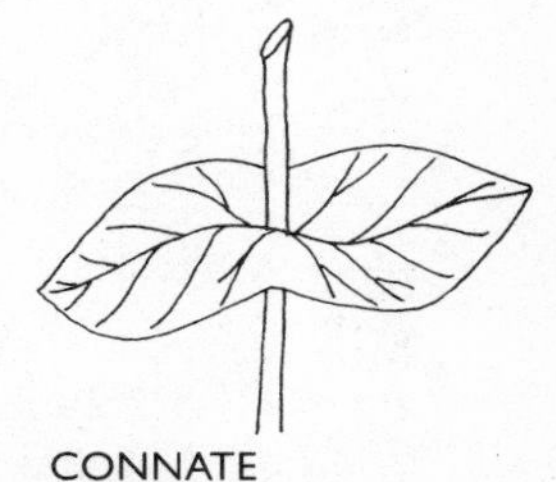
CONNATE PERFOLIATE

Connective

The tissue connecting an anther with the pollen sacs in stamens.

Connivent

Touching but not joining.

Conoid

Cone-shaped.

Conservatory

A term first used in the seventeenth century to describe a greenhouse but now denoting a garden room, usually attached to the house, equipped with table and chairs in addition to plants.

Consortism

See Symbiosis.

Conspecific

Of the same species.

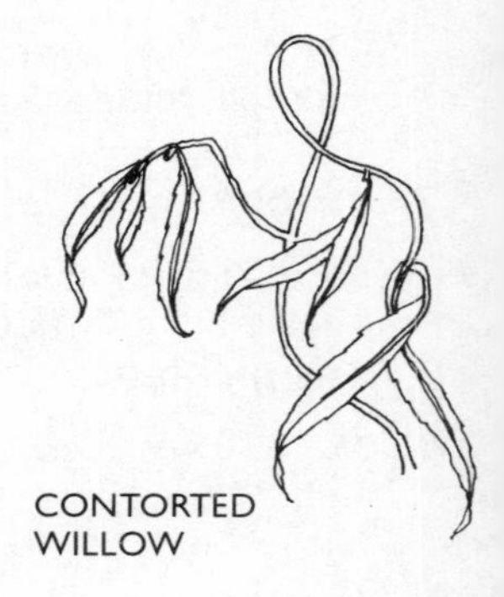

CONTORTED WILLOW

Contact

Characteristic of a type of pesticide which kills by contact, as distinct from a systemic pesticide.

Container Plant

A plant that is grown and sold in a pot or similar container and which can be planted out with the minimum of root disturbance.

CONTRACTILE ROOTS ON THE CORM OF WILD ARUM

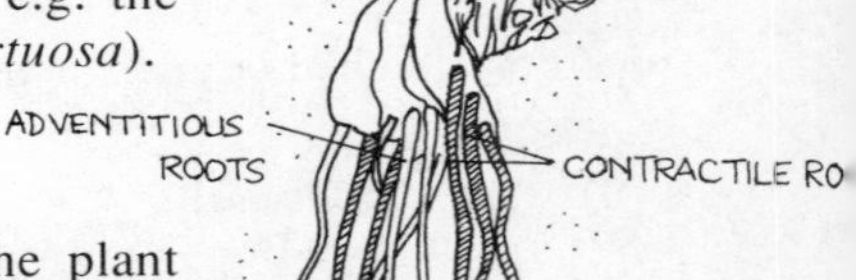

Contorted

Spirally twisted, usually in the bud; sometimes said of sepals and petals but also applicable to stems, e.g. the contorted or twisted willow (*Salix matsudana tortuosa*).

Contractile

Shrinking in length, as of roots which draw the plant further down into the soil, e.g. crocus. Each year the new corm grows on top of the old, hence the need to pull the plant down to compensate.

Controlled Release

Characteristic of a type of fertilizer which releases its nutrients slowly over a period of time.

Convariety

Old term now replaced by 'group'.

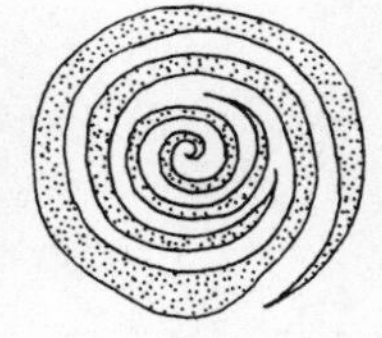

CONVOLUTE CROSS-SECTION

Convolute

Rolled up lengthwise, coiled or twisted laterally, usually referring to petals and leaves.

Coppice

Small wooded area, mainly of small trees, grown for periodical cutting. *Same as* Copse.

Coppicing

The cutting of trees back to ground level in order to promote the growth of several new stems from one root-stock.

CORDATE

Copse

See Coppice.

Cordate

Heart-shaped – usually applied to leaves.

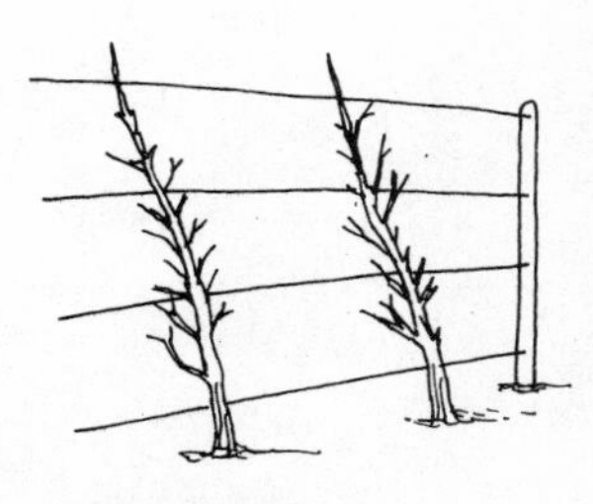
CORDON TREES

Cordon

Usually describes a fruit tree which has been trained to a single stem, such as cordon apples, but it can also apply to other plants such as sweet peas, etc.: *see also* U cordon.

Core

Horny capsule surrounded by pulp and containing seeds, as in apple, etc.

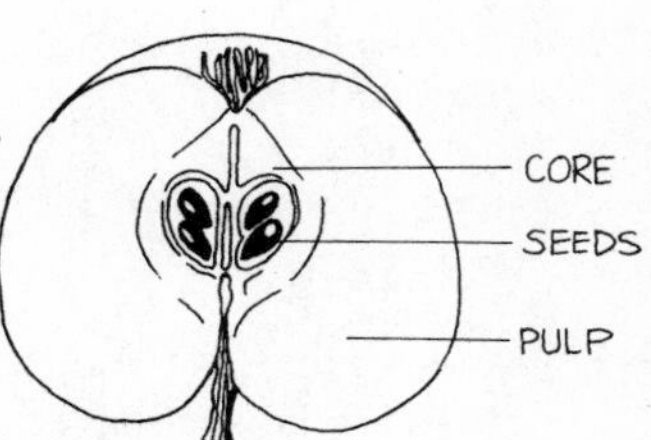

CORE OF AN APPLE

Coriaceous

Leathery, tough, as the leaves of laurel.

Cork

(1) The bark of the cork-oak (*Quercus suber*).
(2) The tissue forming the outer skin of the bark in higher plants, consisting of near-waterproof, close-packed, air-containing cells.

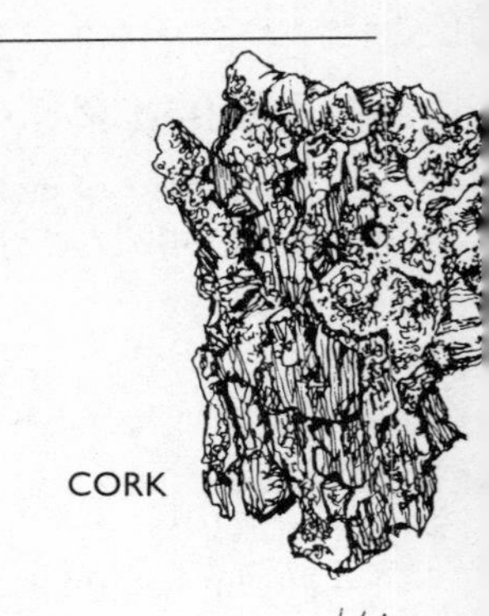

Corm

The swollen, solid, usually subterranean base of a stem, without scales, e.g. gladiolus and crocus. Tiny corms which are produced around the base of the corm are known as cormels or cormlets.

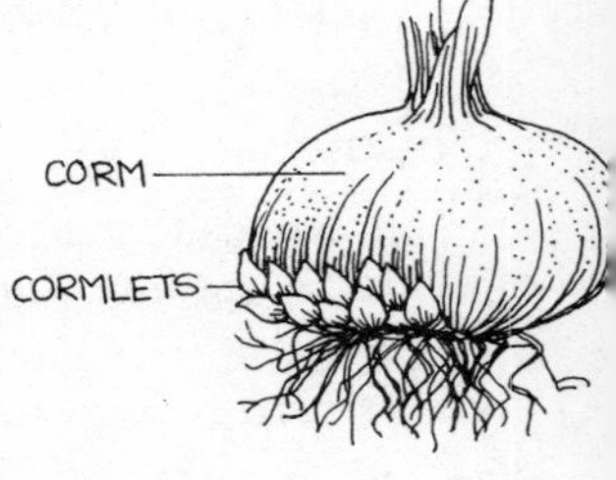

Corniculate

Bearing a small horn-like attachment.

Corolla

A general term applied usually to the inner whorl of a flower or floral leaves as distinct from the sepals. It may be of several petals or sometimes in one piece.

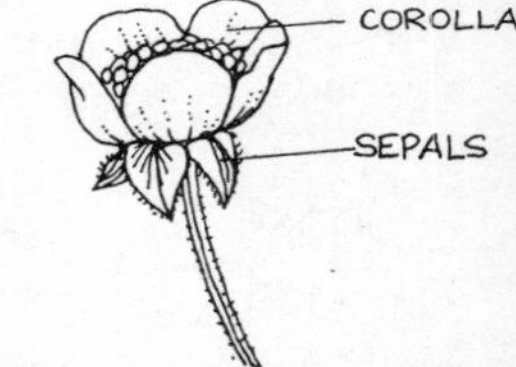

COROLLA OF A BUTTERCU

Corona

A crown. In a flower, the outgrowth of the perianth, e.g. the trumpet of a narcissus.

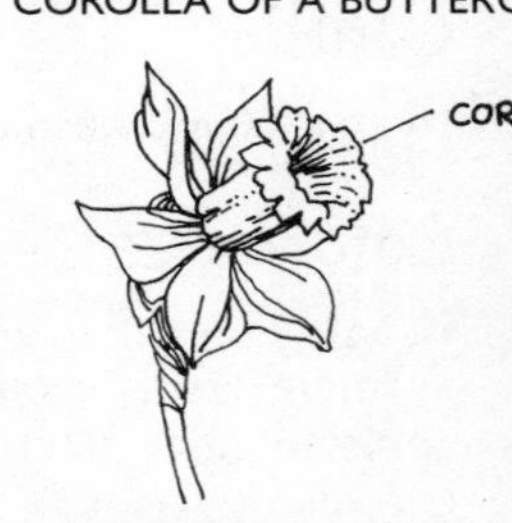

Coronate

Having a corona.

Cortex

The inner bark of a tree.

Cortical Cells

Those cells contained in the cortex.

Corticate

Having bark.

Corymb

A more or less flat-topped flower cluster in which the flower stalks emanate from different parts of the main stem as distinct from an umbel where they radiate from a single point. The inner stalks are shorter than the outer ones, thus producing the flattish type of head.

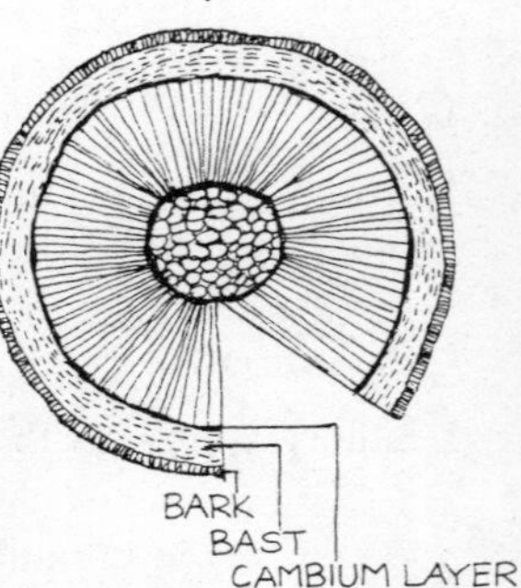

CORTEX

Corymbose

Corymb-like, arranged in a corymb.

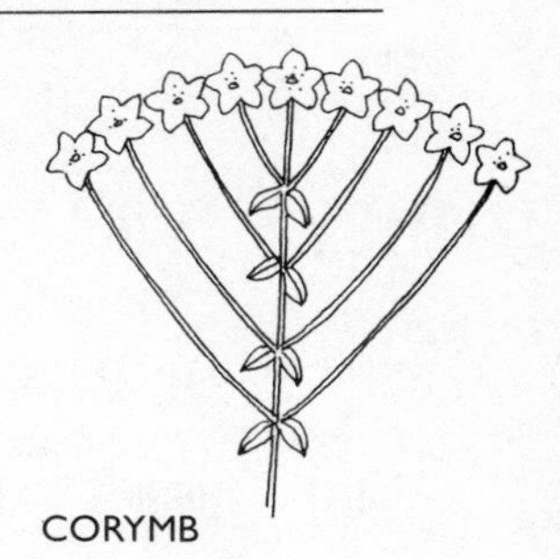
CORYMB

Costa

The central vein or rib of a leaf.

Costate

(1) Ribbed or having ribs.
(2) With the central rib on a leaf well pronounced.

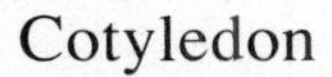

Cotyledon

A primary leaf or seed leaf. Quite different from the true leaves which develop later, they are usually dicotyledons (two seed leaves) or monocotyledons (single seed leaves); gymnosperm seedlings, such as pine, may have several cotyledons. In some plants, such as peas, the cotyledons remain underground.

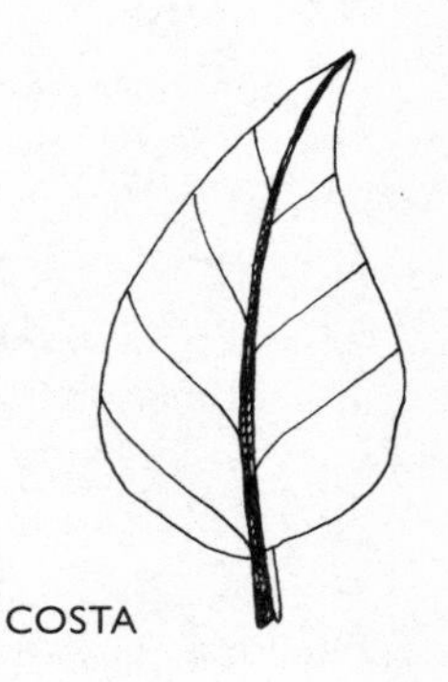
COSTA

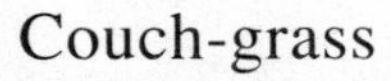

Couch-grass

See Twitch.

Covert

Small man-made thicket originally created as a shelter for game.

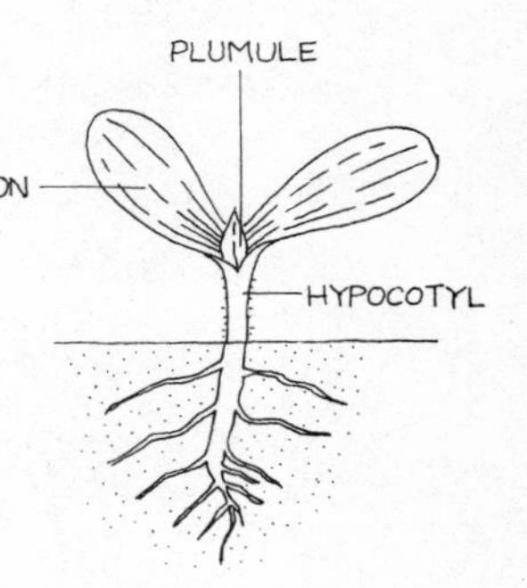

Creeper

Plant that creeps along the ground or, more rarely, up a support of its own accord.

Creeping

Running or trailing over or under the ground and rooting at intervals, as in strawberry, creeping buttercup, etc.

Crenate

Having margins scalloped, with obtuse or rounded teeth, as on some leaves.

Crenulate

Having margins with very small rounded teeth, as on some leaves; diminutive of crenate.

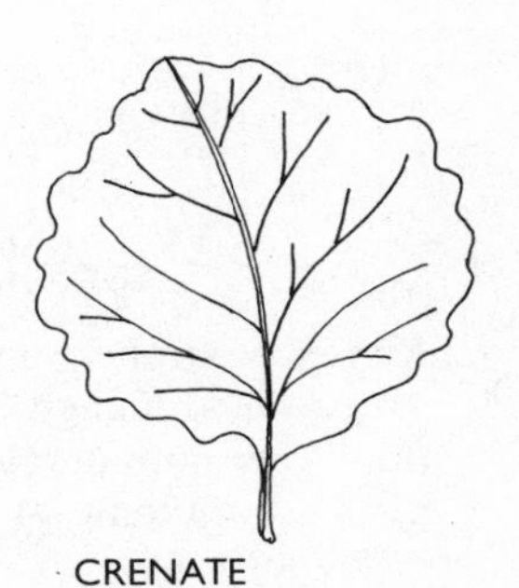
CRENATE

Crested

Having upright fan-like toothed ridges, as on some fern fronds.

CRESTED FERN
(*Asplenium 'cristatum'*)

Cribriform

Perforated with many tiny holes, as on cabbage leaves after an attack by flea beetles.

Crispate

Having curled or ruffled margins, as in leaf of curly kale, but can apply equally to leaves or flowers.

CRISPATE

Crisped

Minutely wavy-edged or closely curled, usually applied to leaf margins but also to some petals.

Cristate

Crested or cockscomb-shaped; a form of fasciation often encouraged, particularly on cacti and ferns.

Croceate

Saffron-coloured, as from the dried stigmas of the crocus.

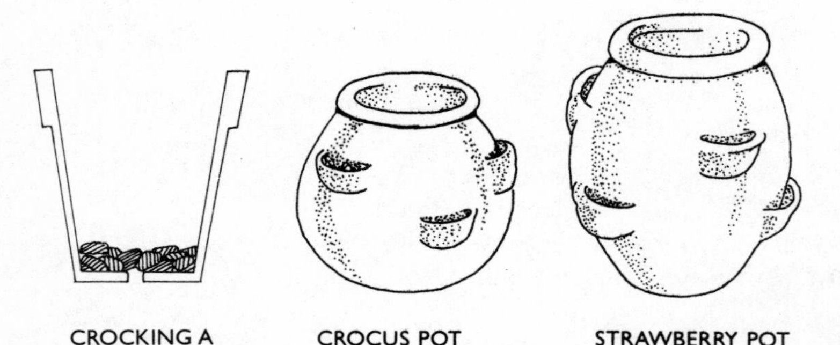
CROCKING A CLAY POT　CROCUS POT　STRAWBERRY POT

Crocks, Crocking

Pieces of broken pot, stones, etc. formerly used to improve the drainage of clay pots. The method was abandoned by commercial growers upon the introduction of plastic pots some thirty years ago but is still practised by a few amateurs.

A CRISTATE VARIETY OF CACTUS

Crocus Pot

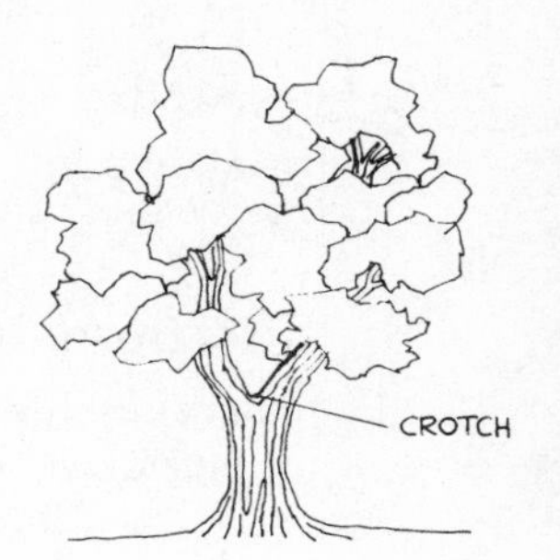

A small fancy container having several apertures in the walls in which crocus bulbs are planted. The pots can also be used for other small plants such as aubretia, etc. Larger sizes are available with more holes in the sides for planting with strawberries.

Crop Rotation

A system by which a crop is not grown in the same place in successive years. A three-year rotation is the norm but two-, four- or more may be operated according to the condition of the soil and the type of crop to be planted. This method improves the fertility of the soil and, to some extent, controls the population of the soil-pests.

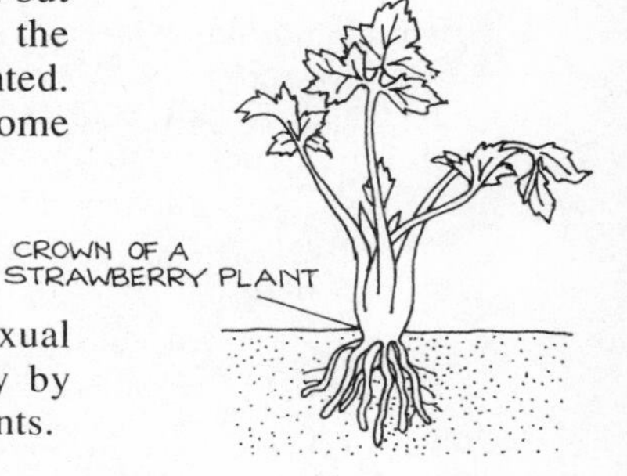

Cross

The progeny of two distinct parents, the result of sexual pollination. Fertilization can be performed manually by human agency, by insects, or sometimes by air currents.

Cross-fertilize, Cross-pollinate

To transfer pollen between different plants of the same species.

Crotch

The fork of a tree where the bole divides to produce branches.

Crown

(1) The entire rootstock, as in rhubarb.
(2) The upper part of a rootstock from which shoots arise, dying back in autumn as in peonies, etc.
(3) The base of a plant where stem and root join.
(4) Part of a rhizome with a large bud, used for propagating.
(5) The corona of a flower, as in narcissi.
(6) The first chrysanthemum buds to appear after initial stopping.

Crown Thinning

Removing the inward-growing branches of a tree to reduce overcrowding and to enable more light and air to penetrate.

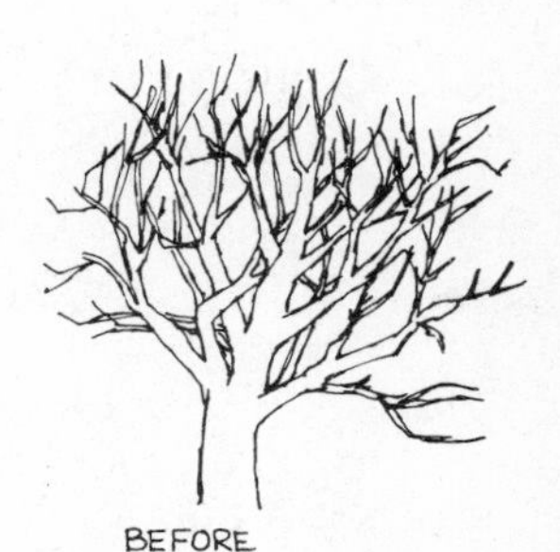

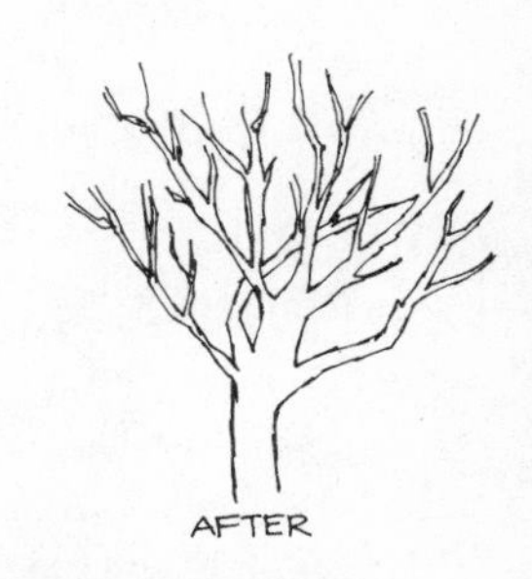

CROWN THINNING

Cruciate

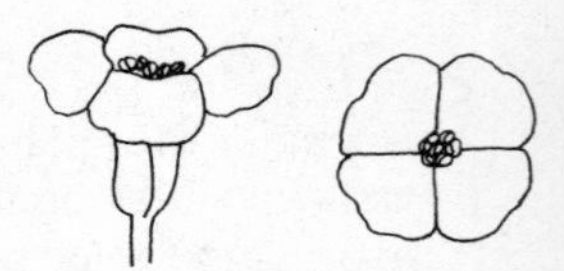
CRUCIATE PETALS OF A WALLFLOWER

Shaped like a cross, as the petals of watercress, wallflower, etc.

Cruciferous

Relating to a member of the mustard family Cruciferae. The name derives from the cross-like arrangement of the four petals. Stocks, alyssum and the brassicas are all cruciferous.

Crumb

The ideal condition of the soil when it crumbles in the hand, not too dry and not sticky. Breadcrumbs would be a reasonable comparison.

Crustaceous

Brittle, hard, gritty.

Cryptogam

A plant that reproduces by spores instead of seeds as in fungi, ferns, etc.; it has no stamens or pistils and therefore no real flowers.

Crytogamia

The class of cryptogams which includes algae, ferns, mosses, fungi, etc. An outdated term since these plants are so diverse that they are no longer grouped together.

Cuckoo-spit

The froth which surrounds the nymphs of the froghopper, a sap-sucking insect which attacks many types of plant. In appearance the nymphs resemble rather bloated greenfly. If sprayed with soapy water the froth will be washed away and the exposed insect will perish.

CUCULLATE
(*Aconitum anglicum*)

Cucullate

Hooded or shaped like a hood.

Cucurbits

Plants of the family Cucurbitaceae such as melon and cucumber.

Culm

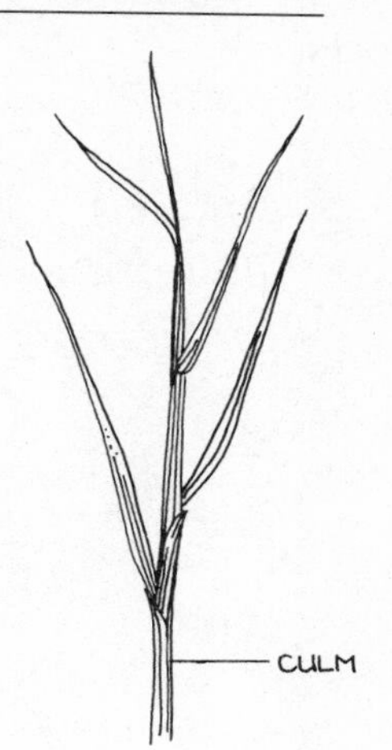

The stem of a plant, usually hollow except at the nodes, such as the stems of grasses, e.g. bamboos. The term is also applied to sedges, although their stems are not normally hollow.

Cultivar

A hybrid name derived from CULTIvated VARiety and applied to a plant that has originated and persisted under cultivation (as distinct from a species). They are usually distinguished typographically by being set in roman, not italic, type and contained within single quotation marks as in *Pelargonium domesticum* 'Grand Slam'.

Cultivate

(1) To break up the soil with or without a mechanical implement.
(2) To till the ground with a mechanical cultivator.
(3) To tend carefully, to nurture (plants).

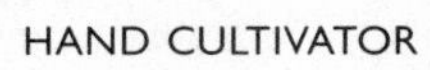

HAND CULTIVATOR

Culture

Correctly called tissue-culture, the artificial development of cells, etc. in specially prepared media under laboratory conditions.

Cuneate

Wedge-shaped. Can apply to both leaves and petals.

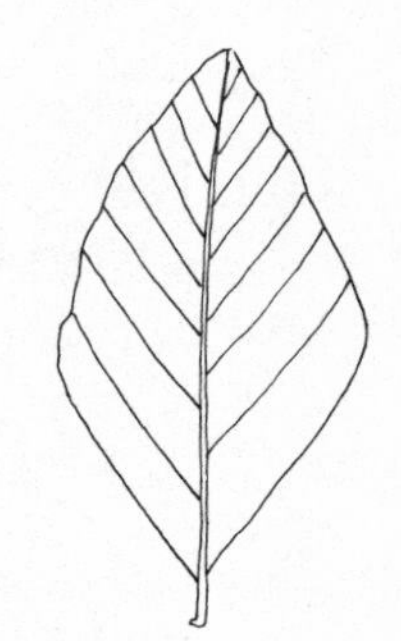

CUNEATE

Cup

(1) The central corona of narcissi when it is shorter than the outer perianth segments; when the corona is longer than the perianth segments, it is a trumpet.
(2) One of various types of fungi.
(3) The receptacle which holds the acorn, the fruit of the oak tree – its botanical name is cupule.

Cup-and-Saucer

See Hose-in-hose.

Cupulate

Having a cup-shaped sheath enclosing or partially surrounding a fruit, as in the acorn.

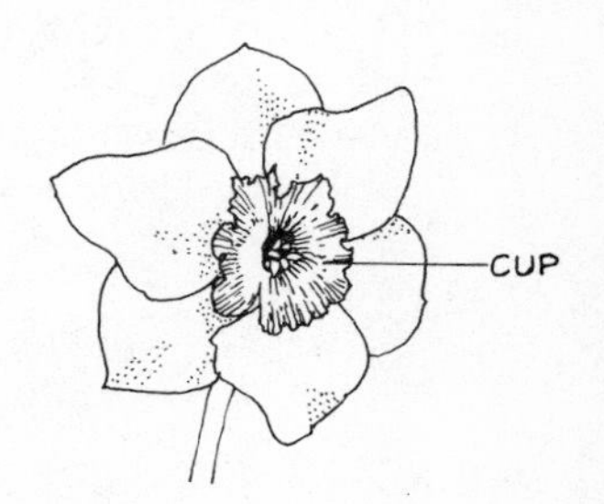

Cupule

See Cup.

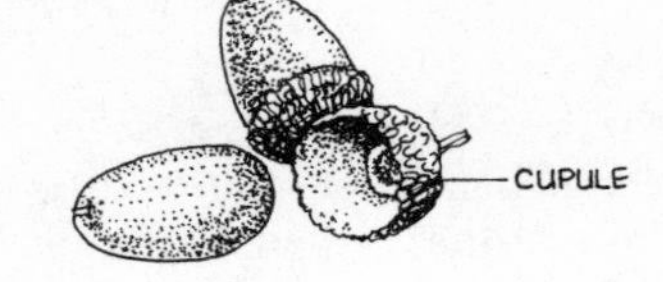

Cupuliform

Cup-shaped.

Curd

Dense cluster of immature flowerbuds that combine to form the head of cauliflowers and broccoli.

CURD

Cushion

Usually a small type of plant with dense, close-packed flowers or foliage, forming a dome-shaped pad.

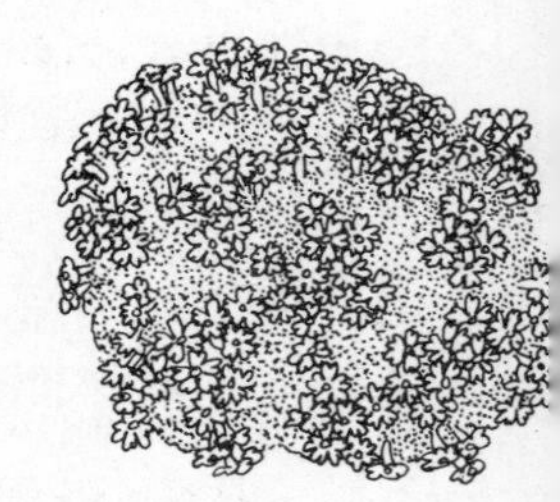

CUSHION PLANT
(*Dionysia curviflora*)

Cusp

A sharp point formed from the extended margin of a leaf: the thorns on holly leaves are cusps.

Cuspidate

Having a cusp or thorn at its apex.

Cuticle

The waxy surface of a leaf, particularly noticeable in laurel, etc. It is a protection against insect damage and also reduces water loss.

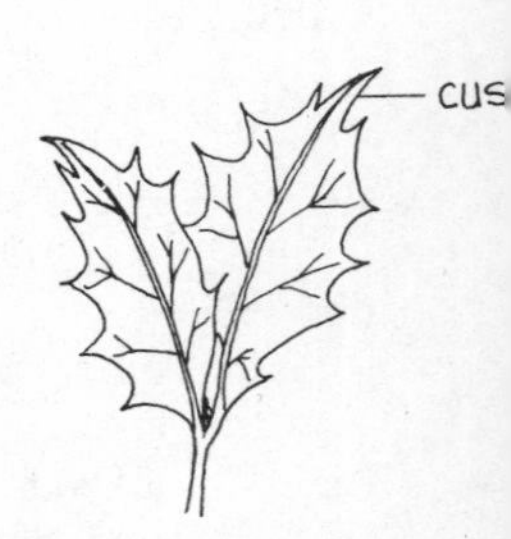

Cutin

The material that forms the plant cuticle; the waxy or corky layer on the epidermis.

Cutting

A portion of stem, root or leaf removed from the parent plant and encouraged to develop into a new plant. There are several types of cutting in general use including hardwood, semi-hardwood, softwood and herbaceous. Hardwood and semi-hardwood cuttings root best under mist, but most herbaceous cuttings will root in peat, sand, soil or water. Begonias and saintpaulias, etc. can be propagated by leaf cuttings, and other species are best propagated by root cuttings, e.g. horse-radish, hollyhocks, etc.

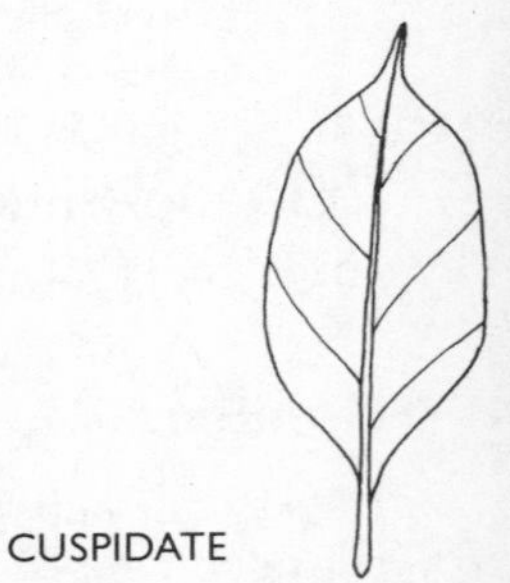

CUSPIDATE

Cyclic

A circular arrangement of petals, flowers or leaves.

SOFTWOOD CUTTING

Cyclic Bearing

A term used to describe crops that produce at longer than annual intervals, from two to five years or longer. It applies, for example, to the biennial cropping of certain fruit trees.

Cylindrical, Cylindric

Circular and hollow in cross-section, and elongated.

MOST MEMBERS OF THE DAISY FAMILY ARE CYCLIC

Cymbiform

Boat-shaped, applicable to the flowers of some orchids.

Cyme

Cymes have many forms but the term broadly covers a more or less flat-topped or dome-shaped flowerhead where the growing point ends in a flower and further flowers occur on a succession of side-shoots. The central flower usually opens before the others.

Cymose

Cyme-like or having flowers carried in cymes.

LEAF CUTTINGS (BEGONIA)

Cymule

A small cyme.

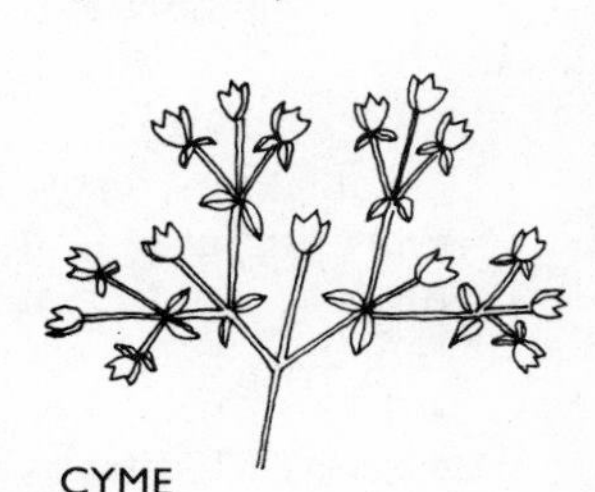

CYME

Cypsela

A single-seeded fruit produced by a unilocular, inferior ovary, e.g. dandelion.

Cyst

Hollow organ in a plant containing a liquid secretion.

Cystolith

A limy concretion or crystal found in some plant cells.

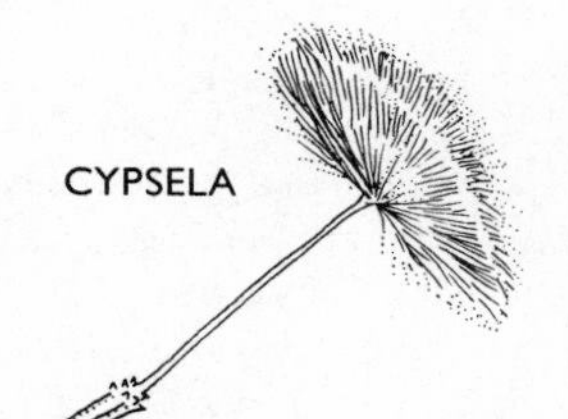

CYPSELA

D

Daddock

The heart of a rotten tree.

Daisy Grubber

Small hand tool having a narrow blade with a V-shaped notch at the end and cranked in the middle to provide leverage. Originally used for grubbing-out daisies in lawns but now more widely used for docks, dandelions, etc.

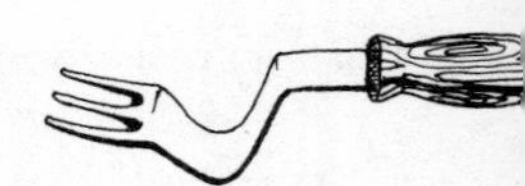

VICTORIAN
DAISY GRUBBER

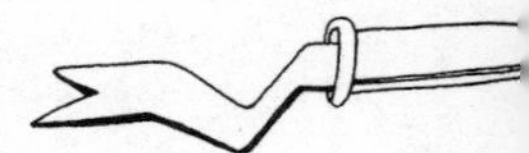

MODERN
DAISY GRUBBER

Damping Down

The spraying of paths and benches in a greenhouse with clean water in hot weather to reduce the temperature and improve the relative humidity of the atmosphere.

Damping Off

The collapse of seedlings at or just below soil level, usually caused by bad handling when pricking out, over-crowding or poor drainage – conditions which encourage botrytis and fungal attack. The term covers a variety of fungal diseases and some control can be obtained by using only sterilized compost and watering with a fungicide such as Cheshunt Compound but it is best prevented by good growing practice.

Damping Overhead

Spraying plants in warm weather to increase the humidity and, in the case of some crops, e.g. tomatoes, to reduce excessive transpiration. Best performed in the morning to give the foliage time to dry off before nightfall, or botrytis may be encouraged.

Dandelion Clock

The seed head of the dandelion flower from which children once professed to be able to tell the time.

DANDELION CLOCK

Dappled

Variegated, blotched, marked unevenly.

Dard

A lateral shoot, not more than 3 in (7.5 cm) long, on a pear or apple tree, carrying a fruit bud at the tip. The variety Bramley's Seedling is noted for its proliferation of dards.

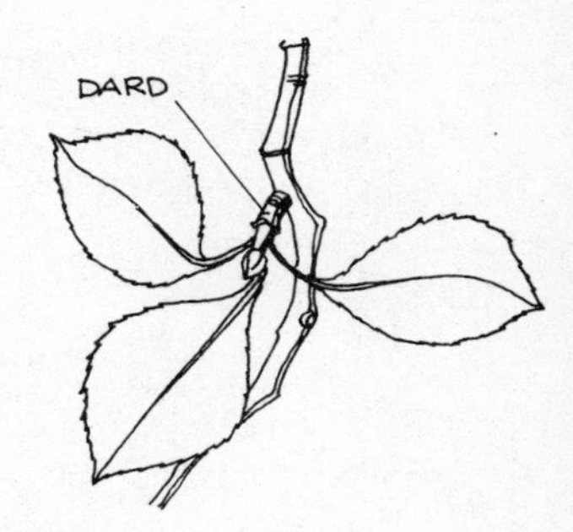

Day-length

The length of daylight hours has a controlling effect upon the initiation of the flowerbuds of many plants. Those which flower when days are long are called 'long-day plants', and those flowering when daylight is shorter are termed 'short-day plants'. 'Day-neutral' plants are not affected by the day-length and are called photoperiodismic. Both long-day and short-day plants are encouraged commercially to flower outside their normal season by manipulating the illumination of the subjects to produce artificial lengthening or shortening of the daylight hours.

Day-neutral

See Day-length and Photoperiodismic.

Dead-heading

The removal of dead flowerheads from plants, essential if further flowers are desired. Dead-heading prevents the development of the seed heads, quite apart from keeping the garden tidy, for if seed is allowed to develop, the production of further flowers will cease or certainly be reduced; it is especially important on annuals which normally die after producing seed.

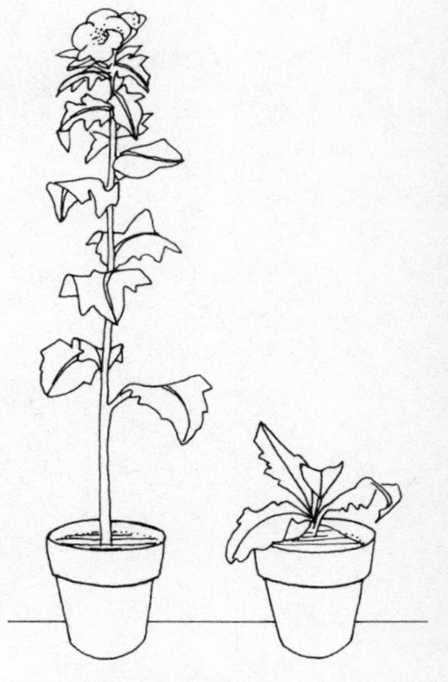

LONG-DAY PLANT (HENBANE)
LEFT: GROWN IN FULL DAYLIGHT
RIGHT: DAYLIGHT RESTRICTED TO 8 HOURS IN EACH 24

Deblossoming

The removal, either partially or totally, of blossom from top fruit trees. Total deblossoming is usually applied to fruit trees in their first season after planting to enable the tree to establish itself before fruit production begins; if a tree is allowed to fruit immediately after it is planted, its growth will be inhibited and it will be several years before it recovers and gets into full production. Partial deblossoming is practised when there is an excessive amount of blossom, which, if allowed to develop into fruit, would seriously strain the tree, with a possible risk of the tree developing the habit of biannual bearing.

Decalcify

Deprive of lime or calcareous matter.

Deciduous

Applied mainly to trees which shed their leaves in winter, although in some cases, e.g. young beech, the leaves are retained (known as narcescent): *cf.* Evergreen.

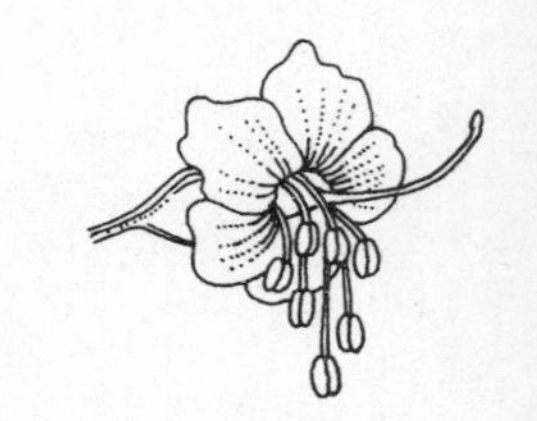

UNRIPE STAMENS OF THE HORSE-CHESTNUT ARE DECLINATE

Declinate

Curved forward or downward, as of stamens.

Decompose

Break up into constituent elements; rot; disintegrate.

Decorative

(1) A class of dahlia with broad, flattish petals.
(2) Groups of chrysanthemums classed as reflexed decorative, incurve decorative and intermediate decorative.

DECORATIVE DAHLIA DECORATIVE CHRYSANTHEMUMS

Decumbent

Lying on the ground but tending to turn up at the growing tip, applicable to stems, as of some mints.

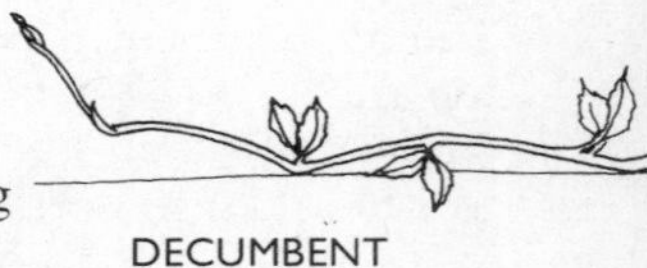

DECUMBENT

Decurrent

Extending down and along, as of a leaf, attached to the stem for almost its entire length.

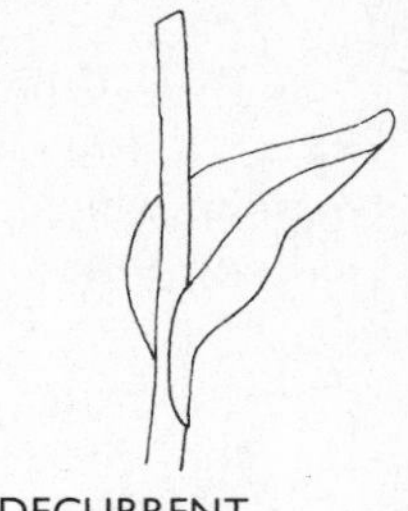

DECURRENT

Decussate

Arranged in pairs on opposite sides of the stem, as in the leaves of carnations; x-shaped when viewed from above.

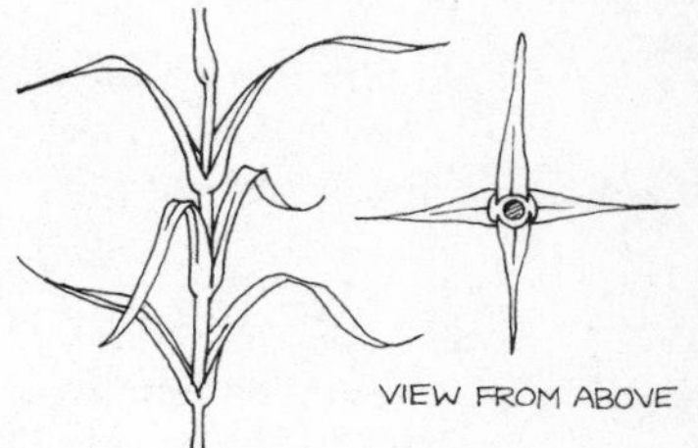

DECUSSATE

Deficient

Lacking all or some of the essential elements in the soil, the results of which may often be seen in the plants. It can apply to the major minerals or to the trace elements, but one of the main causes of deficiency is an excess of lime which can lock up certain minerals such as iron and cause chlorosis (yellowing) of the foliage.

Deflexed

Bent sharply downward or backward. Branches of several species of willow are deflexed.

Defloration

The act of stripping off flowers or removing flowerheads: usually applied to the gathering of healthy blooms as opposed to dead-heading – removal of dead flowerheads.

DEFLEXED (GOLDEN WEEPING WILLOW)

Defoliation

The removal of unwanted leaves from tomato plants to expedite ripening, increase air movement and thus reduce the risk of botrytis; removal of the lower yellowing leaves helps to control disease generally. Whether defoliation actually speeds up ripening is disputed, but some growers still remove all the lower leaves, leaving only those carried on the top third of the plant.

Deforestation

Artificial clearance of all or part of a forest area. Total deforestation rarely occurs naturally. *Same as* Disforestation.

DEFOLIATION

Defruiting

The removal of all immature fruit from a tree to allow it to build up its strength. Usually done when deblossoming has been overlooked or forgotten at the correct time.

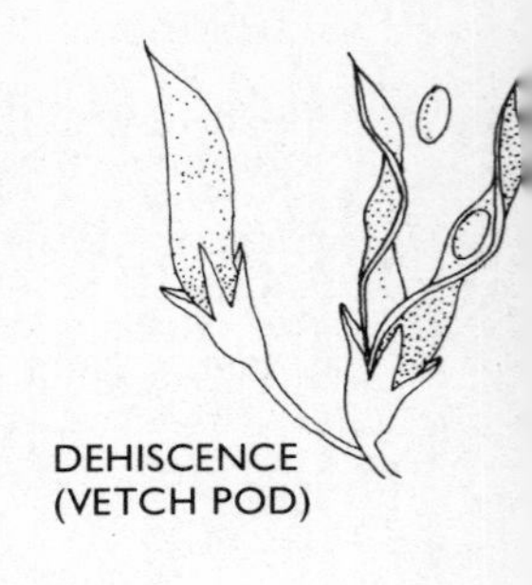

DEHISCENCE (VETCH POD)

Dehiscence

The opening or splitting of a seed capsule and disgorging of the contents.

Dehiscent

Applied to fruits which split open to release their seeds when ripe, such as peas, lupins, etc.; also to anthers which split to release their ripe pollen: *cf.* Indehiscent.

APPLE TREE BEFORE AND AFTER DEHORNING

Dehorning

A severe method of pruning elderly fruit trees to bring them back into production or to reduce the overall size. Some of the larger branches are totally removed, and all dead and decaying wood cut away, together with any inward-growing branches, to open up the centre and permit more air movement and light penetration. The remaining branches are often cut back to about half of their length. All cuts should be carefully trimmed and covered with a protective coating.

Deliquescent

(1) Dissolving by absorbing water.
(2) Breaking up into branches, as the veins of a leaf.

Deltoid

Shaped roughly like an equilateral triangle, as of a leaf with the base at the point of attachment to the petiole.

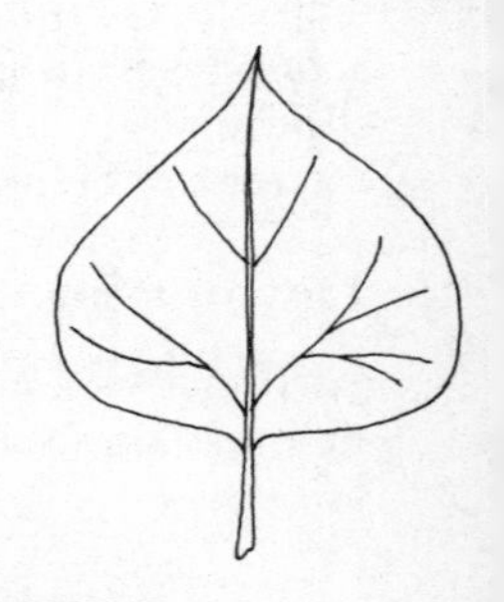

DELTOID

Dendriform, Dendritic, Dendroid

Tree-like.

Dendrochronology

Measurement of tree rings and inference of the tree's age and past climatic conditions.

Dendrology

The study of trees and the natural history of trees.

Dene

Steep-sided, heavily wooded small valley.

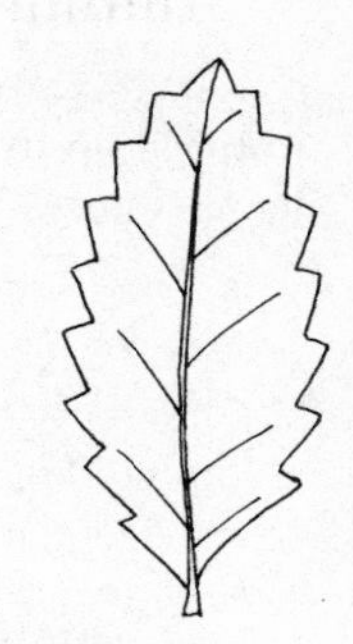
DENTATE

Denitrification

Severe shortage of nitrogen in the soil caused by unbalanced bacterial activity.

Dense

Crowded together; compacted; thick.

Dentate

Coarsely toothed, with the indentations rather spreading and regularly divided, as applied to a leaf margin.

Denticulate

Diminutive of dentate – having smaller teeth.

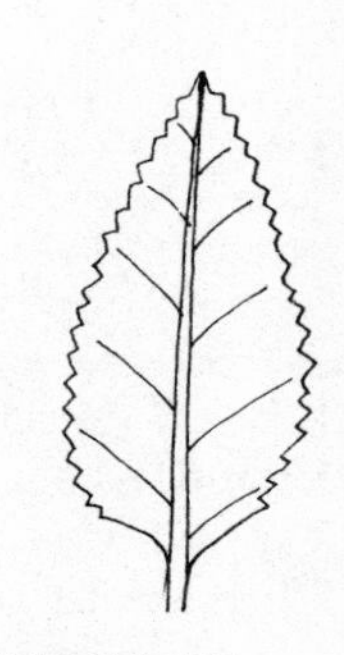
DENTICULATE

Dentition

The arrangement of teeth at the edges of leaves, and occasionally applied to some petals.

Depauperate

Starved; poorly developed or distorted due to unfavourable conditions.

Depressed

(1) Sunken, as in some necrotic spots on foliage.
(2) Flattened.

Deracinate

To pull up by the roots instead of digging or lifting with a fork or spade.

Derived

Originating from an earlier type or form, as applicable to plants.

Derris

An insecticide manufactured from the tuberous root of a tropical plant. It is harmless to humans and animals but toxic to fish.

Deshooting

The removal of very young shoots from a trained fruit tree in order to maintain its pattern or shape. Best done when the shoots are soft enough to be pinched out with the fingers.

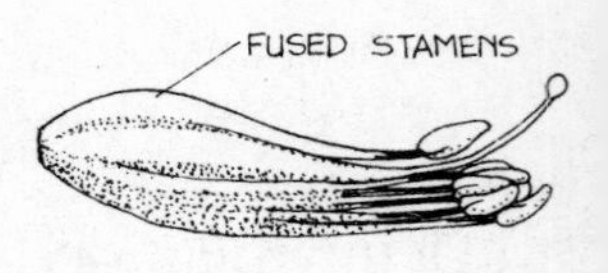

DIADELPHOUS

Determinate

Where the main stem ends in a terminal flower or truss, thereby arresting the further development of the main axis, as in bush tomatoes.

Dewpoint

Air temperature at which dew begins to be deposited. After a warm day in the garden the temperature falls, the air becomes so humid that it is saturated and condensation is visible on smooth surfaces. The foliage will usually show dew first because the transpiration process cools the leaf surfaces and thus lowers the temperature.

DIALYPETALOUS (BRASSICA)

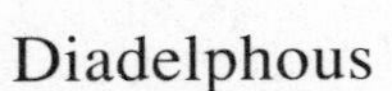

Diadelphous

Having stamens which are united by the filaments in two bundles, as in peas and beans: *cf.* Monadelphous and Polyadelphous.

Dialypetalous

Having free and separate petals.

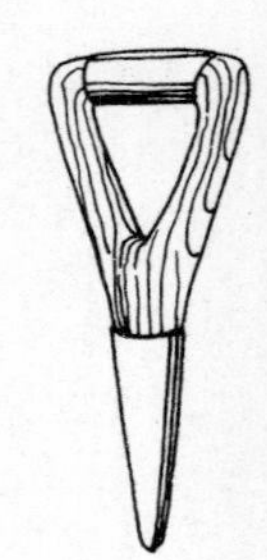
DIBBER

Diandrous

Having two stamens.

Diastase

Enzyme produced in germinating seeds which converts starch into sugar.

Diatom

Type of algae found in fresh water and the sea, forming large part of plankton.

DICHASIAL CYME

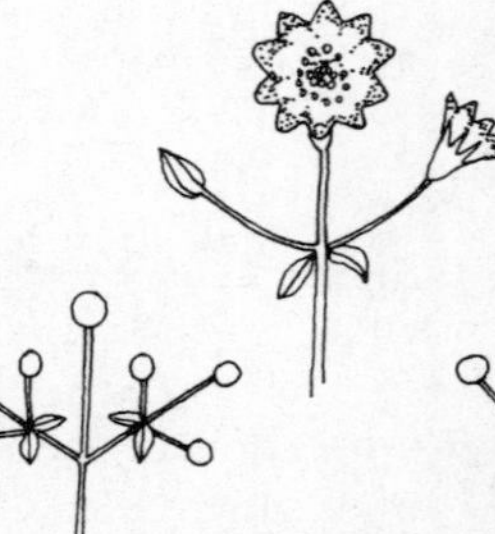

Dibber

Instrument used for making a hole in the soil into which a seed, seedling, cutting or tuber can be inserted. Garden shops sell specially made tools, but anything can be used, such as a pencil, garden cane or fork handle, according to the diameter of hole required. To dibble is to use a dibber.

Dichasial Cyme

Cyme in which the axis carries a terminal flower between two roughly equal branches.

Dichotomous

Divided repeatedly and regularly in pairs with the two branches of each fork roughly equal.

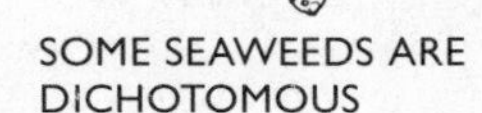

SOME SEAWEEDS ARE DICHOTOMOUS

Diclinous

Having the stamens and pistils on separate flowers, which may be on the same or different plants.

Dicotyledon

A plant bearing two cotyledons or seed leaves. The term is often abbreviated to dicot.

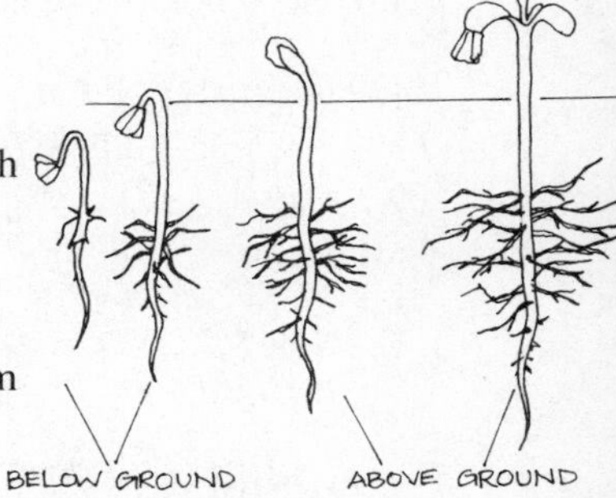

DICOTYLEDONS

Didymous

Having stamens in pairs.

Didynamous

Having four stamens in two pairs, one pair being longer than the others.

DIDYNAMOUS

Die-back

Any of several fungus diseases which cause shoots or branches gradually to die back from the growing tips. It is a much slower process than wilt, where infected shoots collapse rapidly and entirely. Diseased shoots should be cut back to sound wood and the infected material burned.

DIE-BACK

Diffuse

Open habit; loosely spreading or branching.

Digging

Cultivation of the soil by turning over with a spade or similar implement.
Single spit digging. The spade is pushed down to its full working depth into the soil (one spit deep); then the soil

is lifted, turned over and deposited back into the same hole. In autumn the clods are left unbroken and uneven so that the winter frosts can break them down to a fine tilth by the spring.

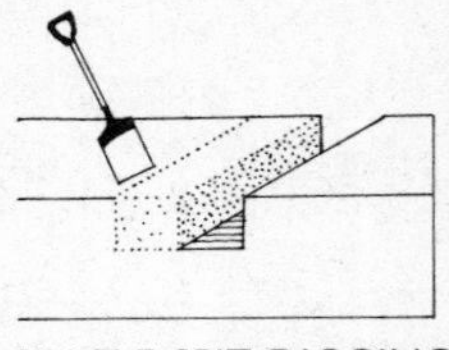
SINGLE SPIT DIGGING

Double-digging. Otherwise known as 'bastard trenching', this system is used to break up the lower spit which has become compacted and therefore resistant to root penetration. Take soil out of the first trench to the depth of one spit and barrow to the end of the plot. Dig the floor of the trench and fork over to the full depth of the tines; then remove the next top spit, invert and place it in position over the forked trench. Fork over the bottom of the new trench as before and continue digging in this manner until the end of the plot is reached; then fill the final trench from the soil in the barrow.

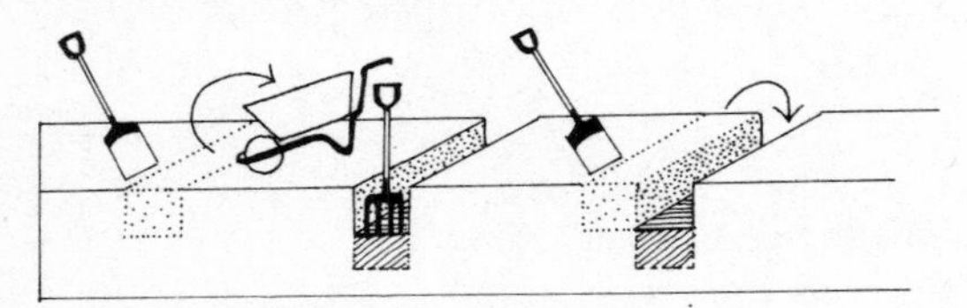
DOUBLE DIGGING (ALSO KNOWN AS BASTARD TRENCHING OR MOCK TRENCHING)

Trenching. This is a method used to cultivate the soil to a depth of three spits – about 30 in (75 cm), an operation only necessary when bad drainage caused by a hard pan is a problem. It is essential that these lower strata of soil are not brought to the surface because their composition is quite different to the top layer and deficient in many of the nutrient requirements. Begin by taking out a trench across the plot 3 ft (90 cm) wide and one spit deep, barrowing the topsoil to the far end of the plot. Divide this trench lengthwise with a line, and remove half of it to a depth of one spit, thereby making a further trench two spits deep; barrow the soil from this trench, too, to the end

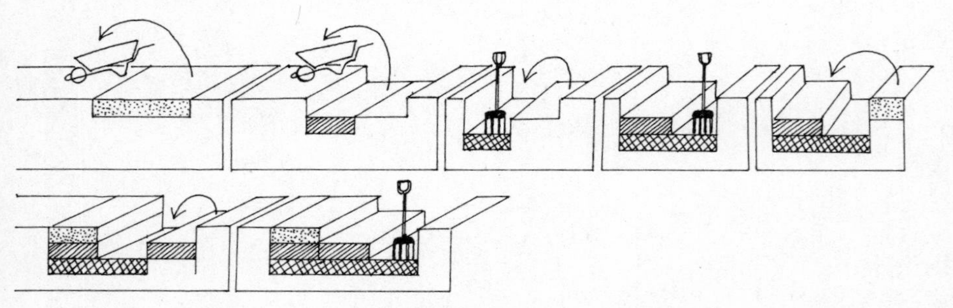
TRENCHING

of the plot but stack separately. Fork and break up the base of this trench to the depth of another spit, then turn over the second half of the upper trench on to the cultivated lower trench. Return now to the surface and start a new row by marking out another 18 in (45 cm) strip and throwing the topsoil forward to fill the far side of the original trench. Continue in this way until the end of the plot, then replace the barrowed spits at their correct levels.

Digging Hoe

The digging or heavy hoe is mainly used for heavy work on stony ground. It is used in much the same way as a pick and has been a basic agricultural implement for more than 2000 years.

DIGGING HOE

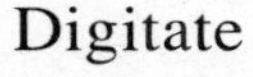

Digitate

Resembling fingers, applicable to a compound leaf with leaflets attached to the petiole at a single point, as in horse-chestnut.

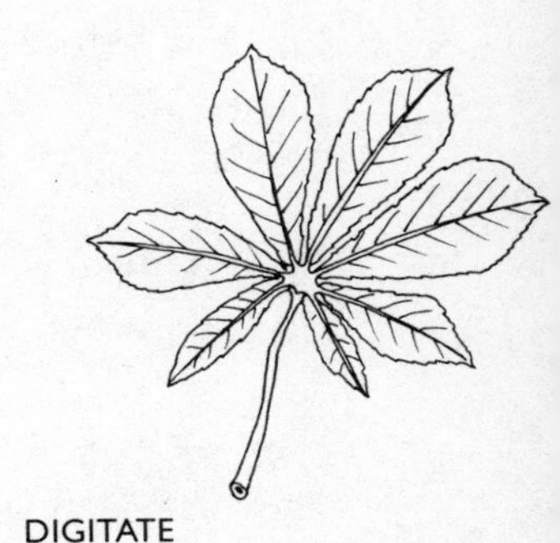

DIGITATE

Dilated

Roughly flattened and expanded into a blade.

Diluter

A measuring device which can be assembled into a pipeline to deliver liquid fertilizer at a predetermined rate of dilution.

Dimorphic

Two different forms in the same species, as in the adult and juvenile leaves of ivy.

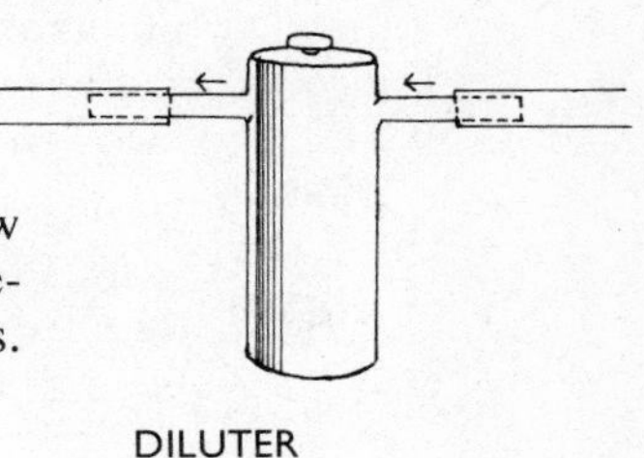

DILUTER

Dimple

Old term used to describe the virus disease of pears, now called ‘stony pit’. The fruits show small dimples or depressions on the skin, below which are hard gritty lumps.

Dingle

Deep hollow surrounded by or shaded with trees; a dell.

Dioecious

Having male and female flowers on separate plants. For fertilization purposes a male plant is usually set among a group of females. Some fruit trees are dioecious, as are skimmias and many hollies: *cf.* Monoecious.

DIPTEROUS SYCAMORE SEEDS

Diplococcus

Type of micro-organism which forms pairs.

Diploid

Having the full or unreduced number of chromosomes for its species; this condition is usual in most organisms, the chromosomes being in pairs.

Diplostemonous

Applied to flowers which have stamens in two alternating whorls, the outer whorl alternating with the petals.

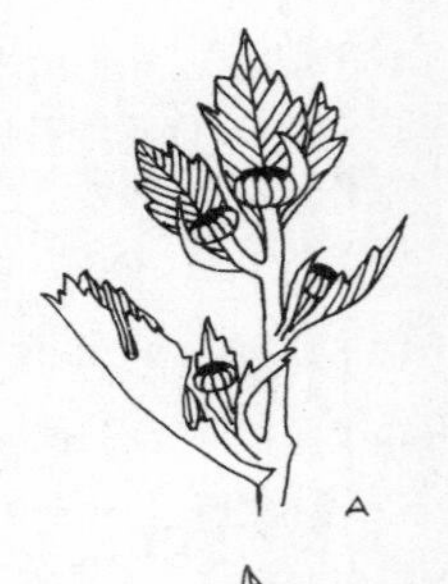

Dipterous

Having a wing-like appendage, as of certain seeds.

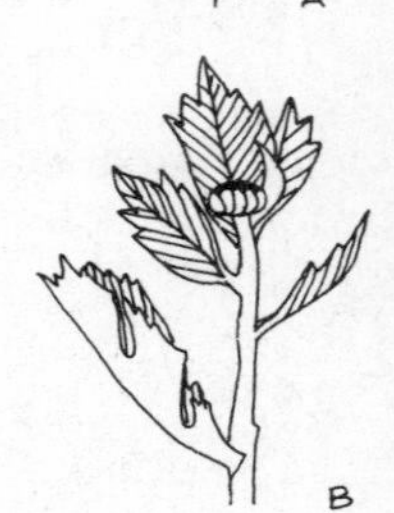

CHRYSANTHEMUM
A. BEFORE DISBUDDING
B. AFTER DISBUDDING

Disbudding

The removal of side shoots and surplus buds to force the plant to concentrate its energies in the remaining single bud. Exhibition carnations, dahlias, chrysanthemums, etc. are usually disbudded.

Disc, DISK (USA)

(1) The centre part of a flower such as the daisy, where the outer florets are attached to a central 'cushion' or disc which often secretes nectar.
(2) The flattened part at the end of a tendril used for adhesion.

Discoid, Discoidal

Disc-shaped; circular and flattened.

DISC OF A DAISY

Discolorous

Not of the same colour. The upper and lower surfaces of a leaf are often discolorous.

Disease

Almost any ailment of plants other than pest damage. Attacks by bacteria, fungi, viruses, and sometimes the results of poor growing conditions, are all grouped (often erroneously) under the general term 'disease'.

Disforestation

See Deforestation.

Disk

See Disc.

Dismember

To divide a plant by pulling stems apart or cutting through the crown.

Disorder

Ailment caused by poor conditions, mineral deficiencies or a hostile environment as distinct from one emanating from attack by pests.

Disperse

To scatter or distribute seed in different directions. Dispersal can be by many different agencies, including wind, water, animals, etc.

Displanting

The word used for transplanting during the early eighteenth century.

Dissected

Deeply cut into narrow segments, as of certain leaves.

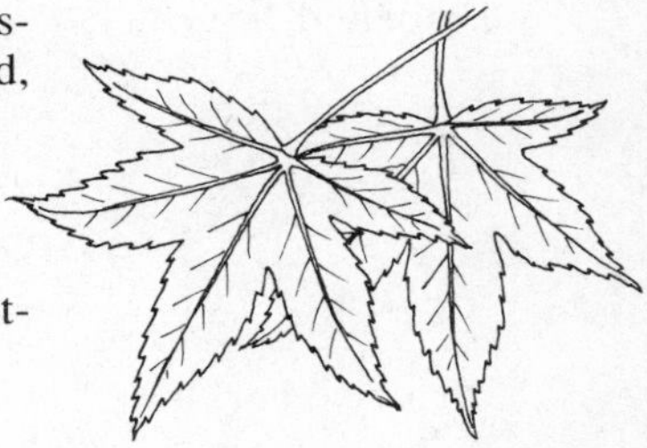

DISSECTED FOLIAGE OF JAPANESE MAPLE

Dissepiment

Partition or dividing membrane.

Distal

Farthest from the point of attachment, usually at or near the apex.

Distichous

Having two opposite, vertical rows, as applied to leaves or branches.

DISTICHOUS

Distinct

Separate, not united (as in petals).

Distorted

Growing into unusual or crooked shapes, as in petals, leaves, stems, etc. Normally the result of viral or insect damage.

MUSHROOMS DISTORTED BY A VIRUS

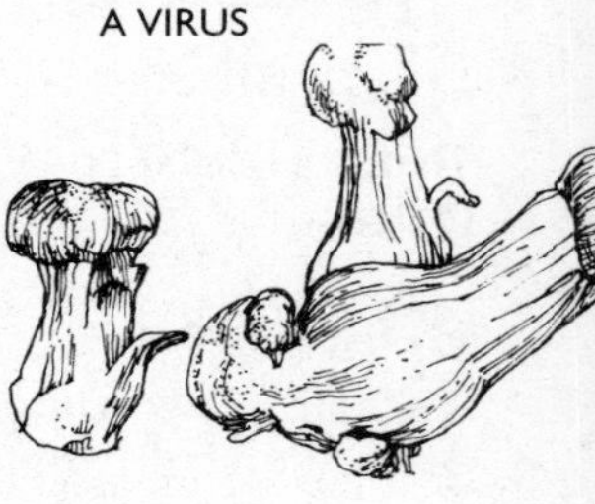

Distributor

Mechanical device for spreading fertilizer and similar materials evenly. Small types can be used on garden lawns, larger ones are towed behind tractors for farm use.

Diurnal

Belonging to the daytime. Usually applied to flowers which open only during daylight hours, e.g. dandelion.

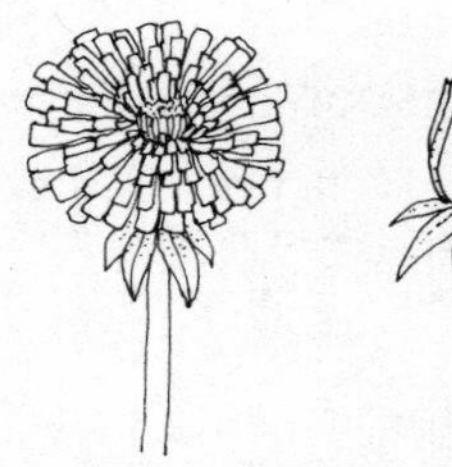

DIURNAL

Divaricate

Widely divergent.

Divergent

Spreading, but to a less extent than divaricate.

Divided

Separated but not completely detached from the midrib of a leaf.

DIVIDED LEAVES OF *Acer palmatum*

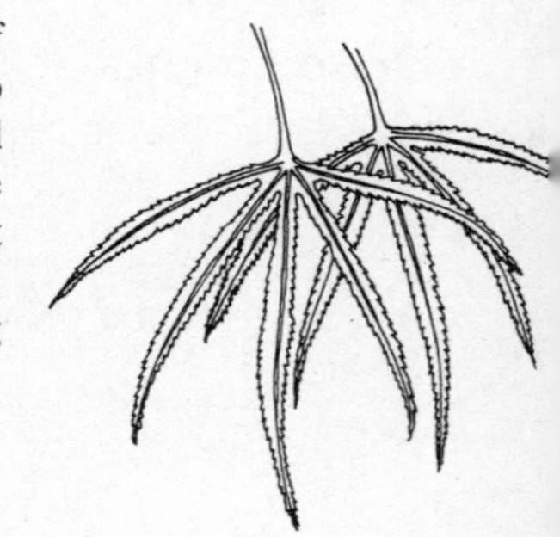

Division

(1) Root division is a method of increasing stock of some herbaceous plants. Two forks are inserted into the clump of roots back-to-back then carefully forced apart to divide the rootstock. In some cases the clump can be pulled apart by hand, but delicate pot plants are best divided with a sharp knife or scalpel. The division of plants is ideally performed during winter when they are dormant.

(2) A major plant group, such as the Spermatophyta.

DNA

Deoxyribosenucleic (alternatively desoxyribonucleic) acid, a molecule found in chromosomes of animals and plants, now known to contain the coded information carried in the genes responsible for the development of individual character.

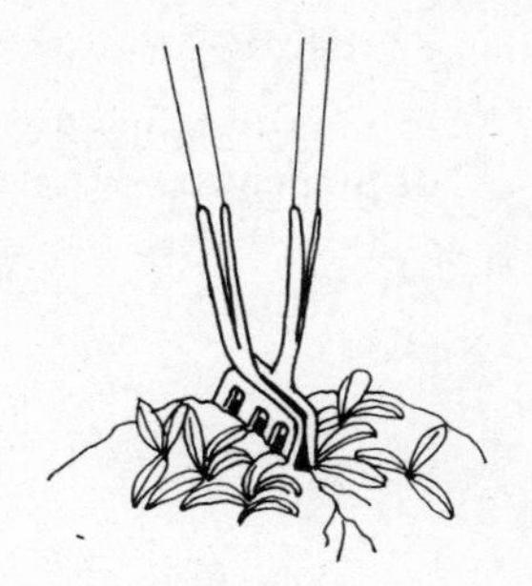

ROOT DIVISION

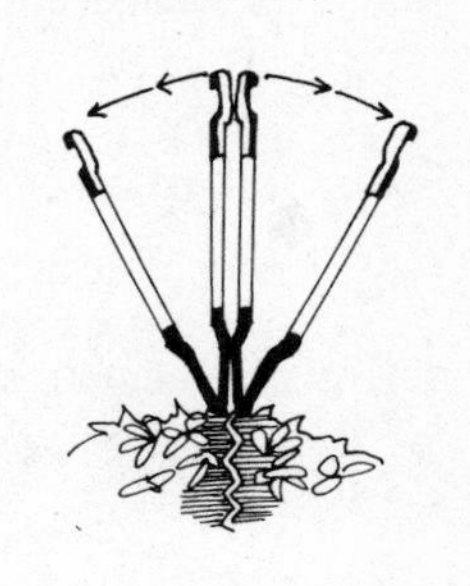

Dollar Spot

A fungal disease of neglected lawns or turf which shows as light-coloured or brownish circular patches about 2 in (5 cm) in diameter. Feeding with a high-nitrogen fertilizer will assist recovery.

Dome

Canopy of trees, vines, etc. formed by the foliage.

Dominant

That one of two alternative characters of a plant which is visible in the offspring when both are inherited.

Dormant

(1) Resting, as of a plant that has temporarily stopped growing, usually in winter.
(2) In a state of suspended growth, as of seeds.

Dormant Oil

Petroleum-based winter wash, mainly used in USA.

Dorsal

(1) Pertaining to the back or outer surface of an organ facing away from the axis.
(2) Ridge-shaped.

Dorsifixed

Of anthers attached to the filament more or less centrally at the back.

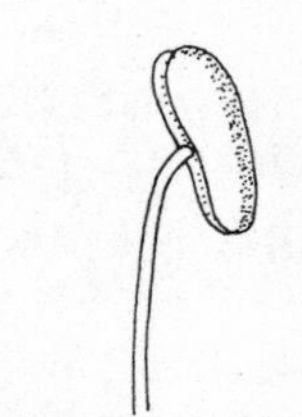

DORSIFIXED ANTHER

Dorsiventral

Having two sides (as a leaf), upper (ventral) and lower (dorsal), especially when the surfaces differ.

Dot Plant

Plant used in bedding schemes to provide height and contrast, usually among shorter varieties or ground cover plants. Standard roses, dwarf trees, standard fuchsias and pelargoniums are among those frequently used as dot plants.

DOT PLANT

Double

Flowers which have the normal complement of petals are termed single and those with more than the normal are called double. A flower with only slightly more than the single amount is termed semi-double although, to be precise, this is a contradiction in terms. Many fully double flowers are sterile because the extra petals have replaced the stamens or pistils.

Double Digging

See Digging.

DOUBLE ROSE SINGLE R

Double Leader

A tree which produces twin shoots at the apex instead of the more normal single spike.

Double-U Cordon

See U cordon.

DOUBLE LEADER

Doubly Serrate

Of leaf margins in which the primary serrations bear smaller, secondary teeth: *cf.* Biserrate.

Downs, Downland

Hilly, treeless open land, widespread in the south of England.

Downy

Usually applied to leaves which are covered with short, stiff hairs.

Draw Hoe

Long-handled tool with metal blade for taking out seed drills, etc.; it is pulled towards the user.

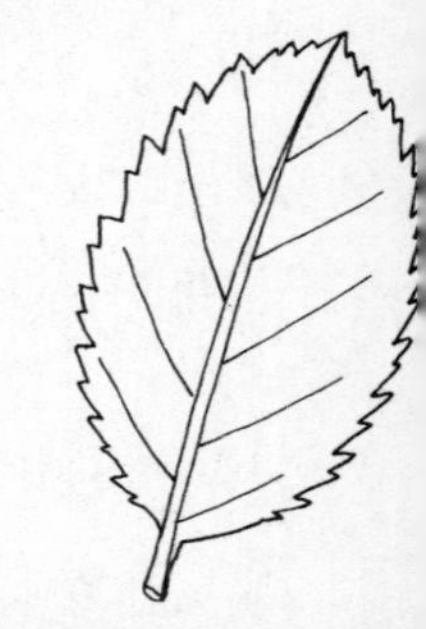

DOUBLY SERRATE

Drawn

Etiolated. Plants grown in poor light or crowded conditions will become tall and spindly, and will often lose much of their green colouring. They can never fully recover from this condition, and the mature plant which develops from a drawn seedling will almost always be inferior in yield, whether of fruit or flower. Drawn plants are often described as 'leggy'.

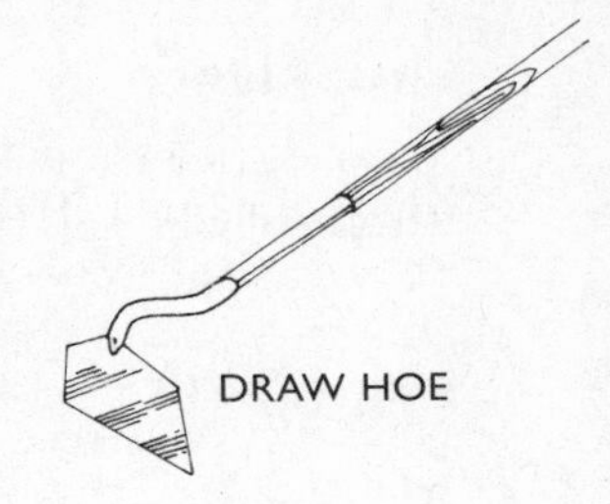
DRAW HOE

Drench

(1) To wet thoroughly; to soak.
(2) To apply liquid to the soil rather than to the foliage of a plant.

BEAN SEEDLING OF SAME AGE
(A) GROWN NORMALLY
(B) GROWN IN POOR LIGHT (DRAWN)

Drepanium

A cymose inflorescence with branches on one side only, and often curved to one side as in some members of the rush family (Juncaceae).

Dressing

Any fertilizer, whether organic or chemical, applied to the soil in a dry as opposed to a liquid from. Base dressings are incorporated into the soil before planting; top dressings are applied to the surface around growing plants. Top dressings of bulky materials such as animal manure are usually called mulches.

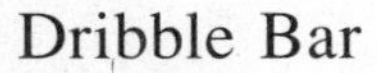

Dribble Bar

A perforated tube of metal or plastic, sometimes called a sprinkle bar, which can be attached to a watering can for the precise application of fertilizers, weedkillers, etc. Wide forms are available for use on lawns, narrow ones for better control around plants.

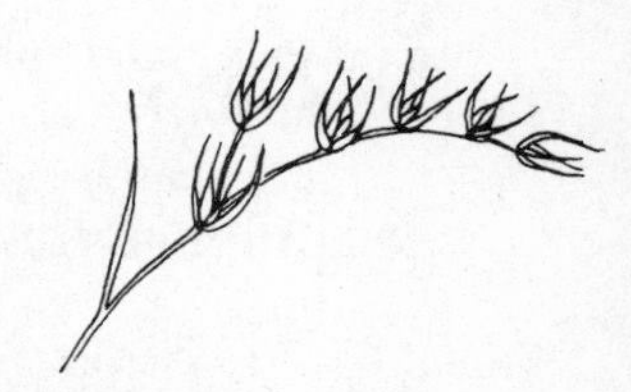
DREPANIUM

Drift

A large, irregular group of plants arranged for effect. Daffodils and other spring bulbs are often planted in drifts, which can also occur naturally in undisturbed woodland.

Drill

(1) A narrow furrow made in the soil for seed sowing.
(2) A row of small plants, the result of sowing seeds in a furrow.

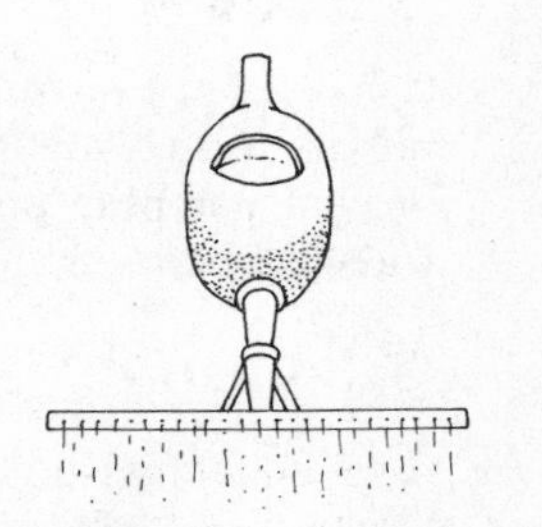
DRIBBLE BAR

Drill Hoe

Normally a long-handled hoe with triangular blade for taking out seed drills, but there are also short-handled versions.

DRILL HOE

Drip-feeding

See Trickle irrigation.

Drip Line

The imaginary line on the soil below a tree where the drips fall from the leaves. Any fertilizer should be applied around this line because that is where the majority of the feeding roots are located; most of those outside the line are simply anchor roots.

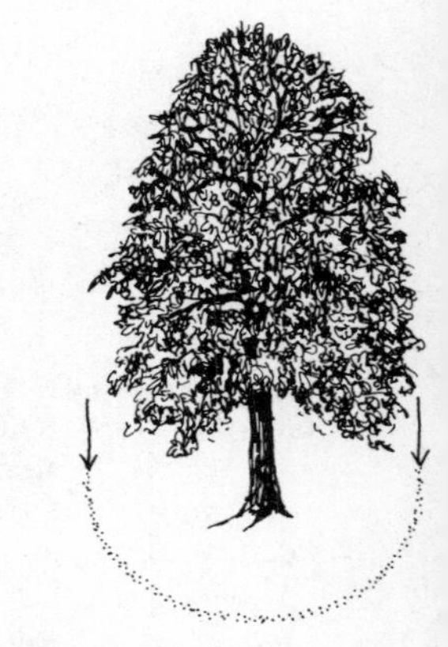
DRIP LINE

Drought

A period of protracted dryness: in Britain, officially a period of 14 days without measurable rainfall.

Drum, Drumhead

Flat-topped variety of cabbage.

Drupe

A stone fruit where the seed is protected by the hard casing (stone) contained in a fleshy layer within an outer skin. Peach and apricot are simple drupes, raspberry and blackberry are compound drupes.

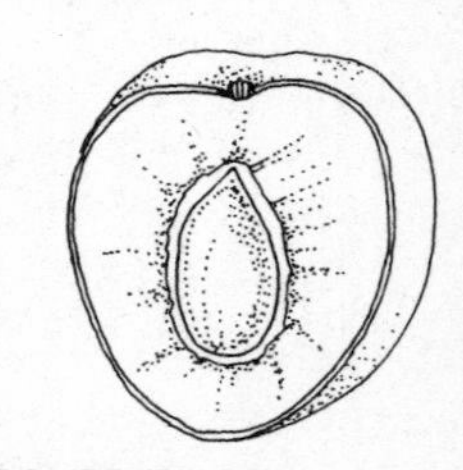
DRUPE (PLUM)

Drupelet, Drupel

A small drupe or the individual component of a compound drupe, as in a blackberry.

COMPOUND DRUPE (BLACKBERRY)

Dry Rot

A loose term covering several diseases in plants such as bulbs, corms, tubers and fruits. Symptoms may include brown patches, shrivelling, areas of dry decay, etc. Infected items should be burned.

DRY STONE WALL

Dry Set

Applied to glasshouse-grown tomatoes which fail to grow larger than pinhead size, usually as a result of an overheated dry atmosphere. There is no cure but the problem can be prevented by regular overhead spraying with clean water or a fruit-setting agent.

SIMPLIFIED DIAGRAM OF THE VASCULAR SYSTEM OF A STEM SHOWING THE DISTRIBUTION OF THE DUCTS

Dry Wall

A stone wall constructed without mortar. Popular in many parts of rural Britain, they are ideal for growing many small plants between the cracks.

Duct

Part of a plant's vascular tissue carrying air, water, etc.

Duramen

The dense centre of a tree trunk; the heartwood.

Dusting

The application of dry powdered fungicides, pesticides, etc. through a mechanical device designed to produce a fine, even fog. Also known as fogging.

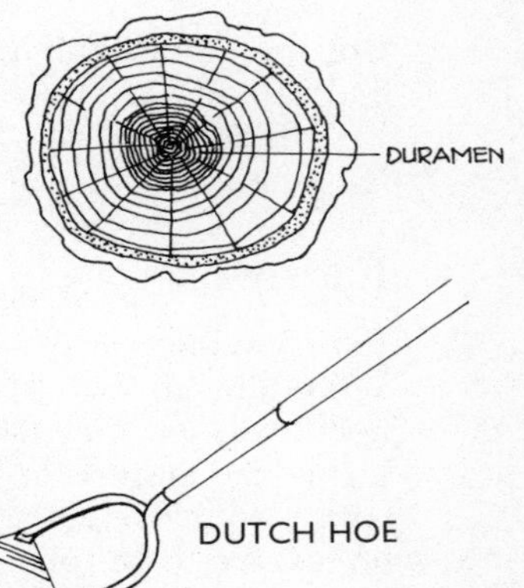

DUTCH HOE

Dutch Hoe

Long-handled tool with metal blade for loosening soil, weeding, etc. It is pushed away from the user.

Dutch Light

A single sheet of glass about 59 × 31 in (150 × 79 cm) enclosed in slots in a wooden edging without putty. They can be used in the construction of garden frames, or bolted together to make greenhouses.

Dwarf

Extremely small. Applied to plants that are naturally tiny; plants which have been artificially dwarfed by root restriction or grafting, and plants which have been chemically dwarfed for one season, after which they revert to their natural form.

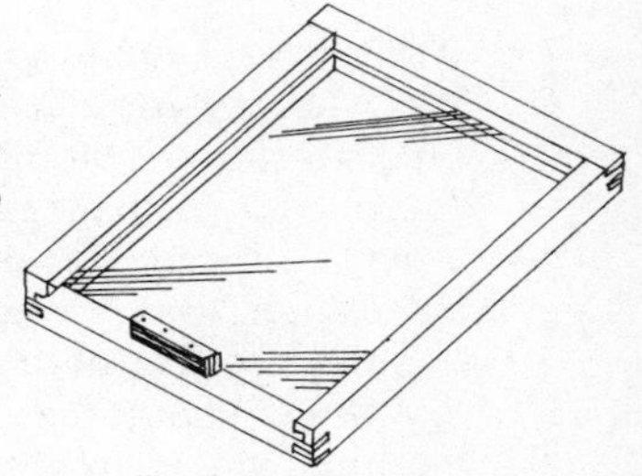
DUTCH LIGHT

E

Ear

Head of corn or ripe corn-spike containing seeds.

EAR OF WHEAT

Earth

The ground or soil, including dust, mould, clay, etc.

Earthing-up

Covering roots with heaped-up soil; drawing soil towards and around plants such as potatoes, to prevent the greening of the tubers, and celery, to blanch the stems. *Same as* Ridging.

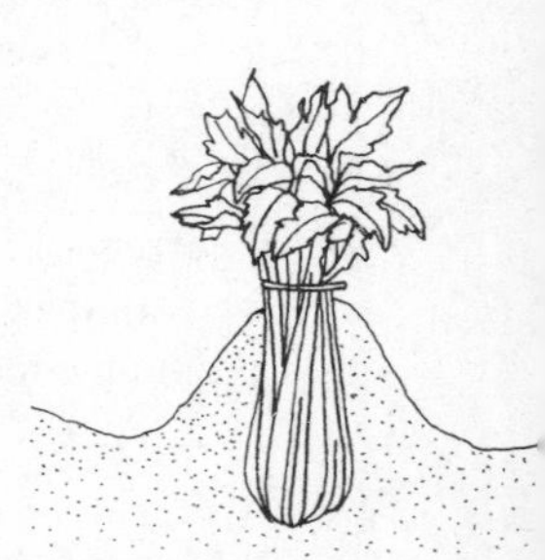

CELERY EARTHED-UP

Earthworm

Beneficial terrestrial animal which lives normally in the soil (although some species can survive in water), mainly nocturnal, spending daylight hours underground and coming to the surface to feed. As a general rule, the more worms to the acre, the more fertile the land. They can be a nuisance to pot plants growing in unsterilized soil.

EARTHWORM

Earwig

An insect related to grasshoppers, cockroaches, etc. having a long, fairly narrow body with a pair of nippers or pincers at the rear. Basically scavengers, earwigs can cause severe damage to some plants, especially those with a dense flowerhead such as dahlias, which provide food and shelter. Mainly nocturnal feeders, they can be trapped by placing paper or cardboard 'dunces' hats' stuffed with dry straw among the flowerheads, inspecting them each morning and destroying the occupants.

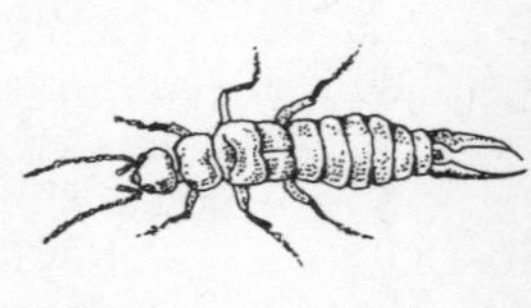

EARWIG

Ebracteate

Without bracts.

Eccentric

One-sided or lop-sided, e.g. the leaves of begonias.

Echinate

Generally, prickly or bristly like a hedgehog, but in horticulture usually applied to a plant having thick, short bristles such as the echinocactus.

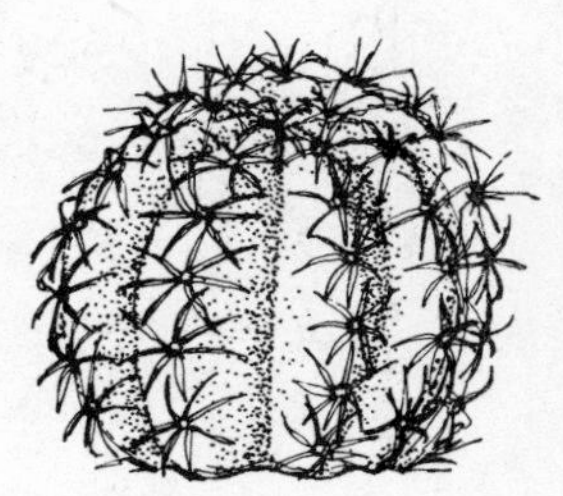

ECHINOCACTUS

Eciliate

Bald, without cilia (hairs).

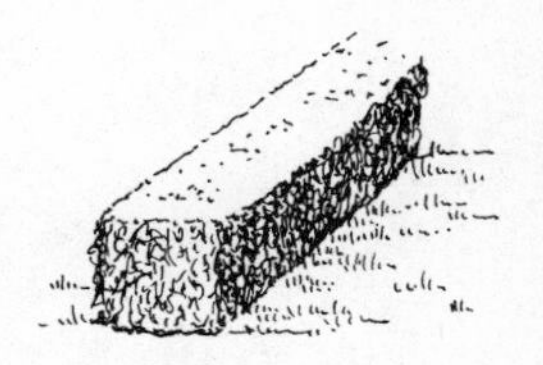

BOX EDGING

Ecology

The study of the relationship between the environment and the living organisms therein.

Edema

See Oedema.

Edge

Border around flowerbeds, ponds, paths, etc. May be of grass, stones, concrete, wood or metal.

Edging

(1) The use of materials such as wood, metal or stone to keep lawn edges and paths adjacent to borders tidy.
(2) Using small, compact plants to outline flowerbeds. In the past box and dwarf lavender were favourite subjects for this purpose.
(3) Trimming lawn edges to give a sharp, clean-cut finish after mowing.

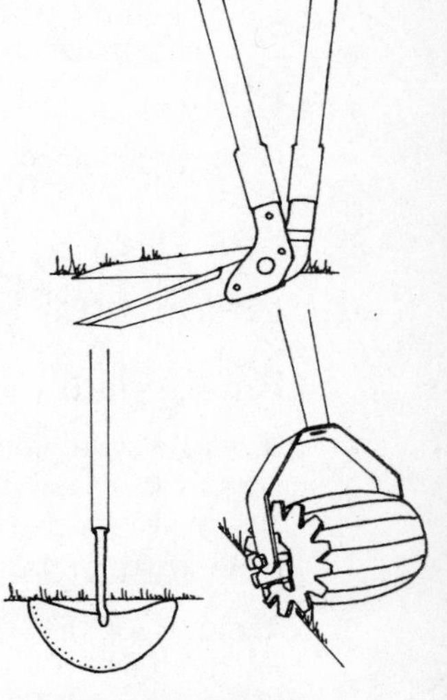

SOME EDGING TOOLS

Edging Tools

A variety of tools used to trim and maintain the edges of lawns such as straight-bladed types of spade, right-angled long-handled shears, electric trimmers, etc.

Eelworm

See Nematode

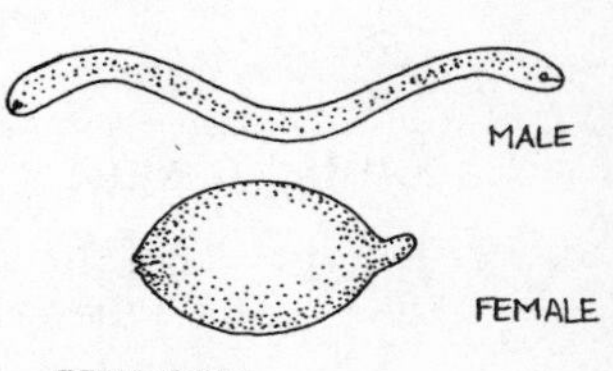

EELWORM

Efflorescence

The bursting or opening of the flowerbud; breaking into flower.

Egg

A female reproductive cell.

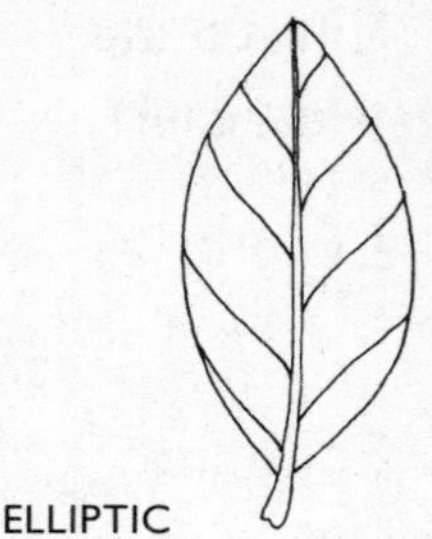

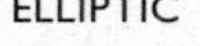

ELLIPTIC

Eglandular

Lacking glands.

Elaiosome

Of seeds which have or contain an oily substance to which ants are attracted. From the Greek *elaion* – olive oil.

Electronic Leaf

Part of a mist-propagation system which controls the supply of liquid to the spray head.

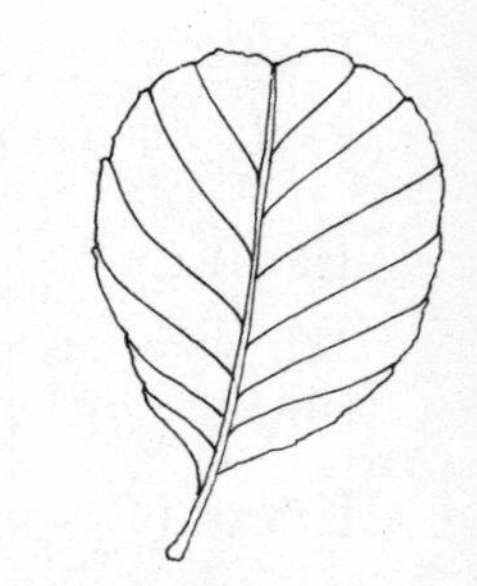

EMARGINATE

Elliptic

Oval, egg-shaped; mainly applied to leaves.

Elongate

Stretched or extended, slender and tapering.

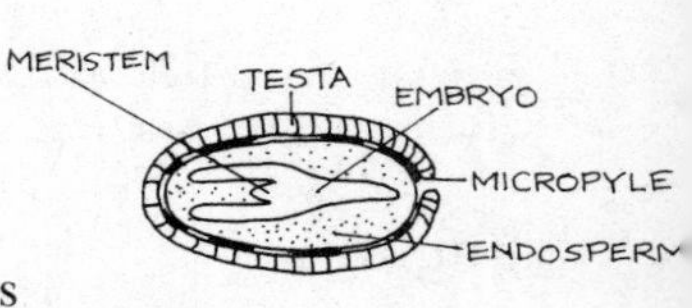

THE ESSENTIAL PARTS OF A SEED

Emarginate

Having a small notch at the apex; usually applied to leaves or petals.

Emasculation

Removal of the anthers of a flower in order to prevent self-pollination.

Embryo

The rudimentary plant in its earliest stages within the seed.

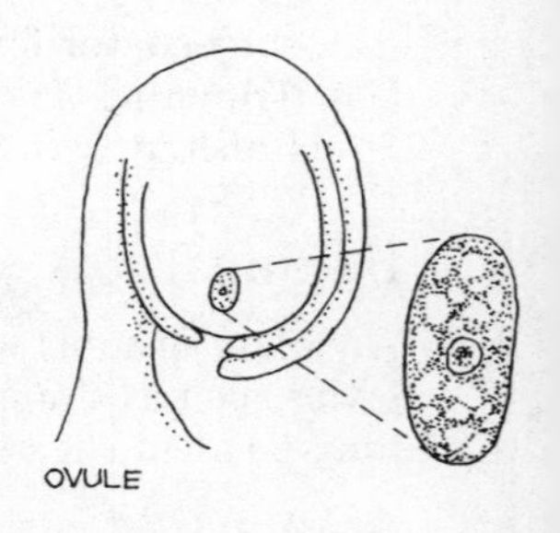

EMBRYO-SAC

Embryo-sac

The megaspore of a flowering plant. One of the cells of the nucellus; the central portion of the ovule.

Emersed

Rising above the surface of water – e.g. leaves of water-lily: *cf.* Immersed.

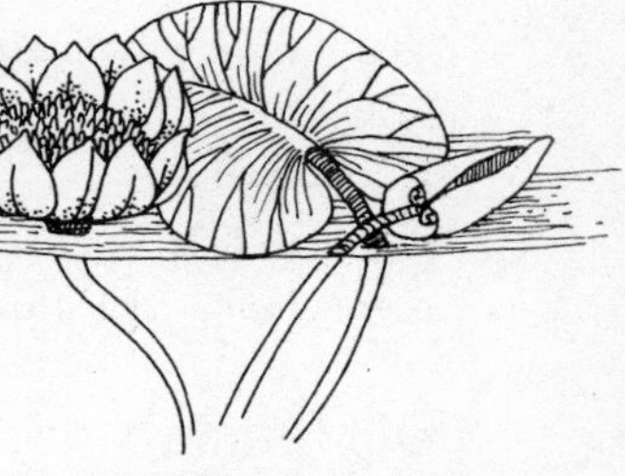
EMERSED LEAVES OF WATER LILY

Emulsion

A mixture of two or more liquids where the particles of one liquid are suspended in another. Several insecticides, etc. are emulsions, e.g. tar oil.

Enation

An outgrowth.

ENATION ON TOMATO

Endemic

Native to or confined to a usually restricted area. Applicable to species occurring naturally nowhere else, not imported or introduced.

Endocarp

The innermost layer of the ovary wall of a fruit, often becoming hard, as a plum stone.

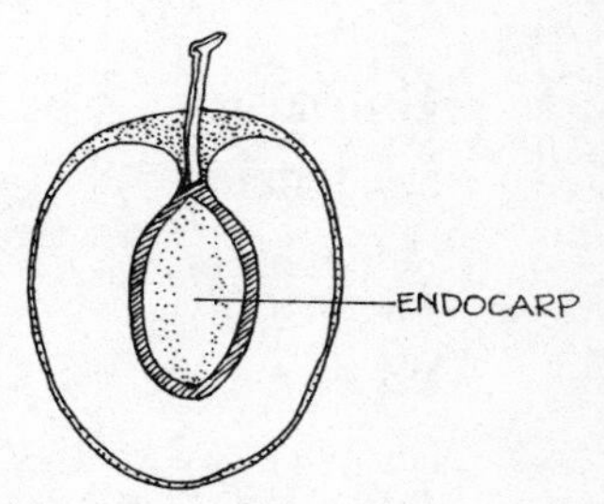

Endodermis

A close-fitting sheath containing the central cylinder of the stem in plants.

Endogen

Any part of a plant regarded as growing from within, such as a rootlet developing from within an old root.

Endophyllous

(1) Formed within a sheath.
(2) Living within a leaf.

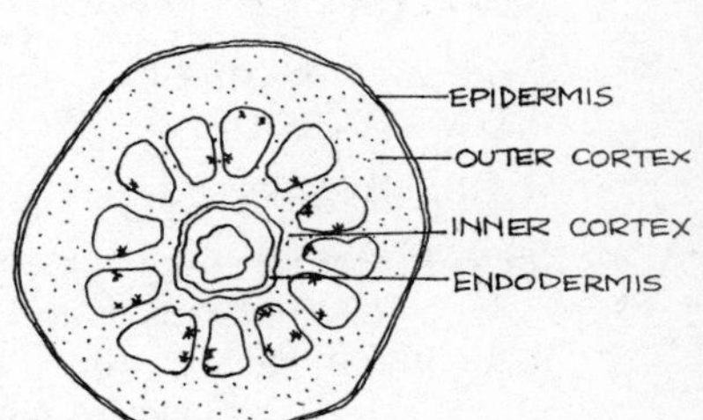

SECTION THROUGH A STEM OF MILFOIL SHOWING THE ENDODERMIS

Endophyte

A plant living within another, not necessarily as a parasite.

Endorhizal

Having the radicle of the embryo enclosed within a sheath.

Endosperm

The albumen or starch-containing tissue of many seeds.

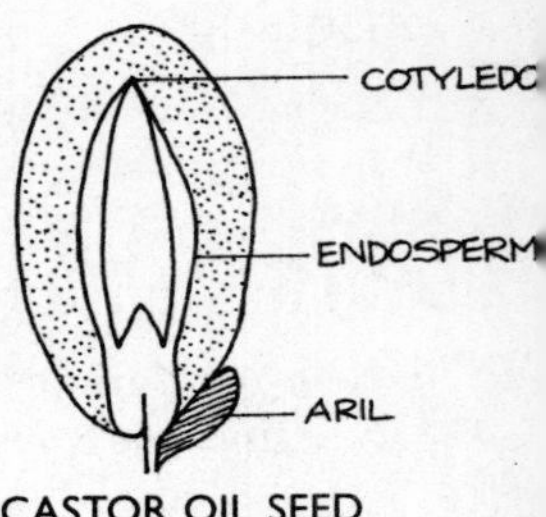

CASTOR OIL SEED

Endozoic

Having seeds dispersed by animals or birds that swallow them. *Same as* Entozoic.

Ensiform

Sword-shaped, as applicable to a leaf.

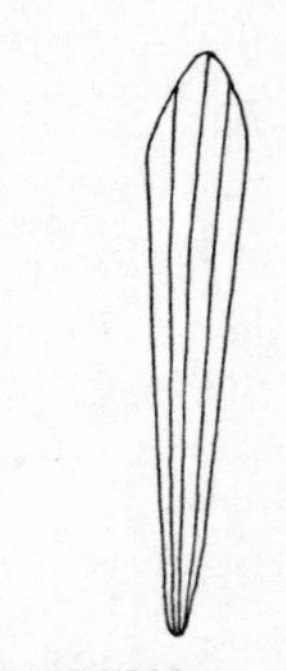
ENSIFORM

Entire

Completely smooth-edged, without indented margins, as applicable to a leaf.

Entomology

The study of insects.

Entomophilous

Pollinated by insects.

Entozoic

See Endozoic.

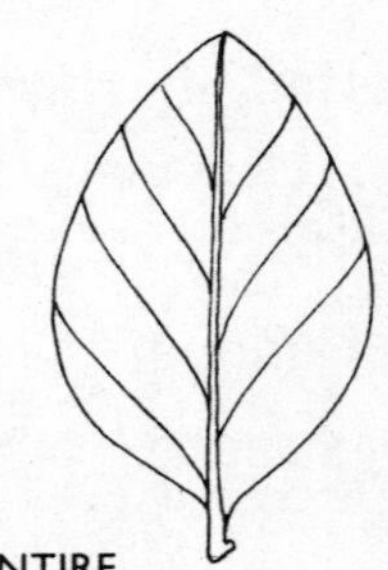
ENTIRE

Enucleate

Without a nucleus.

Envelope

The calyx or corolla (or both) of a flower.

Ephemeral

(1) Flowers lasting for a very short time, usually a day or less; e.g. morning glory (*Ipomoea acuminata*).
(2) Plants such as those annuals which complete their entire life cycle from germination to seed production within a few weeks, e.g. after rain in a desert environment.

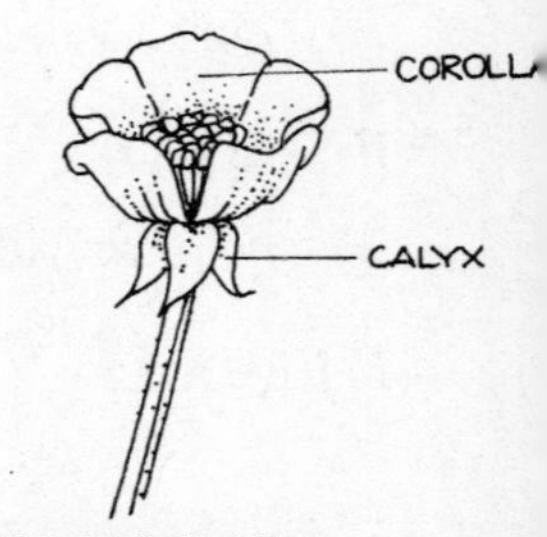

ENVELOPE OF BUTTERCUP

Epicalyx

A subsidiary false calyx outside the true calyx, composed of the bracts or the fused stipules of sepals.

Epicarp

The outermost layer or skin of a fruit. *Same as* Exocarp.

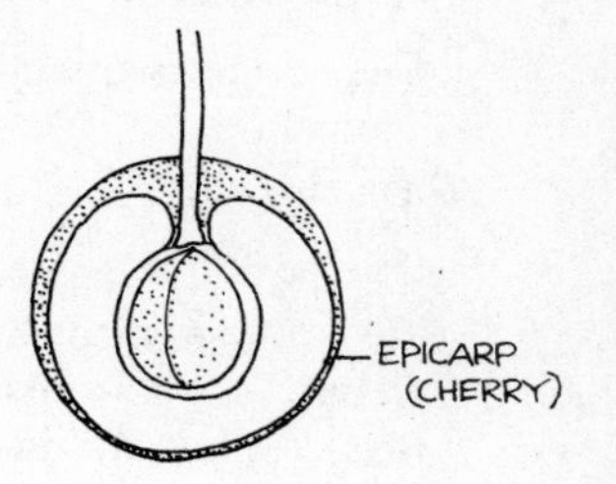

Epichil, Epichile

The termination of the lip in some orchids when distinct from the basal part.

Epicotyl

Seedling shoot above the seed-leaves; the emerging growing-point.

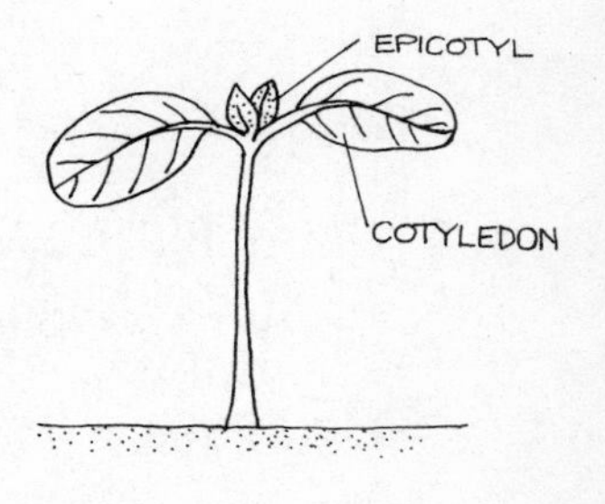

Epidermis

The outermost layer of living cells which forms a protective sheath for many plant organs, including leaves, petals and stems.

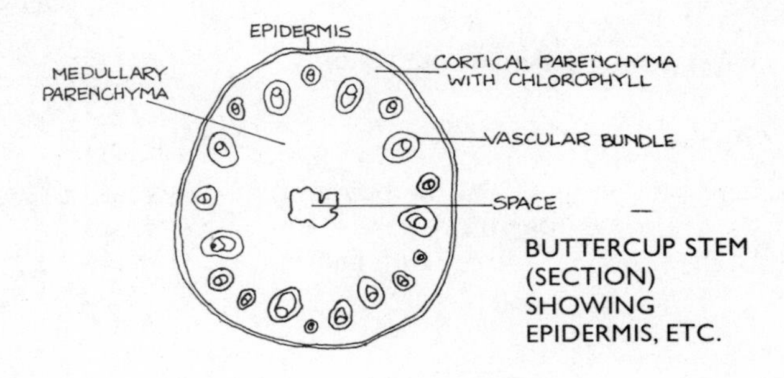

BUTTERCUP STEM (SECTION) SHOWING EPIDERMIS, ETC.

Epigeal

Producing seed-leaves above ground in germination: *cf.* Hypogeal.

Epigynous

Growing on top of the ovary; having calyx, corolla and stamens inserted near the top of the ovary.

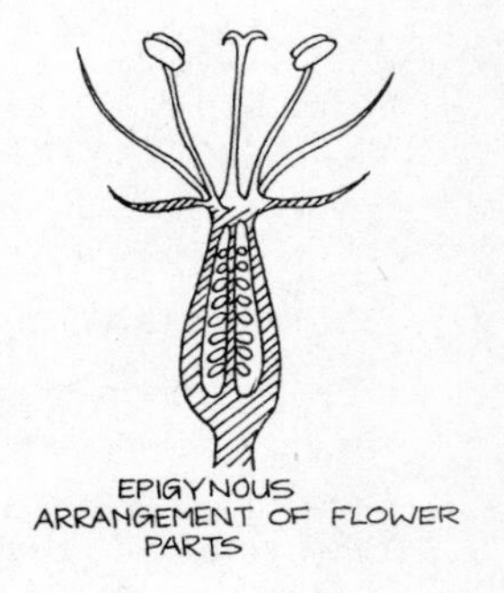

EPIGYNOUS ARRANGEMENT OF FLOWER PARTS

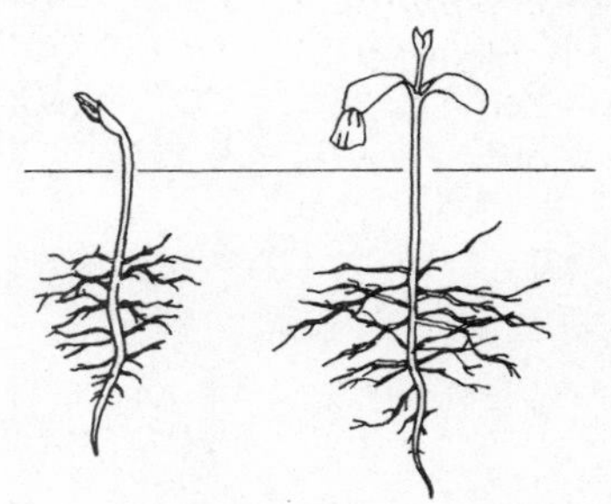

EPIGEAL SUBJECTS PRODUCE SEED LEAVES ABOVE GROUND

Epipetalous

Having stamens arising from or carried on the petals or corolla.

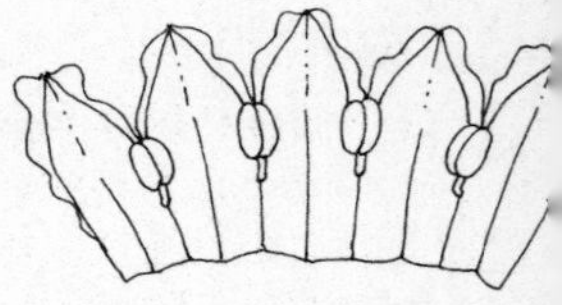

EPIPETALOUS STAMENS

Epiphyte

A non-parasitic plant which uses another as its host, often having aerial roots for the collection of atmospheric moisture. Some orchids and many bromeliads are epiphytic.

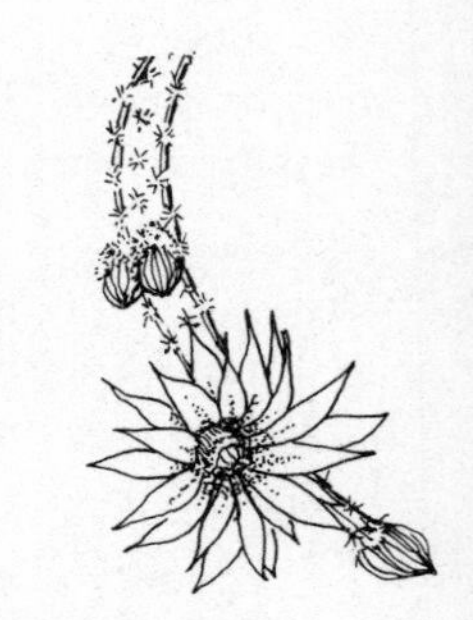

AN EPIPHYTE CACTUS (*Wilcoxia albiflora*)

Epiterranean

On or above the surface of the ground: *cf.* Subterranean.

Epithelial Layer

The superficial cell-tissue of plants; the outermost layer.

Epithem

The group of cells which in some leaves exude water.

Epithet

The second part – the 'specific epithet' – of the binomial (two-word name of a species), or the third part – the 'subspecific epithet' – of the trinomial (three-word name).

Epizoic

Having seeds which rely upon animals for dispersal, such as goose grass (*Galium aparine*) which produces seeds with hooked prickles.

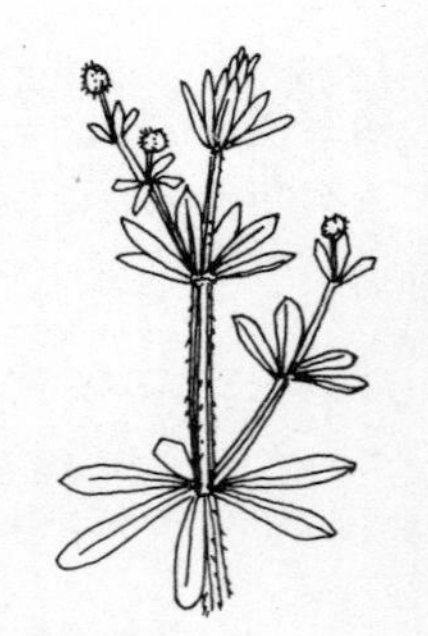

EPIZOIC (GOOSE GRASS)

Epsom Salt

The common name for magnesium sulphate, a vital element in the production of chlorophyll in a plant. A deficiency shows in the discoloration of the leaves between the veins, developing into dead areas if the condition is allowed to persist. The remedy is to spray the foliage with a foliar feed containing magnesium sulphate.

Equatorial Dehiscence

See Circumscissile.

Equitant

Overlapping so as to form a type of fan, applicable to leaves, e.g. the iris.

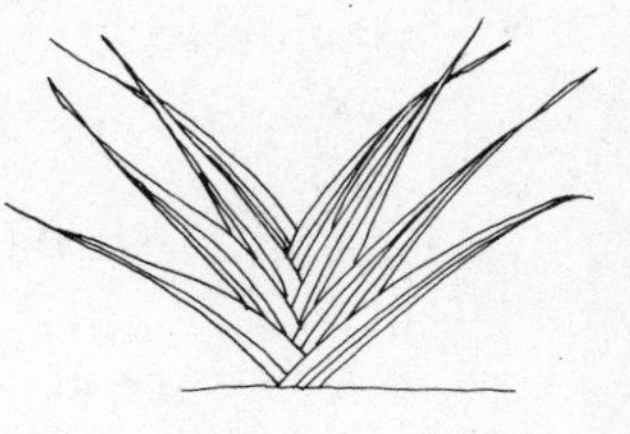

EQUITANT

Erect

(1) An upright ovule with its stalk at the base; directed upward, not decumbent.
(2) The upright growth of stems or of the whole plant.

Ergosterol

A sterol found in yeasts and fungi and present in small amounts in the fats of plants and animals.

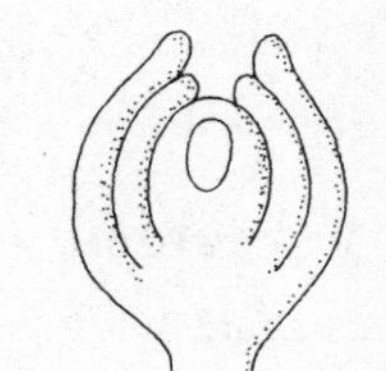

ERECT OVULE

Ergot

A disease of grasses (particularly of rye).

Ericaceous

(1) Relating to plants of the Ericaceae family (heathers), including erica, calluna, rhododendron, etc. Most are calcifuge plants requiring acid soils with a pH of 6.5 or lower.
(2) A type of lime-free potting compost suitable for such plants.

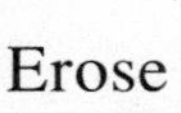

Erose

Irregularly jagged or eroded, notched as if bitten by an insect. Usually applied to margins of leaves or petals.

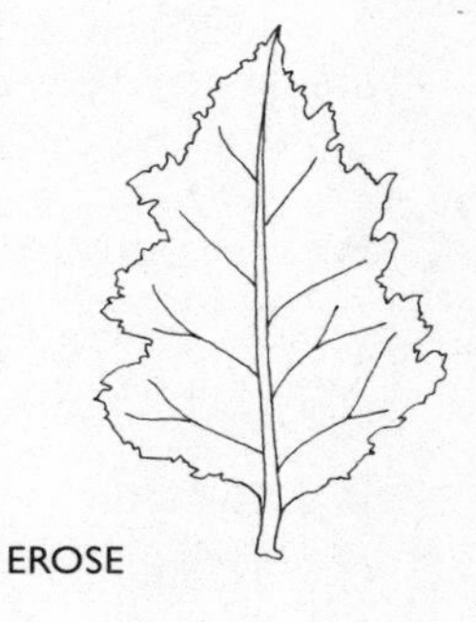

EROSE

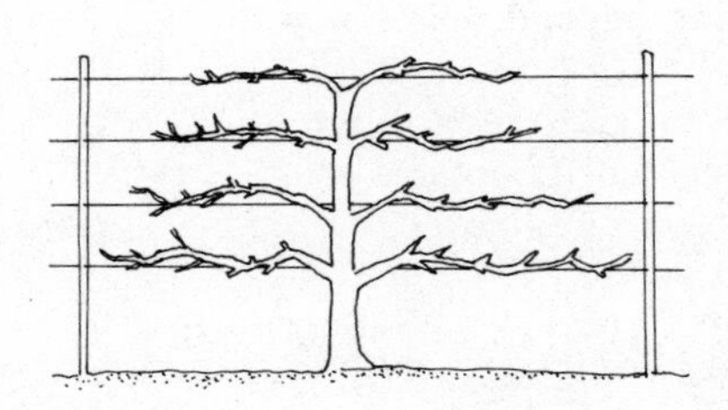

ESPALIER TREE

Espalier

A tree, usually a fruit tree, with a vertical trunk from which lateral branches are trained horizontally. Espalier trees are often attached to walls but they can equally effectively be trained to wires strained between posts.

Estipulate

See Exstipulate.

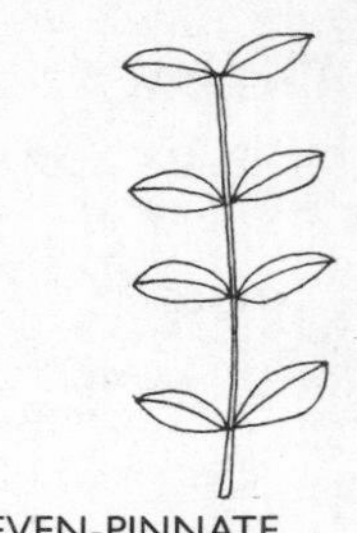

EVEN-PINNATE

Etiolate, Etiolated

Of a plant with pale leaves and extended internodes caused by inferior light conditions: leggy; drawn.

Even-pinnate

Applied to a leaf composed of leaflets in pairs but without a terminal leaflet. When a terminal leaflet is present the leaf is described as 'odd-pinnate'.

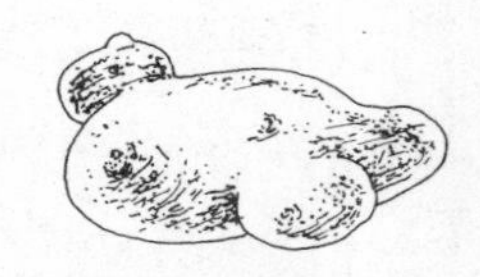

EXCRESCENCE ON POTATO

Evergreen

A plant having leaves that remain green and functional throughout the year: *cf.* Deciduous.

Everlasting

Applicable to some flowers and seed heads which last a long time when carefully dried and stored. Also known as 'immortelles', e.g. statice and teasel.

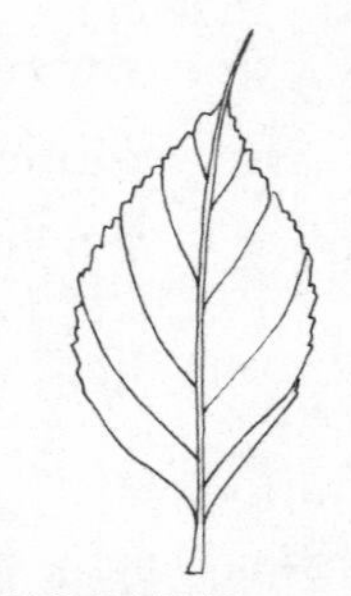

EXCURRENT

Excrescence

Abnormal outgrowth on plants, most often applied to root vegetables.

Excurrent

(1) Having the main stem reaching to the top.
(2) Extending beyond a margin or apex.
(3) Having the midrib extending beyond the lamina or blade of a leaf.

Exfoliate

To peel off in thin layers or shreds, as the bark of some trees.

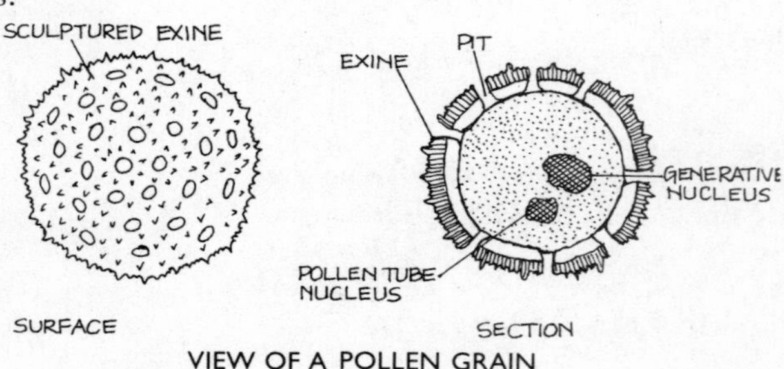

VIEW OF A POLLEN GRAIN

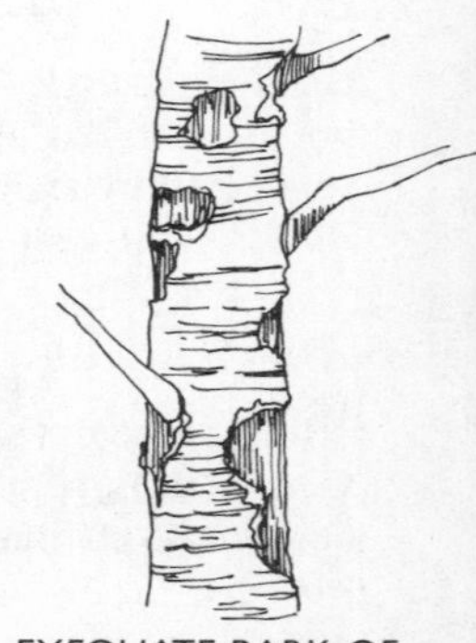

EXFOLIATE BARK OF SILVER BIRCH

Exine, Extine

The outer membrane of a pollen grain or spore.

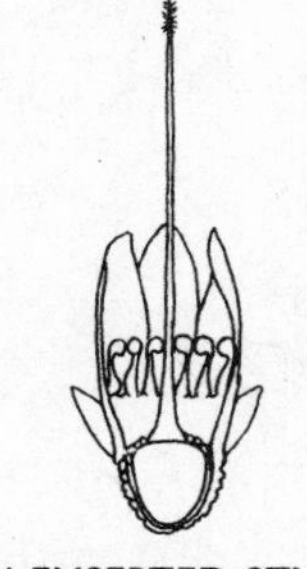

AN EXSERTED STYLE (MYRTLE)

Exocarp

See Epicarp.

Exotic

Introduced from a foreign country. Not native to a country; imported. Sometimes used in sense of showy or gaudy: *cf.* Indigenous.

Explant

A piece of plant tissue used in tissue culture.

Exserted

Protruding or projecting organs such as stamens or pistils which protrude beyond the rest of the flower.

PRIVET IS EXSTIPULATE

Exsiccata, *pl.* Exsiccatae

A dried herbarium specimen.

Exstipulate

Lacking stipules, as in privet. *Same as* Estipulate.

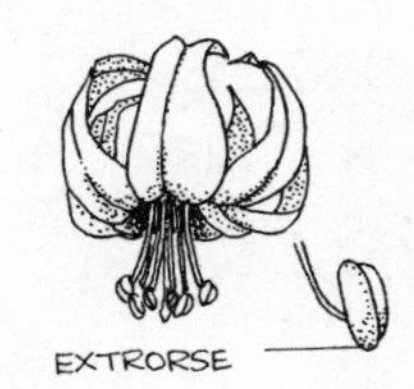

EXTRORSE

Extrorse

Applicable to anthers that turn and shed their pollen outward, opening toward the outside of the flower: *cf.* Introrse.

Eye

Usually an undeveloped growth bud, such as the eye of a potato, but can also refer to the centre of a flower, as in some primroses where the eye is a different colour or shade to the rest of the flower.

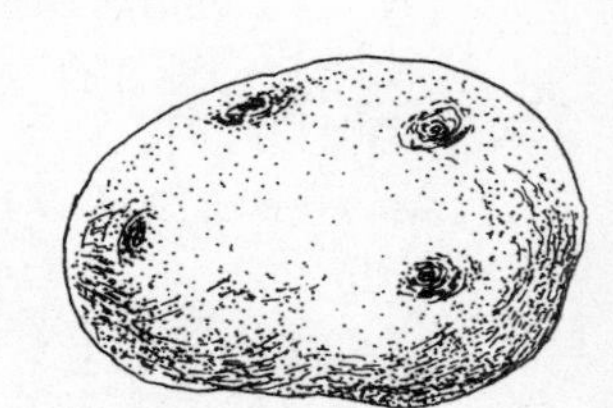

POTATO EYES

Eye Cutting

A short piece of stem containing a growth bud, often used as a means of propagating grape vines, wistarias, etc. during the dormant period.

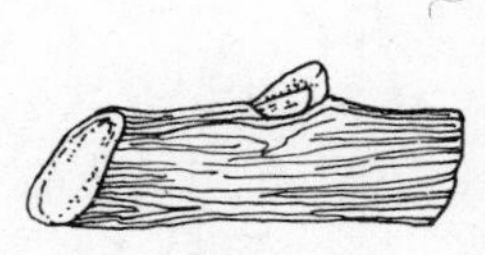

EYE CUTTING

F

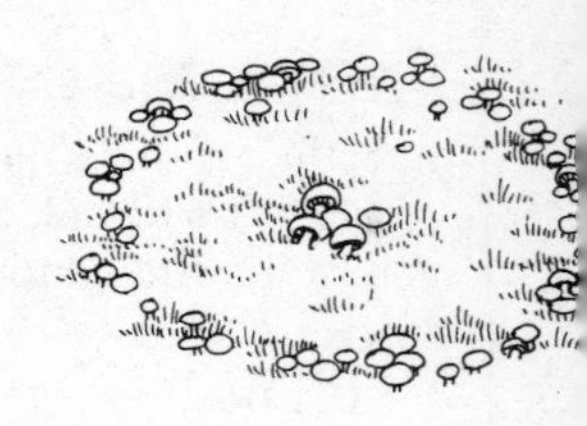

FAIRY RING

Faggot, Fagot

A bundle of sticks or small twigs for fuel or kindling. Called a knitch in some rural areas.

Fairy Ring

A ring of fungi, usually in grassland or meadows. Many toadstool-shaped fungi grow in rings outward from the point of origin and in the past these were attributed to fairies dancing. Fairy rings can be of almost any size from a few inches to a hundred yards or so in diameter.

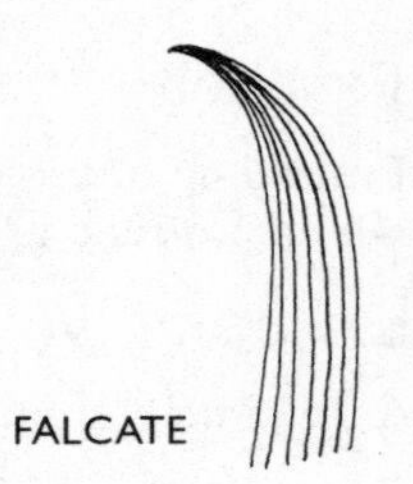

FALCATE

Falcate

Bent like a sickle or claw; hooked; sickle-shaped.

Fall

(1) The drooping outer petal of a flower, as in the iris.
(2) A quantity of felled timber.

Fallow

An area of land which is 'rested' for a season between crops.

Family

A group of related plants, more comprehensive than a genus, with names usually ending in -aceae. *See* Classification.

Family Tree

A fruit tree, usually apple, on to which several different varieties have been grafted. A novelty rather than a utility proposition.

A FANCY CARNATION

Fan Palm

The European fan palm (*Chamaerops humilis*) has palmate rather than pinnate leaves, and is the only palm indigenous to Europe. It is not really hardy in the British climate but will withstand temperatures almost down to freezing point. The Canary Island palm (*Phoenix canariensis*) is indigenous to the Canary Islands – politically but not geographically in Europe.

FAN PALM
(*Chamaerops humilis*)

Fan-shaped

A method of training fruit trees and bushes, often against walls or on wires.

Fancy

Types of flower having variegated petal colouring in contrast to single-coloured varieties (known as 'selfs'). A term mainly used by exhibitors.

Fanging

See Forking.

Farinaceous

(1) Starch-like or containing starch.
(2) Farinose.

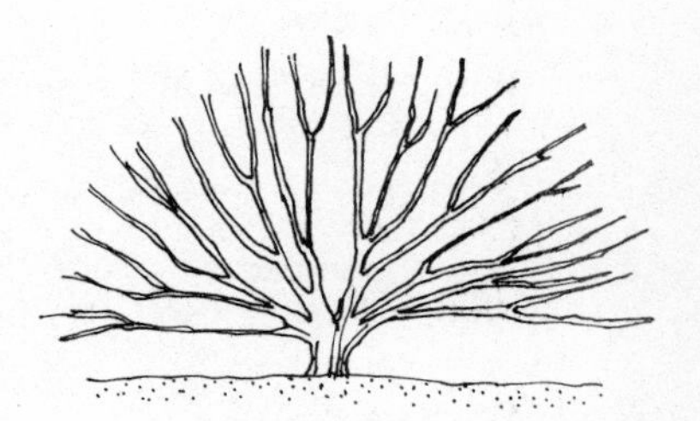

FAN-SHAPED

Farinose

Covered with a mealy or floury coating as a means of protection, as the 'bloom' on grapes and plums, or the foliage of some auriculas.

Farmyard Manure

Sometimes abbreviated to FYM, it contains the excrement of animals such as pig, cow, horse, etc. mixed with the straw or other litter used for their bedding. The nutritional value depends upon the type of animal, how it has been fed, how often the litter is replenished, and the length of time it has been stacked. Horse manure is usually reckoned the best, followed by pig and cow. Poultry manures are normally dried before use, mixed with twice their volume of soil or sand, then scattered as a top dressing.

Primula frondosa HAS FARINOSE FOLIAGE, SILVERY UNDERSIDES

Fasciation

Where two or more stems abnormally grow together lengthwise, or when a single stem is flattened and malformed to give the appearance of several stems joined.

THE CRESTED FERN IS A WELL-KNOWN FORM OF FASCIATION

Fascicle

A cluster or bundle of stems, leaves or flowers, e.g. the flowerhead of sweet william.

Fastigiate

Pointed or sloping to a point. When applied to trees, the branches are erect and somewhat appressed, e.g. the Lombardy poplar.

FASCICLE OF SWEET WILLIAM

Feathered

Applied to flowers with feather-like markings on the petals, such as some tulips and gladioli.

Feathery

Plumose or plumate; resembling a feather.

Fedge

A portmanteau word derived from FEnce and heDGE denoting a hedge used as a fence or vice versa.

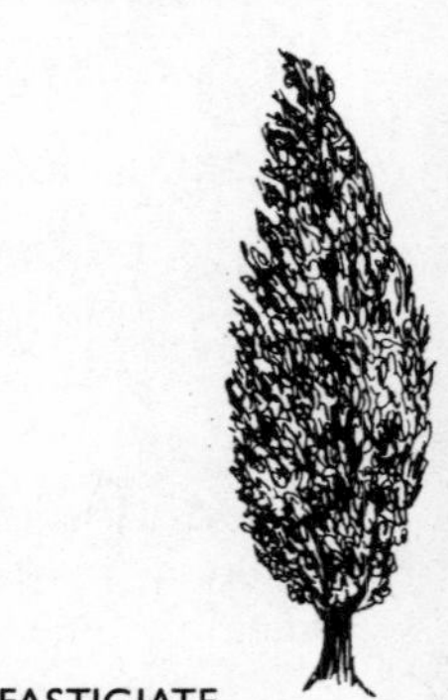

FASTIGIATE (LOMBARDY POPLAR)

Felted

Covered with dense short hairs, as on plant surfaces. Some succulents are felted, and many rhododendrons have leaves which are densely felted on the undersides.

Female Flower

A flower that has no stamens but contains functional carpels. Such a flower can only reproduce after fertilization by pollen from a male flower of the same species.

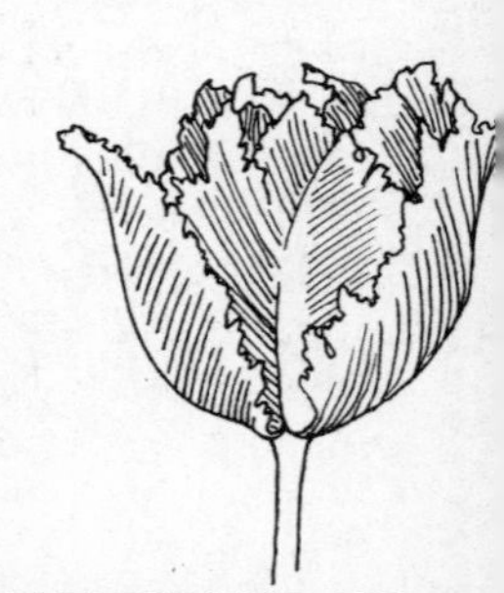

'PARROT' TULIPS ARE FEATHERED

Fence

A barrier made from wood or other natural products, plastic or wire. Fences are distinct from walls which are constructed from brick, stone, etc. There are numerous types and patterns of wooden fencing, also wire fences and electrified wire fences for restricting animals to specified areas. The former comprise lengths of tensioned wire stretched between posts, the latter carry special insulated supports and emit small electric shocks in pulses.

Fenestrate

Perforated with translucent, window-like openings or areas.

Fern

A plant of the order Filicales which reproduces by spores instead of seeds; non-flowering, with fronds instead of leaves.

Elaeagnus ebbingei HAS LEAVES WHICH ARE FELTED ON THE UNDERSIDE

Fernery

Place where ferns are grown. Formerly a type of conservatory set aside for the cultivation of ferns.

Ferruginous

The colour of iron-rust, or impregnated with iron. Some soils are said to be ferruginous.

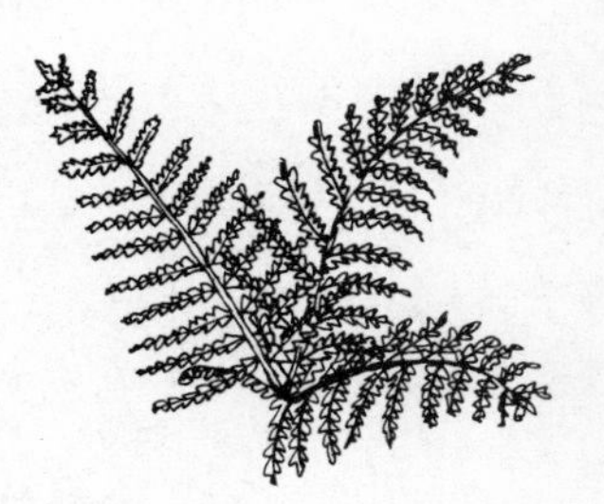

FERN

Ferrugo

Rust disease of plants.

Fertile

(1) Capable of germinating; stamens carrying ripe pollen, flowers with receptive pistils or fruits containing seeds.
(2) Soil in good heart and capable of producing quality crops.

Fertilization

Pollination; impregnation. The fusion of the pollen of the male plant with the ovule of the female to form a seed.

Fertilizer

Natural organic or synthetic substances incorporated into the soil or growing medium to improve its fertility. Manure.

Fescue

A grass of the genus *Festuca*, valuable as pasture.

TALL FESCUE

Festoon

FESTOON

An ornamental chain or necklace of flowers, fruit or leaves, fastened at each end and drooping in the middle rather in the manner of a hammock. *Same as* a Swag.

Fetid, Foetid

Stinking; having or emitting an unpleasant odour.

F_1 Hybrid

First-generation cross between two pure-bred strains. Seed is produced by controlled hand-pollination but the second generation will seldom breed true. The crop is usually very consistent in size, colour, vigour, etc.

F_2 Hybrid

Second-generation cross between two F_1 hybrids. Not so vigorous as F_1 hybrids and usually showing great variation.

Fibre

(1) Thread-like cell or filament, part of vegetable tissue.
(2) Small root or twig.

Fibril

(1) Small fibre or filament.
(2) Ultimate subdivision of root.

Fibrous

Applicable to loam composed of rotted grass and fine grass roots; ideal for incorporating into soil-based composts.

Fibrous Root

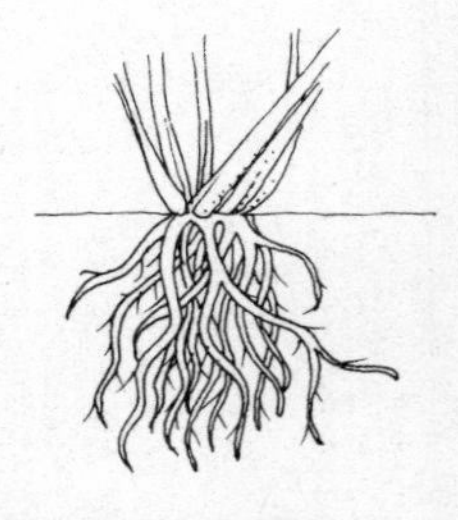
FIBROUS ROOTS (GRASS)

A root that branches in all directions, as in plants that have a mass of fine matted roots as distinct from thicker, fleshy ones.

Field Capacity

Agricultural term sometimes used by gardeners to denote a free-draining soil which is holding its maximum amount of water.

Filament

Thread-like stem that carries the anther in a stamen: *cf.* Fibre.

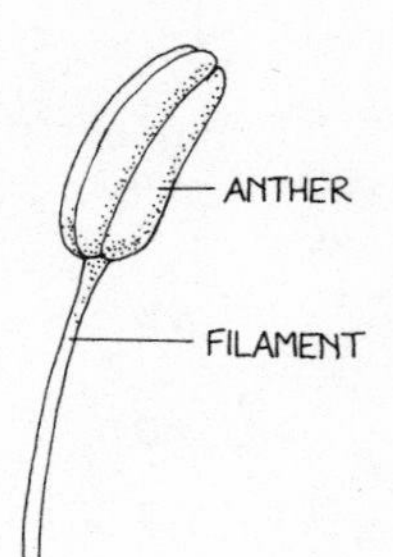

Filiation

The formation of offshoots or branches.

Filiferous

Having thread-like appendages.

Filiform

Very long and thin, thread-like; usually applied to leaves but also to other organs.

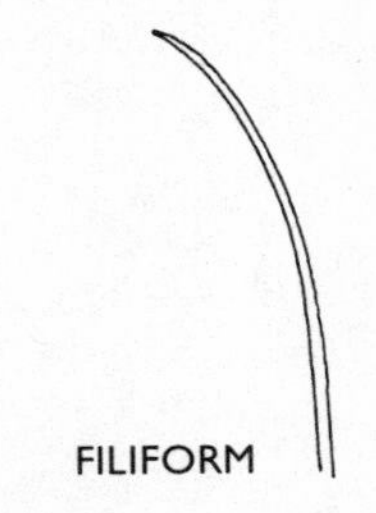
FILIFORM

Fillis

Soft string specially made for gardeners and used to tie plants to supports without damage to the stems. Originally made from natural materials, now more often a plastic product.

Fimbriate

(1) Having fringed margins as in a stigma.
(2) Of flowers with petals divided or split at the edges, e.g. certain varieties of chrysanthemum.

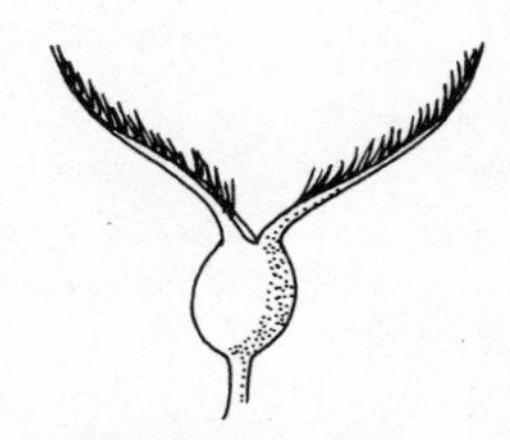
FIMBRIATE STIGMA

Finger and Toe

Victorian gardeners' term for club root disease of brassicas.

Fire

Loose description of any disease or condition which causes the leaves of plants to appear scorched; also an incurable fungus disease of tulips.

FIMBRIATE PETALS OF *Dianthus superbus*

Fire Blight

A bacterial disease of the rose family which can also affect pear and apple trees. It is a notifiable disease in Britain and all infected material must be burned.

Firm

To compact the soil after planting; to fix plants firmly in the soil, especially young trees which can be killed by wind-rocking. Trees are normally staked as well as firmed.

FIRMER

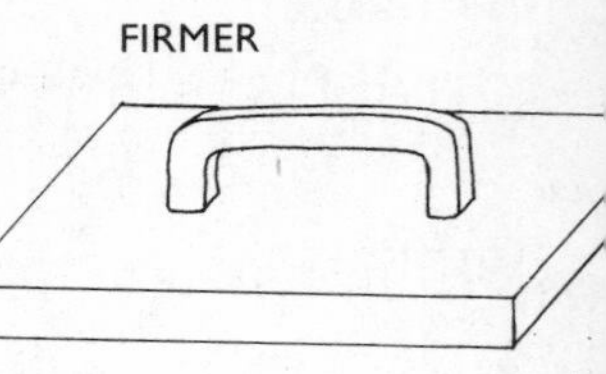

Firmer

A flat-bladed implement (usually of wood) used to firm the soil in seed boxes, etc.

FIRMING A YOUNG TREE

Fish Manure

Made from dried and pulverized fish waste, it should be dug into the soil in autumn, and kept well away from the roots of plants, which may be scorched. Fish waste from the kitchen can be dug in in the same way, but it needs a few months to decay before the nutrients will become available to the plants.

Fission

More commonly called budding. Method of reproduction by division of cells or organisms.

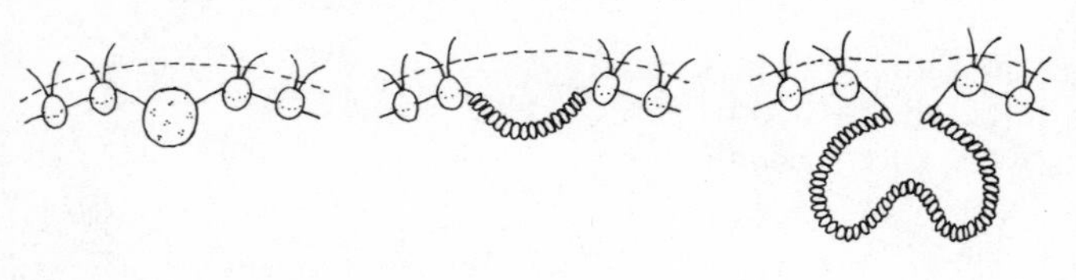

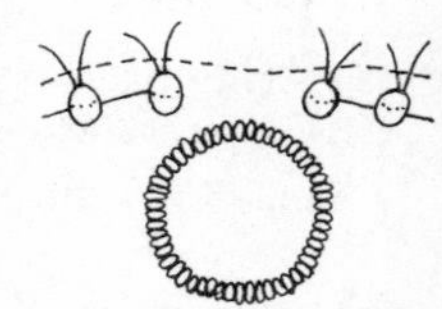

FISSION OR BUDDING: STAGES OF REPRODUCTION BY DIVISION OF THE CELLS OF VOLVOX (ALGAE)

Fissure

Narrow opening or slit in an organ, etc.

Fistulose

Hollow like a tube or pipe. Bamboo stems are fistulose except at the nodes.

Flabellate

Fan-shaped, or a fan-like structure. The foliage of the Chusan palm is flabellate.

FLABELLATE (CHUSAN PALM LEA

Flaccid

Soft and weak; limp; lax.

Flagellum

Microscopic tail-like appendage on protozoan, used for locomotion.

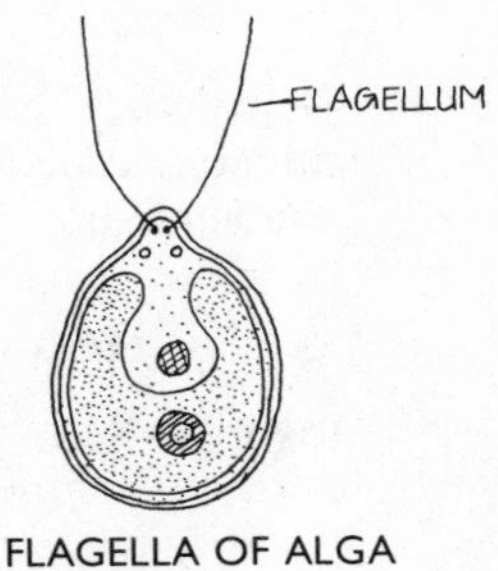

FLAGELLA OF ALGA (SECTION)

Flagging

Wilting, notably applied to the drooping of foliage, mainly on greenhouse plants, in hot weather when the atmosphere becomes too dry and the plants are transpiring faster than fluids can be drawn up from the roots. The remedy is to shade the glass and spray the foliage with clean water, wetting the benches and paths.

Flaked

Former description of bicoloured flowers in which one colour would overlay the base colour in bold splashes.

Flame Gun

A type of pressurized blowtorch on a handle used to burn off surface weeds, etc. or to sterilize small areas of ground.

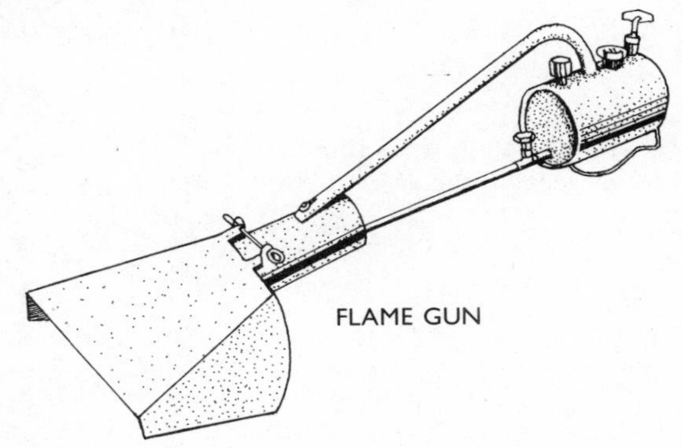
FLAME GUN

FLAKED TULIP

Flamed

Applied to petals which have a solid patch of colour up the centre edged with fine feathering.

Flat

See Seed box.

Flay

To remove the bark from a tree, or strip off turf from a meadow.

FLAMED TULIP

Flea-beetle

A small beetle which feeds mainly on root vegetable crops and brassicas. Seeds and adult plants should be dusted with an insecticide for control.

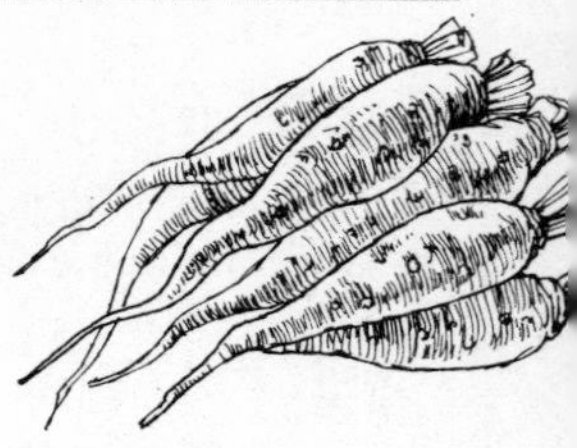

FLESHY SALSIFY ROOTS

Fleshy Rooted

Any plant with thick fleshy roots, e.g. salsify; also applied to fleshy storage organs such as corms, bulbs, etc.

Flexuose

(1) Pliable, easily bent.
(2) Bent or curved alternately in opposite directions.

FLEXUOSE BRANCHES AND TWIGS OF WEEPING WILLOW

Flitch

Slice cut lengthwise from a tree-trunk, often retaining the bark on one side.

Floccose

Woolly; covered with soft woolly tufted hairs that are usually easily rubbed off.

Flocculation

A method of improving soil texture by the use of lime or other soil conditioners, thereby increasing the air content, especially in heavy clay soils.

Flocculent

Like tufts of wool; having woolly tufts.

Flora

The complete range of plants growing in a particular area or climate, or a book that catalogues these.

Floral

Anything pertaining to flowers – petals, sepals, anthers, etc.

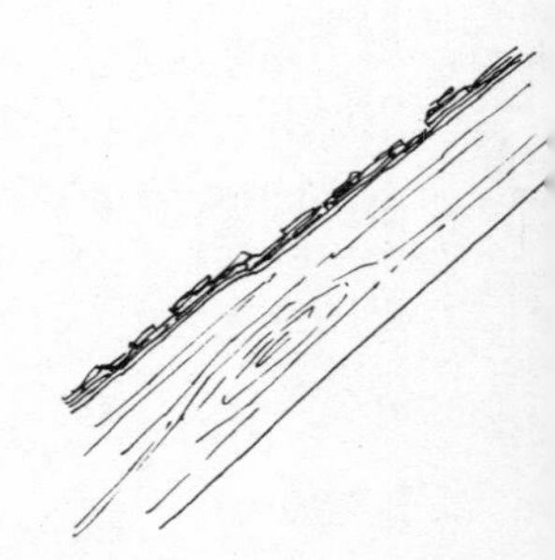

FLITCH

Floral Arrangement

The art or craft of arranging plant material (foliage, flowers, etc.) in a decorative manner.

Flora Pleno, *pl.* Flore Pleno

A term used in the past, but seldom nowadays, to describe an extreme double flower.

Floral Diagram

A diagram showing the arrangement of the parts of a flower in the bud.

Floret

A small flower, or one flower in a close-packed inflorescence, as in a daisy.

Floribunda

Characteristic of a plant whose flowers grow in clusters, usually applicable to a rose but also to other species.

Floricane

The biennial cane of bramble, raspberry, etc. during its second year, upon which flowers and fruit are produced.

Floriculture

Victorian expression for the cultivation of flowers, nowadays seldom used.

Floriferous

Literally 'bearing flowers', but often misused to denote an abundance of flowers.

Floriform

Flower-shaped.

Florilegium

Originally a collection of flowers, now more often a written anthology concerning plants.

Florist

A cultivator or seller of flowers; a student of flowers; a flower specialist.

Floristics

The study of floras.

Floury

Farinose, or having a flour-like coating as on leaves, e.g. the foliage of *Primula farinosa* and *P. frondosa*.

Flower

A specialized shoot of a plant bearing the reproductive organs, often brightly coloured to attract the fertilizing insects; the blossom of a plant.

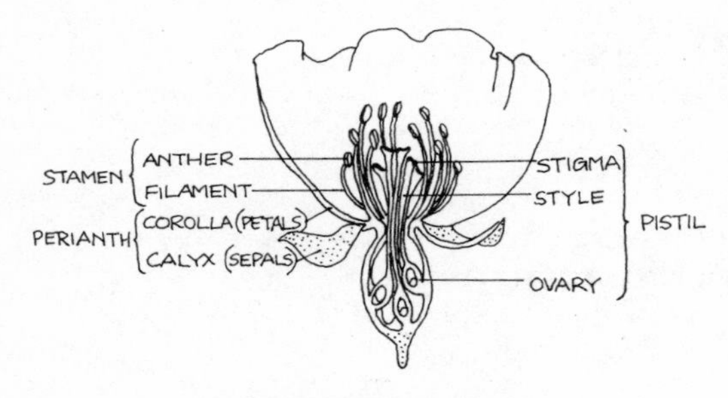

PARTS OF A FLOWER

Flower Box

Another name for a window box, mainly used in USA; a narrow container constructed formerly for use on windowsills but now often a feature of patios and balconies.

Flower Gatherer

(1) One who gathers flowers, e.g. daffodils, from the field to pack for market.
(2) A type of secateur specially designed to hold the stem of a flower after cutting it.

Flower Pot

A container for holding the soil in which a plant can grow. Made of clay, plastic, paper or peat, containers come in various sizes and shapes to suit all subjects at every stage of development.

Flower-stalk

The stem that supports the flowerhead or flower.

Flowerbed

A garden bed for flowers; often a section of open ground cut into a lawn or paved area.

Flowerhead

A close inflorescence in which all the florets are stalkless on the receptacle.

FLOWERHEAD OF YARROW

Flowers of Sulphur

Refined sulphur dust used mainly under glass against mildews, but also to control red spider and other mites.

Flue Dust

Soot, formerly used as a nutritional additive. The sweepings from modern high-temperature boilers have little food value for plants; it bears no comparison with old-fashioned soot from coal-fired boilers or the chimney stacks serving coal fires.

Flush

(1) An eruption of fresh shoots, usually from a previously dormant root or stool.
(2) A burst of production of fruit or flowers which do not produce continually, such as mushrooms and certain types of ivy-leaved pelargoniums, etc.

Fluvial

Found in or excavated from rivers. Often applied to types of sand used in horticulture.

Fly

General term, not always accurately used, to describe many insect parasites found on plants. Greenfly, for instance, is not a true fly.

Fog

(1) Long grass left uncut over winter, also known as after-grass.
(2) Insecticidal dust produced by a machine for coating greenhouse and field crops to control a variety of pests.

Fogging

See Dusting.

Foliaceous

Leaf-like; having organs like leaves.

Foliage

Leaves; leafage.

Foliage Plants

Those plants which are grown mainly for their foliage, most having insignificant flowers but attractive, often brightly coloured leaves.

A FOLIAGE PLANT (*Sansevieria trifasciata laurentii*)

Foliar Feeding

Supplying plant nutrients through the leaves instead of the roots by means of a fine liquid spray.

Foliar Spray

A liquid nutrient applied to the foliage of plants through which it is absorbed.

Foliate

Leaf-like; having leaves or a specified number of leaflets.

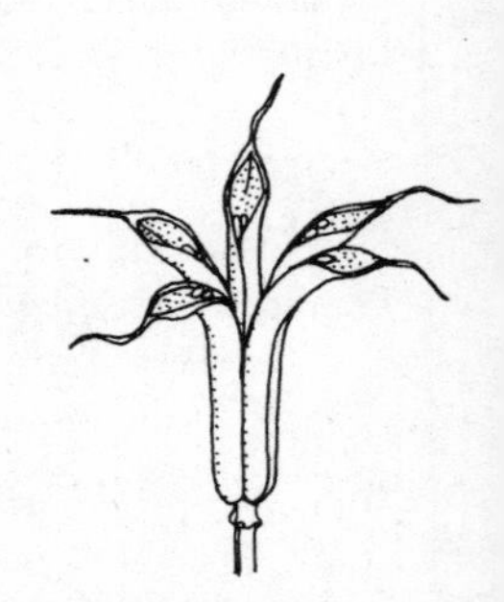

FOLLICLES OF AQUILEGIA

Follicle

A dry fruit formed from a single carpel containing several seeds, splitting along the ventral suture only; often pod-shaped with the seeds attached to the seam, as in peas.

Folly

A non-functional but decorative structure, such as an artificial ruin, erected in a garden as a focal point; a frequent feature of grand late eighteenth-century gardens, and then called a conceit.

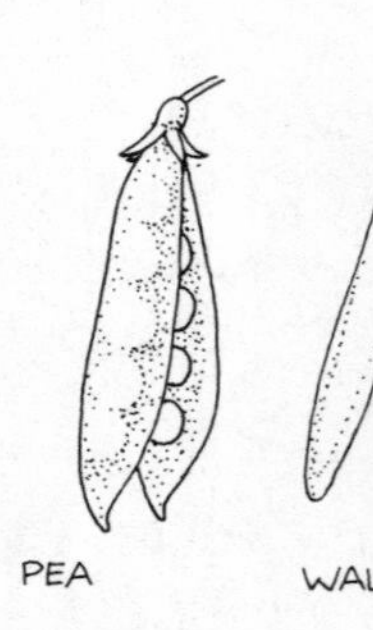

FOLLICLES

Foot Rot

Alternative name for collar rot, a fungal disease of cucumbers and tomatoes which attacks the stem at or well above ground level; it can also affect the fruits.

Footstalk

Petiole of a leaf; peduncle of a flower.

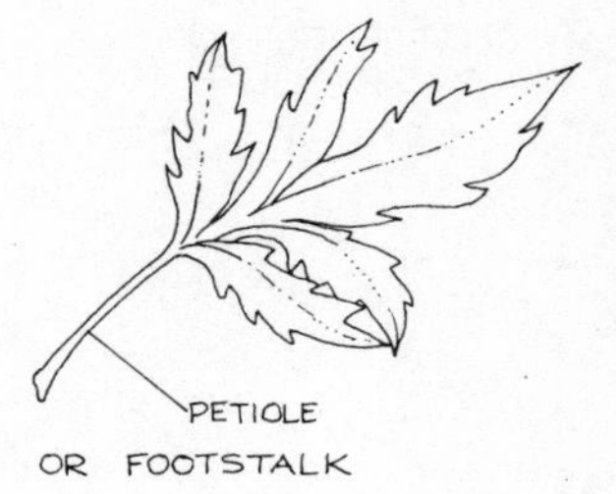

Forcing

Inducing accelerated plant growth by the artificial control of light and heat. In some cases light is increased to encourage early flowering (e.g. chrysanthemums) while in others, including rhubarb, light is almost totally excluded to produce earlier and more tender growth.

Forest

Large area of trees and undergrowth, occasionally interspersed with glades or pasture.

Forester

A person who has charge of a forest; one who has the care of growing trees.

Forestry

The science of forest management; the art of planting, tending and managing forests.

Fork

(1) Pronged implement for digging, lifting, carrying, etc.
(2) The point where two branches of similar age and size in a tree divide, as opposed to a crotch, the point where a main branch joins the trunk.

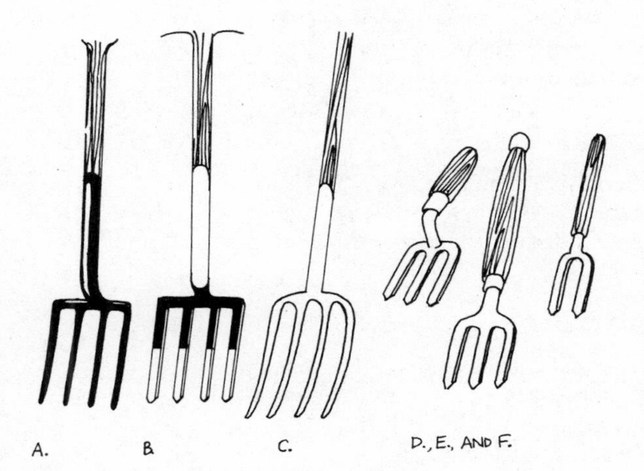

A. DIGGING FORK
B. LIFTING FORK WITH FLAT TINES
C.PITCHFORK
D., E., AND F. HAND FORKS
(NOT TO SCALE)

Fork Hoe

A type of hand-cultivator with two or three prongs instead of the more usual blade.

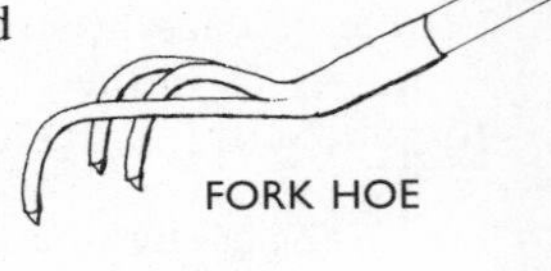
FORK HOE

Forked

Branched, cleft, divergent.

Forking

(1) The division of the root of a tap-rooted vegetable such as carrot. Also known as fanging.
(2) The loosening and aeration of the topsoil with a fork or similar implement.

FORKED BRANCH

Forma, *pl.* Formae

The lowest subdivision of a species ranking after variety; mainly used to indicate a trivial variation such as a difference in coloration. The term is no longer permissible under the 'International Code of Nomenclature for Cultivated Plants', being replaced by cultivar for cultivated plants.

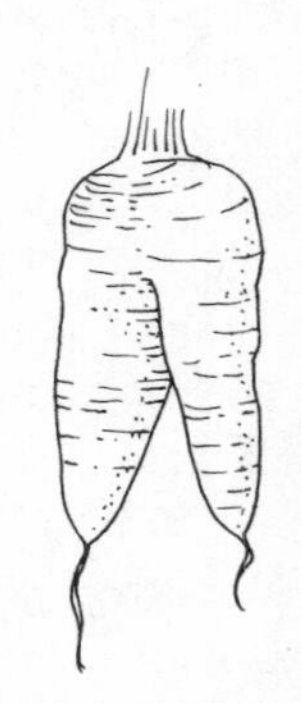
FORKED CARROT

Formal Garden

A garden laid out in geometrical designs such as square and circular flowerbeds, often including clipped hedges and architectural features, fountains, etc.

A FORMAL GARDEN LAYOUT

Formaldehyde, Formalin

Formic aldehyde, a gas in solution used as a disinfectant and sterilant; formalin is the commercial formulation used (diluted) for cleaning tools, pots, etc. and for general sterilization of greenhouses, seed boxes, etc.

Formation Pruning

The pruning of trees and shrubs to desired patterns, either 'natural' as in the shaping of a shrub, or 'artificial' as an espalier fruit bush.

EXAMPLE OF FORMATION PRUNING

Fornix

A small elongation of the corolla.

Forwarding

The advancement of a crop by the use of artificial protection such as cloches. Distinct from forcing, where light is restricted.

FOVEOLATE (APPLE)

Foveolate

Pitted, having one or more small depressions. 'Bitter pit' of apples was once thought to be the result of virus attack but is now confirmed as emanating from poor growing conditions.

Frame

A low structure, covered in glazed lights or similar transparent material, used mainly for hardening-off young plants before planting out in the open ground, when it is unheated. Heated frames can be used for raising seedlings of bedding plants, etc. and for producing early salad crops.

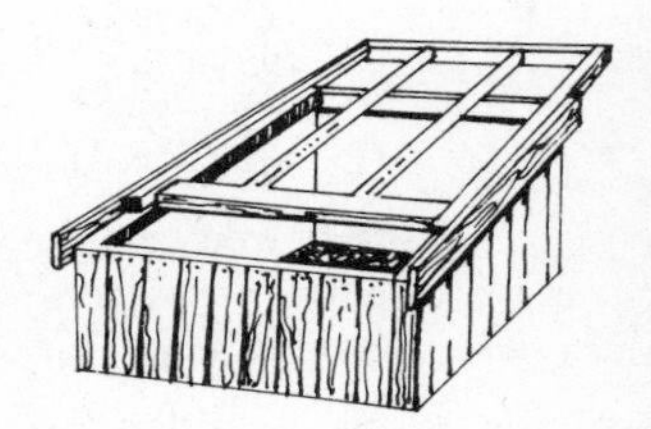

A SIMPLE GARDEN FRAME

Free

Separate, not joined to one another, as when applied to sepals, petals, etc. Free petals are sometimes described as 'primitive'.

FREE PETALS

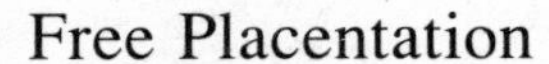

Free Placentation

As applied to the arrangement of ovules in a plant ovary, this can be either free-central or free-basal according to whether the ovules are carried on a stem attached at the base only, or at opposite ends.

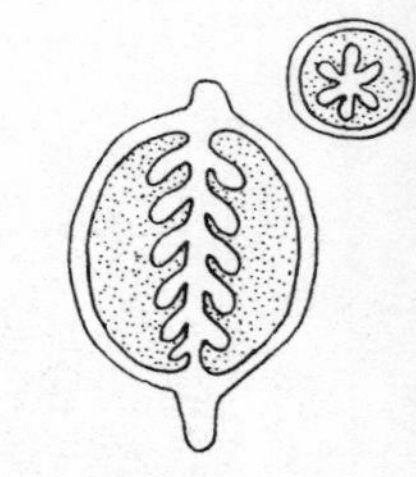

FREE CENTRAL

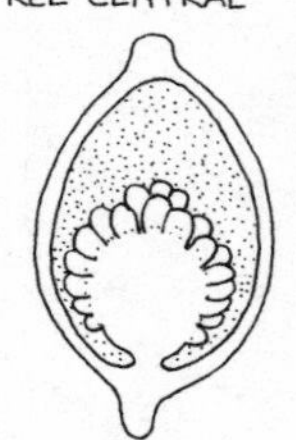

FREE BASAL

PLACENTATION

Friable

Soil which is crumbly and easily worked, neither too wet nor too dry. An ideal tilth, perfect for sowing and planting.

Frill

(1) Natural fringe of hairs, etc. on edges of leaves or petals.
(2) The wavy edge of some petals.

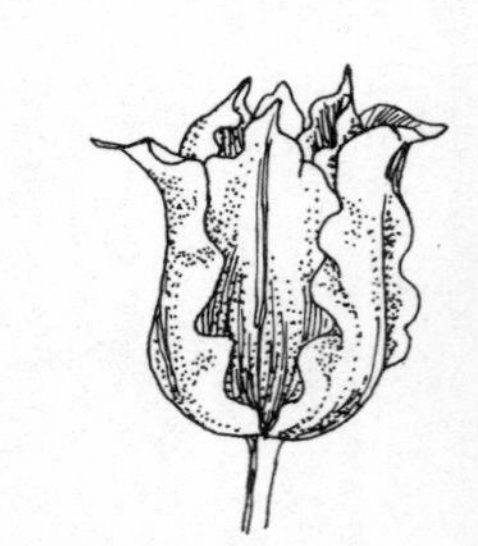

FRILLED TULIP

Frill-girdling

A method of destroying established trees by making downward-sloping, overlapping cuts with a hatchet or similar implement, encircling the trunk at or just above ground level, then filling the cuts with a powerful brushwood killer.

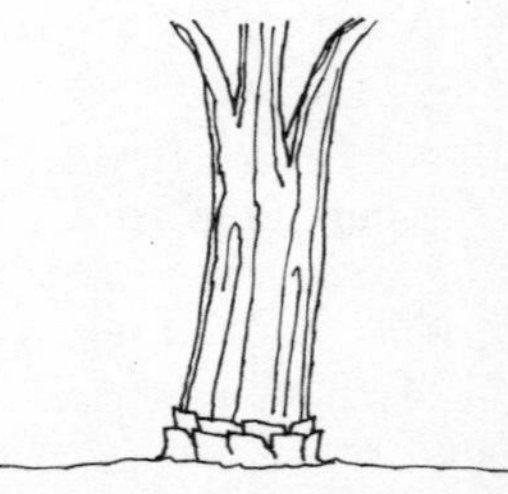

FRILL-GIRDLING

Froghopper, SPITTLEBUG (USA)

A froth fly; the insect which produces cuckoo-spit. A member of the suborder Homoptera.

Frond

The leaf-like part of a palm or fern.

Frost Pocket

Small area subject to frost. If cold air, as it flows downhill, meets an obstacle such as a wall, a frost pocket may form where the air temperature is considerably lower than it is above, presenting a hazard to tender plants.

Frost Resistance

Varying ability of certain plants to withstand frosts. Plants which are totally frost-resistant are termed hardy and those which are not, tender; half-hardy and near-tender convey some degree of frost tolerance.

Fructiferous, Fructuous

Bearing fruit; fruitful.

FERN FROND

Fructification

The action of forming or producing fruit. Reproductive part of plants, mainly applied to fungi and liverworts; the structure that contains the seeds or spores.

STAGES IN THE DEVELOPMENT OF THE FRUCTIFICATION OF THE MUSHROOM

Fructify

To fertilize or impregnate.

Fruit

The ripened ovary; the whole seed-bearing organ whether it be pod, capsule, drupe, etc.

TYPES OF FRUIT

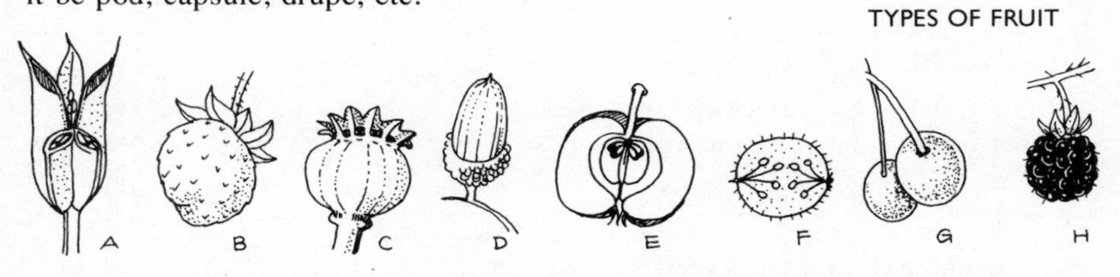

A - FOLLICLE (COLUMBINE); B - ACHENE (STRAWBERRY); C - CAPSULE (POPPY); D - NUT (OAK); E - POME (APPLE); F - BERRY (GOOSEBERRY); G - DRUPE (CHERRY); H - SYNCARP (BLACKBERRY)

Fruit Bud

A bud from which a flower and eventually a fruit will develop.

Fruit Gatherer

A mechanical device for gathering fruit from tall trees without damage, consisting of a pair of clippers on a long pole with a bag or similar receptacle fixed just below the clipper head to catch the fruit. There are several variations to the basic design to suit the particular fruit to be harvested.

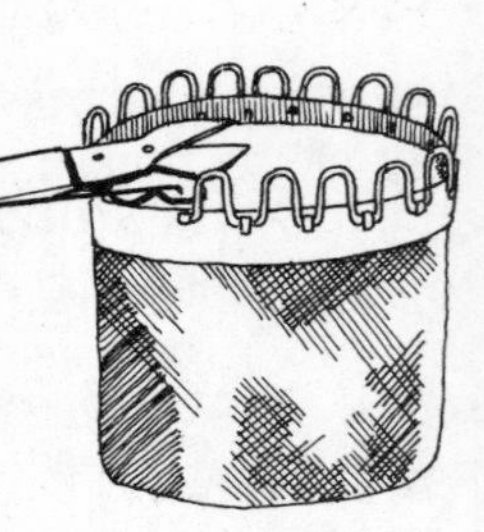

FRUIT GATHERER

Fruiting Body

The spore-bearing organs of the larger fungi, including mushrooms, toadstools, common puffball, deathcap (fatal if eaten), etc.

Fruitless

Barren.

FRUITING BODY OF LAWYER'S WIG FUNGUS

Fruitlet

A small or immature fruit.

Frutescent

Shrubby with woody stems.

Fruticose

Shrubby but having no main trunk.

Fugacious

Easily or quickly shed or withering.

Full Trenching

A method by which the topsoil is replaced by one spit of subsoil: *see also* Mock Trenching and Ridging.

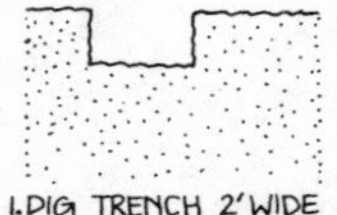

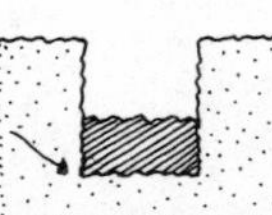

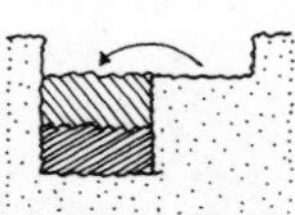

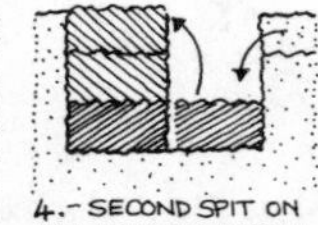

FULL TRENCHING

Fumigation

Applying insecticides or fungicides in the form of gases or smokes.

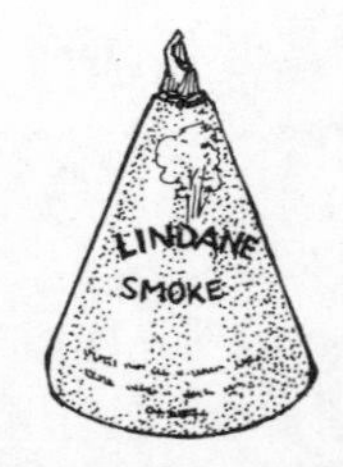

FUMIGATING 'SMOKE'

Fungicide

A chemical to control or kill fungus.

Fungoid

Fungus-like.

Fungus *pl.* Fungi

Cryptogamous plant which, because it lacks chlorophyll, feeds on organic matter. Parasitic fungi, e.g. *Pythium*, rely upon living plants or animals to provide their nourishment, saprophytic fungi, such as toadstools and mushrooms, feed on decaying matter. Many fungi are beneficial but others are the cause of numerous plant diseases and quite a number are poisonous to humans when eaten. Reproduction is by spores.

FIELD MUSHROOM
FUNGI

Fungus Gnats

See Mushroom flies.

Funicle

The stem of an ovule by which it is attached to the placenta in the ovary. Also known as seed stalk.

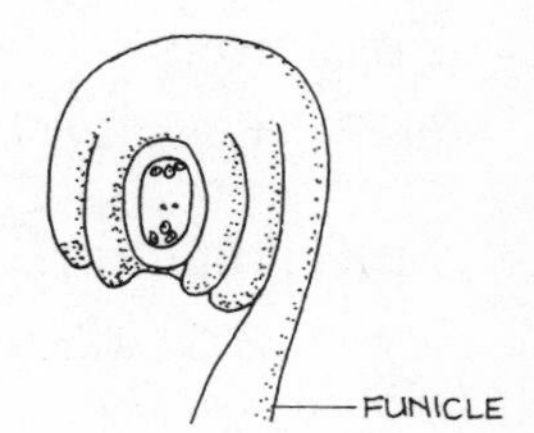

Funnelform

Gradually widening upward, as in the flowers of morning glory, convolvulus, etc.

Furcate

Forked or branched: *cf.* Bifurcate – double forked.

Furfuraceous

Scaly, scurfy, covered with soft bran-like flakes or scales.

Furrow

Narrow trench or rut made in the soil by an implement such as a plough or spade.

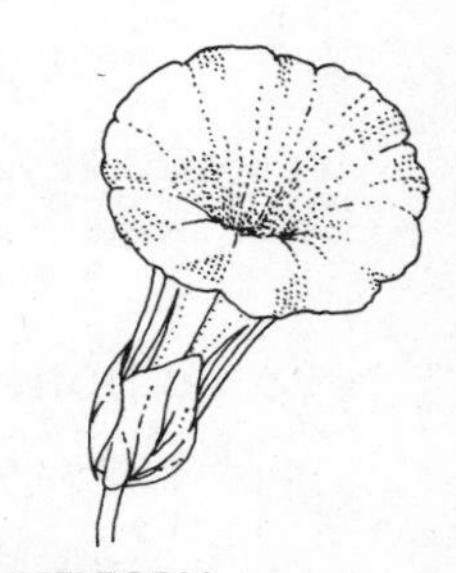

FUNNELFORM
(CONVOLVULUS)

FURROWED FIELD

Furrowed

(1) An area of soil containing several furrows.
(2) Wrinkled or creased, as applied to petals, leaves, etc.

Fusiform

Spindle-shaped, having a swollen middle and tapering to both ends.

SOME SPORES ARE FUSIFORM

Fynbos

Bushy like a thicket; plants having an abundance of fine, hard, closely packed leaves.

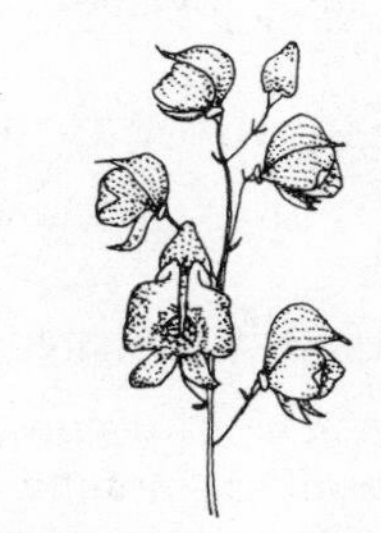
GALEATE
Aconitum anglicum

Galeate

Helmet-shaped or hooded, as in the aconites.

Gall

(1) An abnormal growth on a plant caused by the parasitic attack of fungi, insects or bacteria, e.g. the oak apple.
(2) Bare spot in a coppice or field.

OAK APPLE GALL

Gall Weevil

A beetle whose grubs may attack cabbages and other brassicas, causing pea-sized galls on the stem both above and below ground. They are fairly harmless, not to be confused with club root disease.

GERMINATING POLLEN GRAIN CONTAINING TWO MALE GAMETES

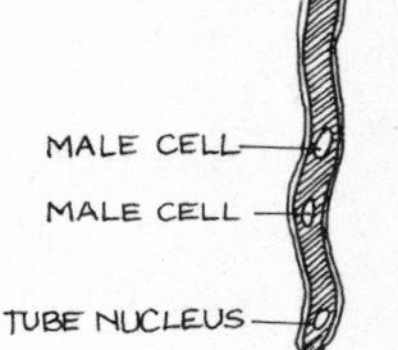

Gamete

A 'sex-cell', either male or female, which when united with another gamete will produce a cell that in turn is capable of developing into a new individual. *Same as* Germ cell.

Gametophyte

A plant producing gametes, the stage (of a plant having alternating generations) which produces gametes, e.g. the prothallus of a fern.

Gamopetalous

Having the petals joined at the edges in the form of a tube. *Same as* Sympetalous.

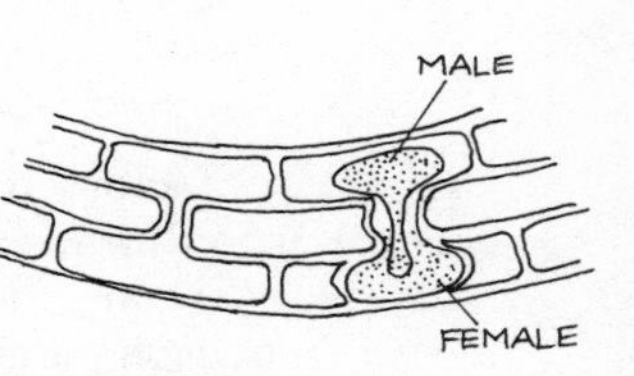

GAMETE

Gamophyllous

Having leaves or similar parts (bracts, petals or sepals) joined together at their margins.

Gamosepalous

Having sepals joined together marginally, at least basally.

GARLAND

Garden Line

A length of cord or string, stretched between two sticks or pegs, forming a straight line as a guide for planting or drilling.

Garland

Wreath or necklace of flowers or foliage, or of a combination of both.

Garotting, Garrotting

A method of encouraging fruiting on fruit trees by encircling the trunk of the tree, or an individual branch, with a tourniquet to restrict the sap flow. Basically similar to ringing but less drastic.

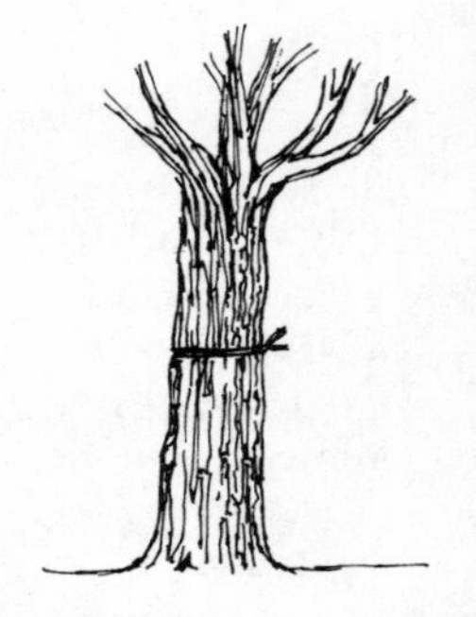
GAROTTING

Gazebo

Originally a garden house of two storeys from which the surrounding area could be viewed. Nowadays the word is applied to any circular, octagonal or similar-shaped structure set in a garden environment, even a summerhouse or teahouse of one storey only.

Gemma, *pl.* Gemmae

A bud-like reproductive body capable of separating and becoming a new individual.

Gemmate

Reproducing by gemmation; having buds.

Gene

One of the factors in the germ cells (gametes) which become paired when two gametes fuse to form a new individual (zygote), one factor of each pair being transmitted from each parent.

Generic

Latin adjective indicating the genus to which a plant belongs and preceding the specific name, e.g. *Hedera helix*. Anything to do with a genus, such as its name or some constant characteristic of the genus, applicable to any member of a group or class.

Genetics

The science of plant breeding and the study of the mode of inheritance of the distinguishing characteristics.

GENICULATE STYLE

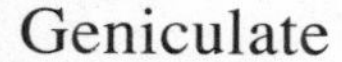

Geniculate

Bent abruptly like a knee.

Geniculum

A knee-like joint, often thickened.

Genus, *pl.* Genera

The smallest natural group containing related but distinct species. Related genera form a family: *see also* Classification.

Geophyte

A plant growing from underground organs such as tubers.

Geotropic

The movement of shoots and roots of plants caused by gravity. Positive geotropism causes roots to grow downward; negative geotropism causes shoots to grow upward: *cf.* Heliotropic, Phototropic.

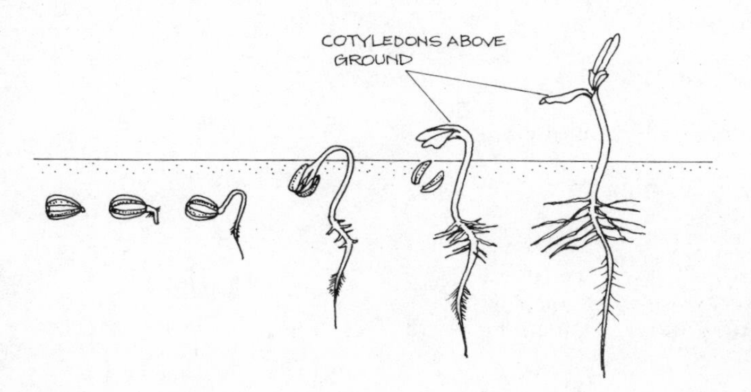

EPIGEAL GERMINATION (SEE NEXT PAGE)

Germ Cell

See Gamete.

Germination

The earliest development of seedlings from fertilized seed; sprouting.

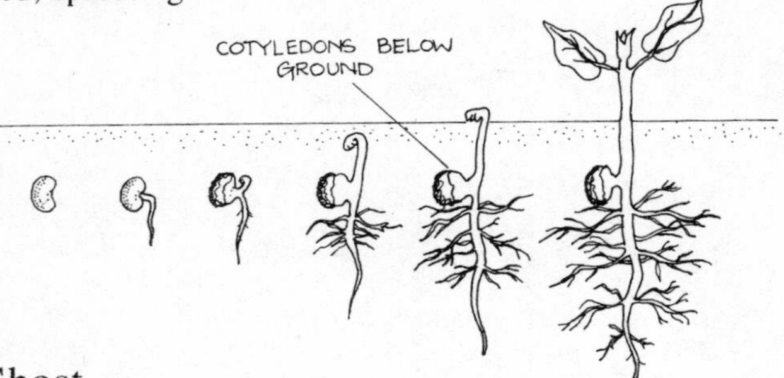

HYPOGEAL GERMINATION (RUNNER BEAN)

Ghost

See Albino.

Gibbous

Swollen or humped. Mainly used botanically to indicate swollen joints as in the 'gouty geranium' (*Pelargonium gibbosum*).

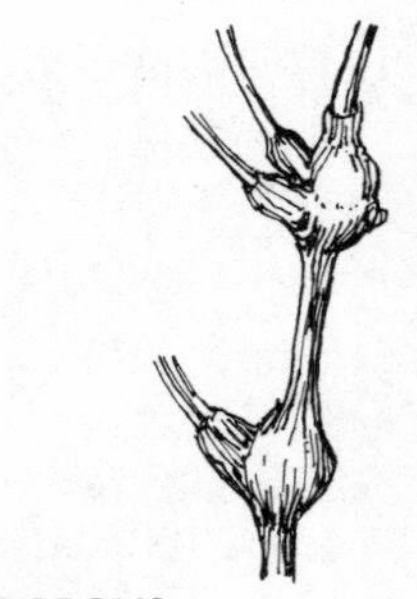

GIBBOUS (PELARGONIUM)

Gill

(1) Vertical radiating plates on the underside of mushrooms and some other fungi.
(2) A deep wooded ravine or small valley.

GILLSON WHICH SPORES ARE PRODUCED

MYCELIUM

Girdling

(1) Killing a tree by removing bark in a ring around the trunk.
(2) Increasing the fertility of a fruit tree by removing bark from the trunk in an incomplete ring, leaving about a third of the bark intact.

Glabrate

Nearly glabrous, or becoming so as it reaches maturity. Often applied to foliage.

Glabrous

Having no hairs or similar projections; bald, but not necessarily smooth.

Glade

Open space or path through a forest or dense woodland.

Gland

A wart-like structure on the surface, protruding from any part of a plant from which oils and resins are often secreted. Some plants produce glandular hairs which sting, such as the common stinging nettle.

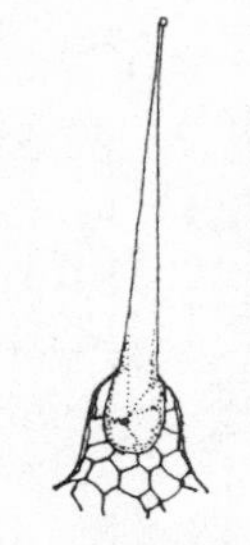

SECTION THROUGH A GLANDULAR HAIR OF STINGING NETTLE

Glandiferous

Bearing acorns, or other nut-like fruits.

Glandular

Consisting of, pertaining to, or containing glands.

FRUITS OF CHERRY PLUM ARE GLOBOSE

Glandular-pubescent

Having glands and hairs intermixed.

Glaucescent

Slightly glaucous.

Glaucous

Covered with a bluish-green bloom, as in grapes and plums.

Glebe

Originally a portion of land attached to a clergyman's benefice; now more often applied to lands owned by and usually close to the church.

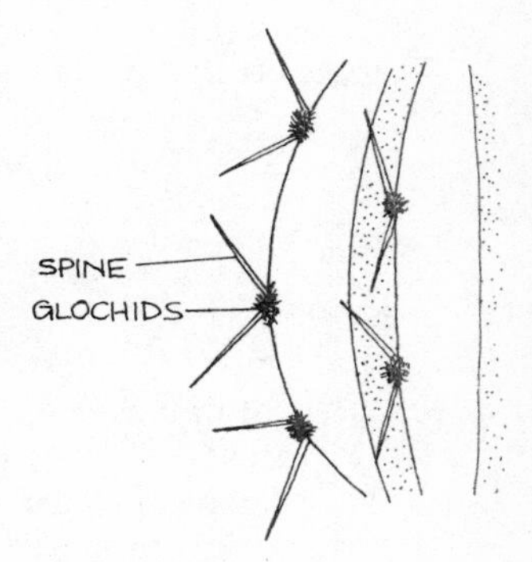

Glen

Narrow valley or ravine, especially in Scotland.

Globose, Globular, Globulous

Circular, round, spherical or nearly so. Applied to plant organs, including fruits such as cherries, rounded shrubs or bushes, and trees with a spherical type of crown.

Glochid

Tiny barbed spine or bristle growing from the areole in many Cactaceae such as the *Opuntia* tribe.

GLOCHIDS

Glomerate

In a dense cluster or clusters.

Glomerule

(1) A small ball or group of spores.
(2) A cluster of short-stalked flowers.
(3) Antiquated word for a clustered flowerhead.

Glucose

Dextrose, grape-sugar, found in certain fruits, honey, etc.

Glumaceous

Resembling a glume – dry and crisp.

Glume

A small crisp bract with a flower in the axil, as in the grass *Agropyron junceiforme*. Glumes occur in pairs in most grasses.

RYE GRASS
(*lolium perenne*)

Glutinous

Sticky, gluey, tenacious.

Gnarled

Twisted, knotted, distorted like an old malformed tree.

Goblet

A method of training a fruit tree in the shape of a goblet or vase. The branches arising from the short trunk are trained outward at first, then upward to create a cup-shaped framework. A wooden or metal hoop is attached to hold the branches in place, and a further hoop added as they grow to maintain the shape. Said to result in increased fruit production, but in any case makes a very attractive and unusual ornamental feature.

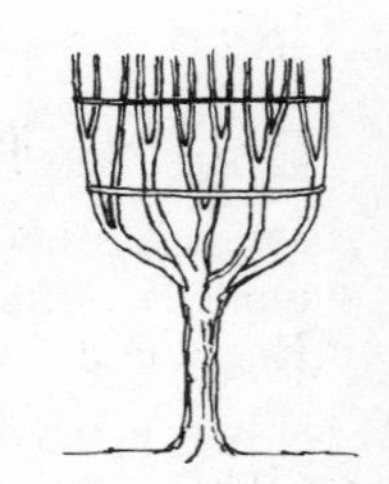
GOBLET

Golgi Bodies

Masses of cytoplasm, once thought to be present only in animal cells, now known to occur in plant cells also, at least in the meristem. They are thought to play a part in primary cell-wall formation but as yet their precise function is not known.

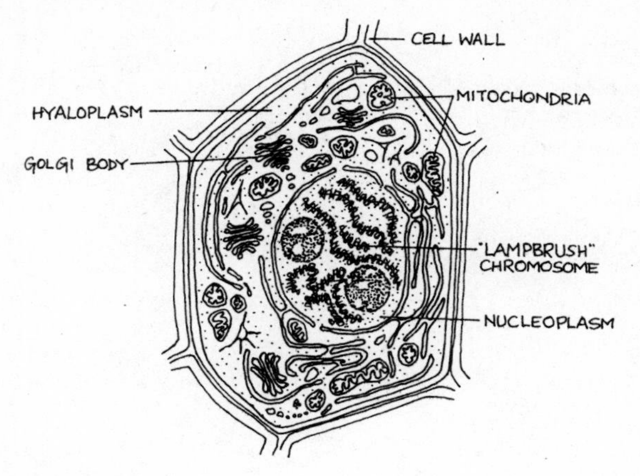

MUCH ENLARGED DIAGRAM OF A PLANT CELL SHOWING GOLGI BODY

Gonad

An organ that produces sex-cells.

Gonidia

The green cells of lichen which make the essential difference between lichens and all other fungi.

Gorge

Narrow opening or cleft between hills.

Gossamer

The filmy webs of small spiders floating in calm air or spread over grass, particularly noticeable in autumn.

Gourmand

Vigorous shoots which develop on the upper surfaces of branches, often trained in pruning systems when arching is required.

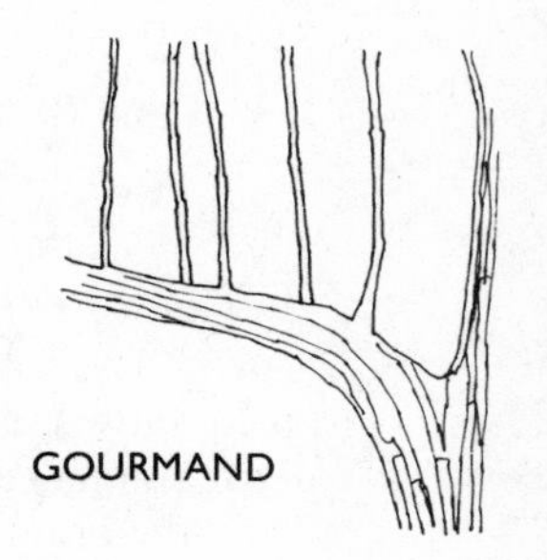

GOURMAND

Graft

(1) A scion or shoot inserted into a stock.
(2) Antiquated word for a spit of earth thrown up by a spade.

Graft Hybrid

The new hybrid plant derived from grafting two different species where the tissues of both species grow to become intermingled, as in a chimera.

Grafting

The artificial union of one part of a plant (scion) with parts of another (stock), e.g. when one variety of apple is grafted on to a different rootstock.

Grain

The fruit of a cereal or species of corn.

Graminaceous

Grass-like or grassy.

Gramineae

One of the largest families of flowering plants; grasses. Species of this family form the most important features of the vegetation in the world's temperate regions and include the cereals and bamboos.

SOME GRASSES (GRAMINEAE) 1. SWEET VERNAL GRASS; 2. TIMOTHY GRASS; 3. QUAKING GRASS; 4. SMOOTH MEADOW GRASS; 5. MEADOW BARLEY; 6. PERENNIAL RYE GRASS; 7. MEADOW FESCUE; 8. COCK'S FOOT; 9. WHITE BENT

Graniferous

Producing grain or grain-like seed.

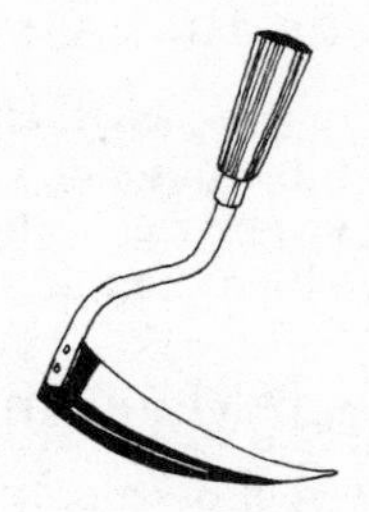

GRASS HOOK

Granular

(1) Small particles or granules.
(2) Covered with small granules.

Grass Hook

Hand-held implement for cutting grass and light brushwood; similar to a sickle but the blade is less curved, nearer to the shape of a scythe.

Grasshopper

Fairly large garden insect with powerful jaws, which feeds on plant material. Grasshoppers cause little damage in Britain but certain migratory species, including locusts, are considered a serious pest in some warmer climates.

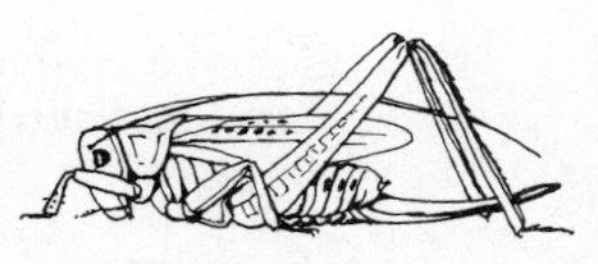

LONG-HORNED GRASSHOPPER OR BUSH-CRICKET

Grassing Down

A practice common in commercial orchards. Grass is sown around the fruit trees to encourage fruitfulness by depriving the trees of some of the nutrients and water to which they would otherwise have access.

Grease Banding

A method of trapping the wingless females of certain moths during the winter months by preventing them climbing up the bole of the tree and laying their eggs among the young shoots. A 4 in (10 cm) band of sticky material is applied to the tree about 3 ft (90 cm) above the ground so that the creatures become stuck to it. When spring arrives, it can be removed and burned.

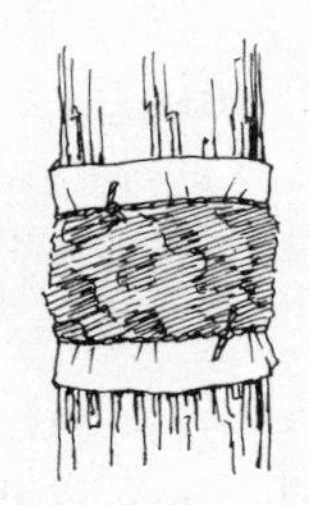

GREASE BANDING

Grecian Saw

A small handsaw with a narrow curved blade, used for removing small tree branches. The teeth are so arranged to cut as the operator pulls the tool toward him.

Green

(1) In leaf, covered with herbage.
(2) Unripe fruit, immature.
(3) Young and tender, not dried or withered.

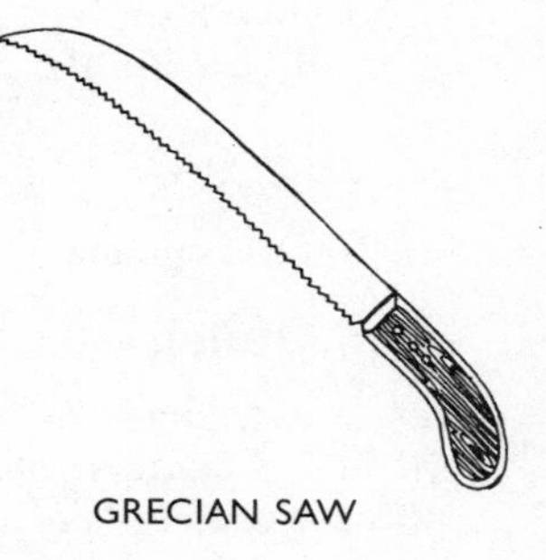

GRECIAN SAW

Green Bud, Green Cluster

On apples, pears, cherries and plums: when the clustered blossom buds can be clearly seen and the outer scales fall. On raspberries: when the leaves expand and the green flowerbuds can be seen.

Green Manuring

A system of improving the fertility of soils by growing a quick-maturing crop such as lucerne, clover, etc. to plough in while still green, followed by a leguminous crop such as peas or beans.

APPLE BUDS

Green Tip

Definition of a certain stage in bud development in apple, pear, plum and cherry trees. *See* Green bud.

Greenery

Green branches or foliage used for decoration or in flower arrangements.

Greenfly

See Aphid.

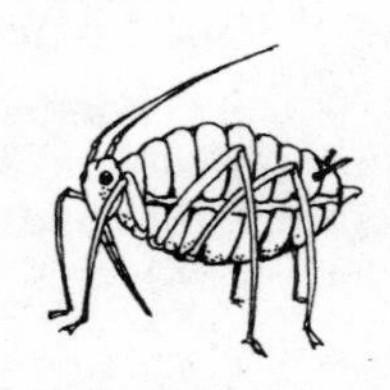

GREENFLY

Greening

(1) Type of apple, still green when ripe.
(2) Algae which grow on sand benches, clay pots and stagnant soil of pot plants.

Greensward

Turf or an area of level, short grass such as a village green.

Grex

A collective term for cultivars of the same hybrid origin, especially used in the case of orchids.

Grey Mould

See Botrytis cinerea.

Grip Trench

A slit trench made by inserting a spade vertically into the soil and rocking it slightly to produce a slit in which cuttings (usually hardwood) can be set.

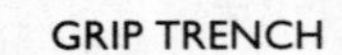

GRIP TRENCH

Ground Cover

Low-growing plants sited between larger subjects such as trees, etc., mainly to suppress weeds.

Groundling

Creeping or very dwarf plant.

Group

Botanically, an assemblage of similar cultivars within a species, also applied at other levels, e.g. species-group.

Grove

Small wood or cluster of trees, usually with an open space in the centre.

Growbag

A plastic sack with sealed ends, usually containing a peat-based compost, in which cuts are made and plants inserted. A popular modern method of growing several subjects including tomatoes, cucumbers and peppers.

GROWBAG

Growing Point

The tip of a stem where further growth will occur. The removal of this point by cutting or pinching will encourage the development of side shoots (laterals) lower down the stem.

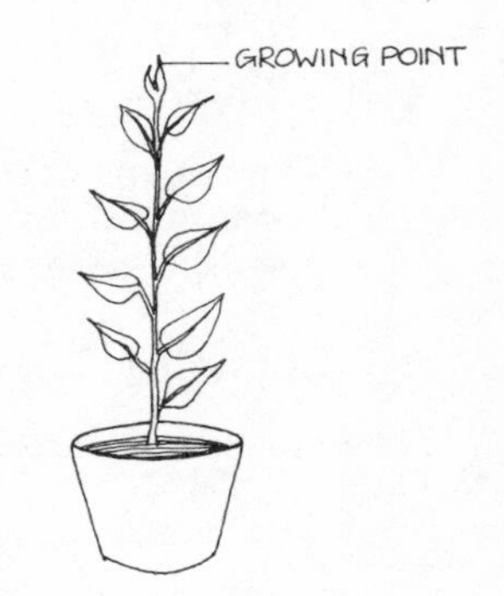

Growmore

See National Growmore.

Growth Regulators

Chemical substances which can alter the normal characteristics of a plant by making tall plants shorter or short plants taller. Widely used commercially for dwarfing chrysanthemums, kalanchoes and poinsettias, and for compacting seed-raised plants (F_1 hybrids) of pelargonium.

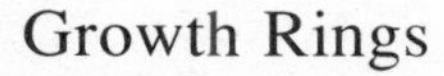

Growth Rings

The concentric rings seen across the severed trunk of a tree caused by the growth of cells at different rates during each season, from which the age of the tree can be calculated.

GROWTH RINGS

Grub

(1) Larva of insect, maggot, caterpillar, etc.
(2) To clear ground of roots, stubs, etc.

Grub Fell

In forestry, to cut a tree down at root level.

Guano

(1) Manure rich in phosphates and ammonia, originally the excrement of Peruvian sea-birds but now includes the droppings of other birds.
(2) Artificial manure manufactured from fish.

Guard Cells

Of a stoma, the apparatus in a leaf-surface by which respiration is controlled to meet varying atmospheric and environmental conditions.

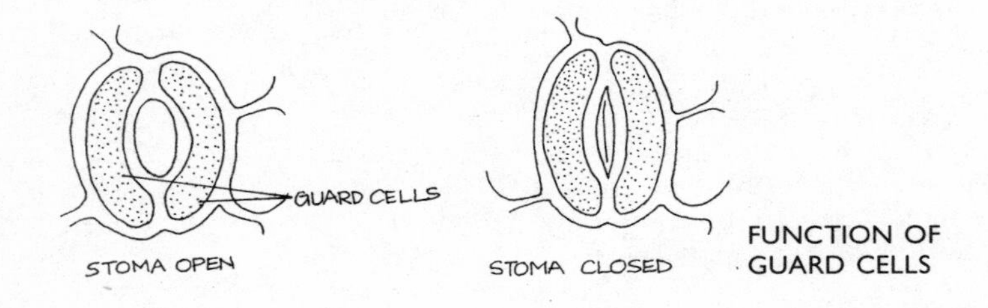

FUNCTION OF GUARD CELLS

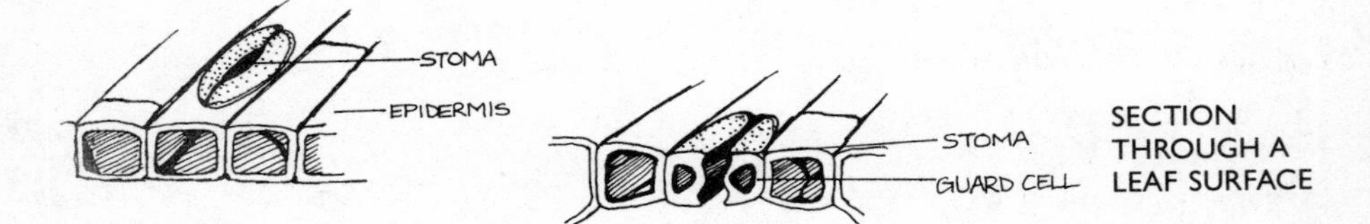

SECTION THROUGH A LEAF SURFACE

Gumming

The gum-like resins exuded by many plants when cut or damaged; mainly by stone fruit trees such as plums, peaches, etc., but also by certain conifers. The gum is harmless.

Gummosis

An ailment of cucumbers caused by a fungus.

Guttation

Droplets of water that appear at the tips and margins of leaves, usually on greenhouse or indoor plants. It is usually a sign that the compost is being watered excessively.

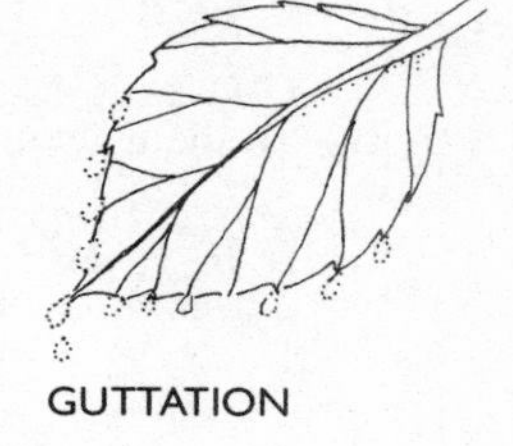

GUTTATION

Guying

A system of supporting a newly planted tree by securing it with guy ropes to pegs in a circle about 4 ft (1.2 m) away from the stem. Care must be taken to ensure that the guys, whether they be of rope or wire, are not allowed to come into direct contact with the stem or severe damage can be caused and, in extreme cases, the tree killed.

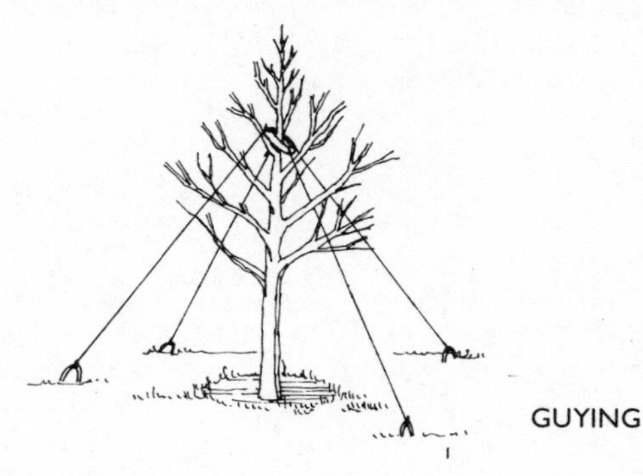

GUYING

Gymnosperm

Plant whose seeds are not protected by enclosure in an ovary, e.g. a conifer.

Gynaeceum, Gynoecium

The female parts of a flower.

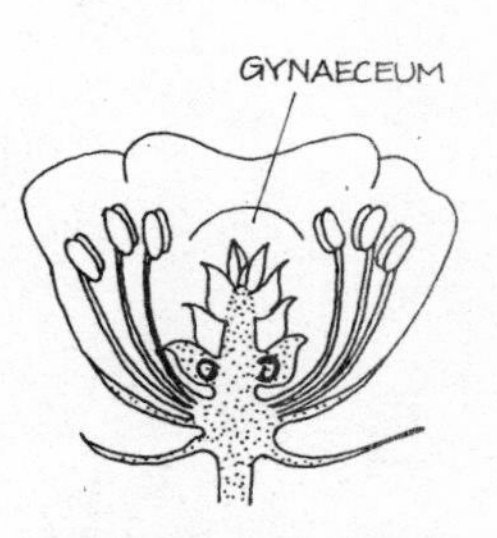

SECTION THROUGH BUTTERCUP FLOWER

Gynandrous

Hermaphroditic; stamens growing together with the carpels, as in orchids.

Gynobasic Style

A style produced at the base of a deeply lobed ovary.

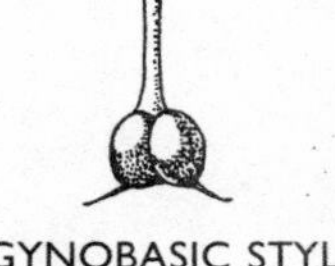

GYNOBASIC STYLE

Gynodioecious

Having female and hermaphrodite flowers on different plants.

Gynoecium

See Gynaeceum.

Gynomonoecious

Having female and hermaphrodite flowers on the same plant.

Gynophore

A stalk carrying the pistil above the receptacle of the flower.

Gypsum

Sulphate of lime, used horticulturally for improving the texture of clay soils without increasing the alkalinity.

H

Ha-ha

A deep, wide ditch (occasionally fenced) constructed in such a manner as to be invisible from the house and therefore not interrupt the view of the distant landscape. Intended to constrain stock and prevent them from entering the cultivated area around the house. Very popular in the eighteenth and nineteenth centuries around some of the great manor houses of the time; a few still exist.

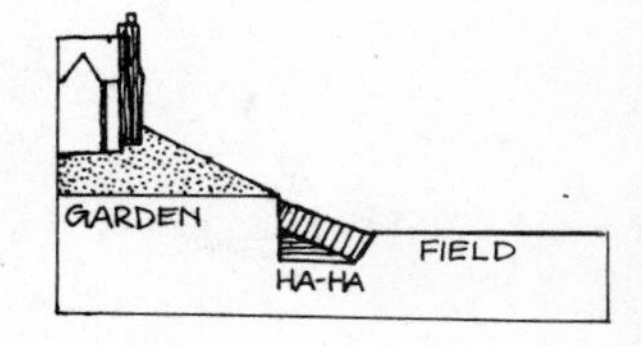

Habit

The general appearance of a plant; the way in which it grows – trailing, spreading, upright, etc.

Habitat

The natural area in which a wild plant grows; the native environment of a plant.

HACK

Hack

(1) To cut or notch.
(2) Mattock or type of pick used for working heavy or stony land.

Haft

(1) A winged leaf-stalk or the narrow stalk-like base to some petals, e.g. irises.
(2) Handle of a tool such as a trowel or spade, more often called a shaft.

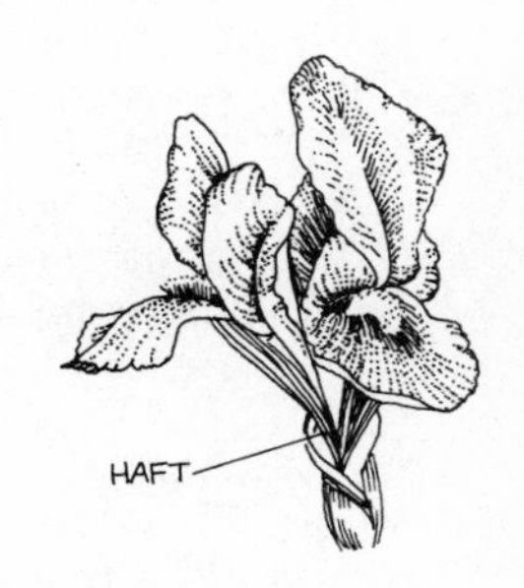

Hag

Part of a peat-bog which is firmer than the surrounding area, often heathery or grassy ground.

Hairy

Degree of hairiness, the opposite to glabrous: *see also* Hirsute, Hispid, Hoary, Lanate, Lanuginose, Lanulose, Multiciliate, Pilose, Puberulent, Sericeous, Setose, Silky, Strigose, Tomentose, Tomentulose, Villous, Woolly.

HALF-STANDARD APPLE TREE

Half-hardy

(1) A plant which can be grown outdoors in the summer months but which will not survive frost, e.g. runner beans, tagetes, etc.
(2) Shrubs and herbaceous subjects that will survive an average winter outdoors but only in sheltered positions, or in favoured areas of the country such as the western counties of England and parts of the western isles of Scotland.

Half-shrub

A perennial plant, only the lower half of which is woody.

HALF-STANDARD PELARGONIUM

Half-standard

A tree having a trunk which is unbranched up to about 4 ft (1.2 m) from ground level, at which point it produces laterals and makes little more growth. Also applicable to plants such as pelargoniums, roses, fuchsias, etc. but in these cases a half-standard would be shorter, as defined by the relevant Societies.

Halophyte

A plant which will tolerate an abnormal amount of salt in the soil. Plants which grow on the seashore and sand dunes are halophytes, such as the common salt-marsh grass (*Puccinellia maritima*).

HALOPHYTE (*Puccinellia maritima*)

Handlight

A small cloche for siting over individual plants, popular in Victorian times: usually square with a removable pyramidal top, but there were other shapes including oblong ones to cover two or perhaps three choice plants.

Hanger

Woodland on the side of a steep hill.

HANDLIGHT

Hanging Basket

A plant container designed for hanging in a porch, under eaves, on walls, in greenhouses, etc. They are manufactured from galvanized wire, wood, cane, plastic, pottery and wrought iron in a variety of shapes, patterns and designs.

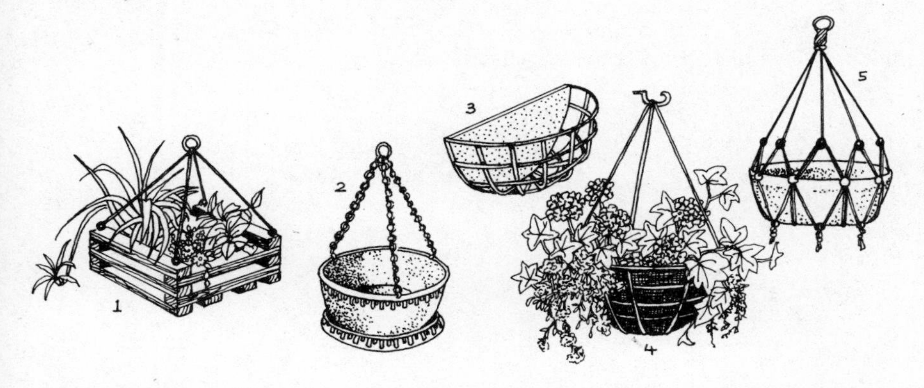

1 WOODEN SLATS 2 PLASTIC WITH DRIP TRAY 3 WIRE WALL BASKET 4 CIRCULAR WIRE CONSTRUCTION 5 CLAY POT IN STRING HANGER

TYPES OF HANGING BASKET

Haploid

Germ cell having half the normal number of chromosomes.

Hardening Off

Adjusting plants to lower temperatures, a process of acclimatization, usually when they have been raised in a greenhouse and prior to planting out in the open ground. Plants which have not been accustomed to the new and harsher environment are liable to fail or at least receive a severe setback.

Hardwood

Timber from broadleaved trees such as oak and elm; the timber from conifers is softwood.

Hardwood Cuttings

Method of propagation used for most hardy subjects such as trees and shrubs. Hardwood cuttings generally take much longer to root than soft-tissue cuttings, and are often best struck under mist.

Hardy

A plant which will survive out of doors all year round without protection. Geographical location, however, can affect hardiness, and a plant which is hardy in Devon may well not be so in Scotland.

Harvest

The mature crop; may be of fruits, cereals, roots, flowers, etc.

Harvesting

Gathering or collecting the mature crop.

Hassock

Tuft of matted grass.

Hastate

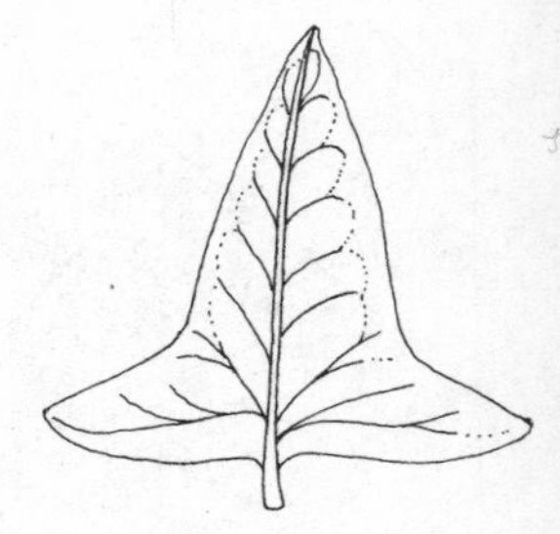

HASTATE

Of leaves which are spear-shaped with the basal lobes facing outward.

Haulm

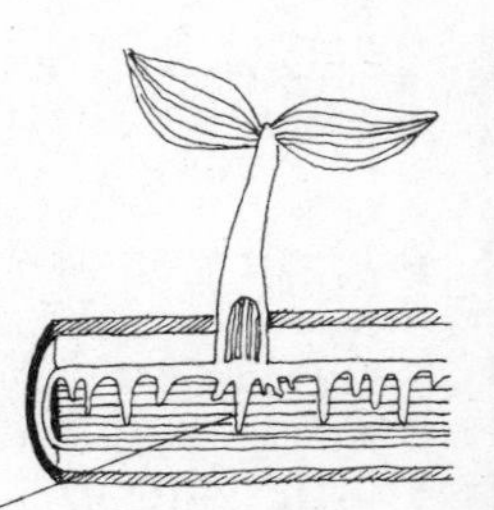

HAUSTORIA PENETRATING XYLEM OF HOST

(1) The foliage and stems of a vegetable crop such as potatoes, peas, beans, etc; sometimes known as the bine.
(2) Straw, or the strawy stems of some plants.

Haustorium, *pl.* Haustoria

The peg-like part of a parasitic plant by which it is attached to and derives nourishment from its host.

Haw

Fruits of *Crataegus* species such as the hawthorn.

Hay

Grass grown and harvested as a crop to cure or dry for fodder.

Hay Band

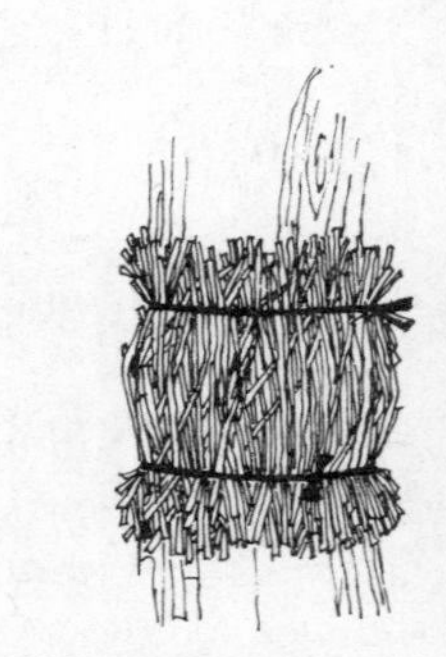

HAY BAND

A band of hay tied around the trunk and main branches of an apple tree as a trap for the caterpillars of the codling moth. Applied in early summer, the bands should be removed and burned in autumn, when the caterpillars descend to find shelter. There may be no effect upon the damage to the fruit in the first year but the numbers of pests will be greatly reduced in subsequent years.

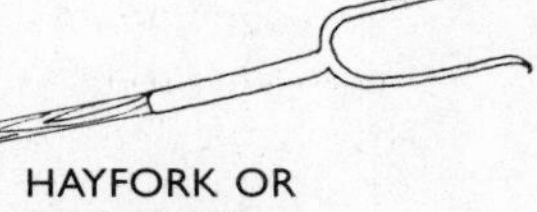

HAYFORK OR PITCHFORK

Hayfork

A hand-fork for turning or lifting hay, sometimes called a pitchfork.

HEADS OF *Bellis perennis*

Head

(1) A cluster of flowers. Usually a dense inflorescence of small, often stalkless flowers, as in the daisy family.
(2) Rounded or compact part of a plant such as lettuce, cabbage, etc.
(3) That part of a tree growing above the main trunk.

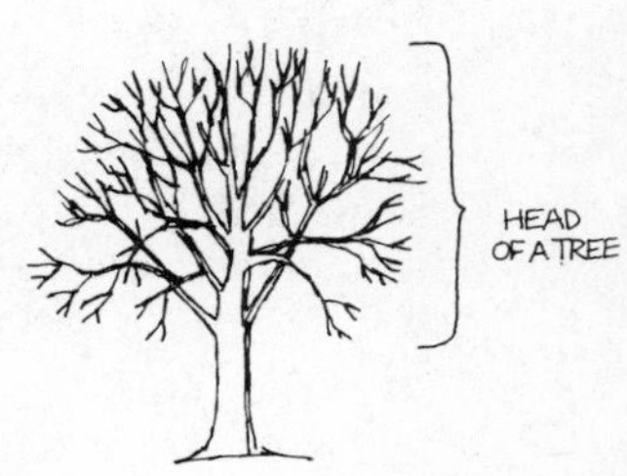

Heading

The compacting of the leaves of a lettuce or cabbage into a dense ball. *Same as* Hearting.

Heading Back

Severe method of pruning fruit trees by shortening all or most of the main branches to about half their length.

HEADING BACK

Headland

(1) A high spot or viewpoint, especially on the coast.
(2) An area of a field at either end of the ploughed furrows used for turning of implements or animals.

Heart-rot

Decay of the internal plant tissues often seen in root vegetables that appear to be quite sound until cut open; often caused initially by slug damage which becomes infected by bacteria. Heart-rot in trees is the result of fungus attack.

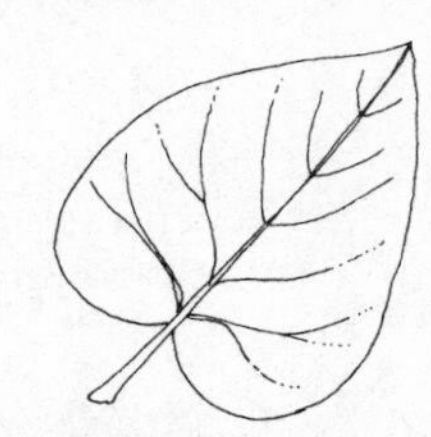

HEART-SHAPED (CORDATE)

Heart-shaped

Cordate; of leaves that are pointed at one end and have a pair of rounded lobes at the other.

Hearting

See Heading.

Heartwood

The dead wood at the centre of a tree-trunk or branch, still providing support although no longer water-conducting.

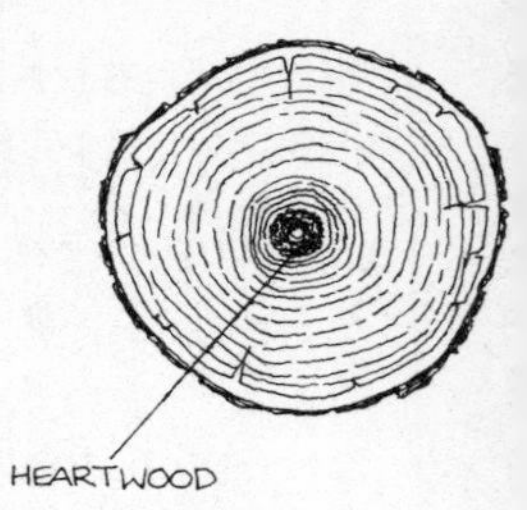

Heath

Open tract of land, often treeless or sparsely wooded and covered with low herbage and scrub including, in many areas, ericaceous subjects.

Heaving

(1) The alternate freezing and thawing of soil by which rough-dug soils are broken down to a workable tilth by winter weather.
(2) Root damage caused by alternate freezing and thawing when frost penetration is severe.

TYPICAL GARDEN HEDGE

Hedge

A boundary or dividing barrier that consists of shrubs, bushes or small trees, close-planted in a continuous line. Some hedges are purely decorative but others function as windbreaks, shelterbelts, etc.

Hedge Maze

See Maze.

Hedging

Planting bushes or small trees close together to form a boundary or fence; a boundary or barrier: *cf.* Leaching, Plashing.

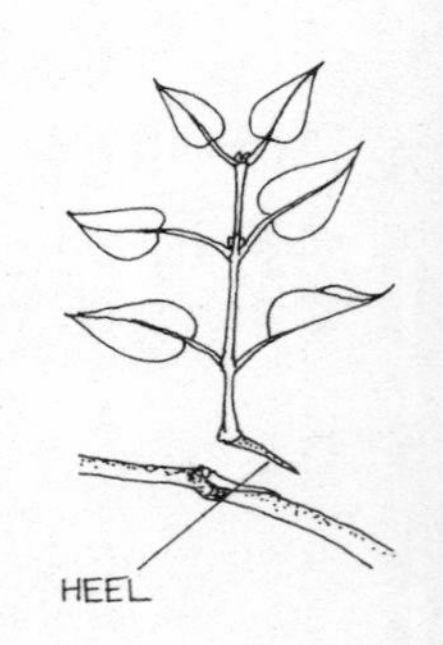

Heel

The small piece of bark and wood attached when a shoot is pulled from the parent plant instead of cutting.

Heeling-in

Temporary planting, usually to prevent plants from drying out when they cannot be set immediately into their permanent positions.

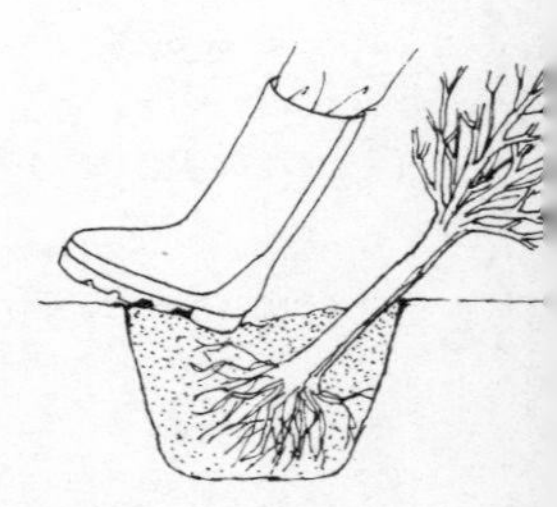

HEELING-IN

Helicoid

Spiral, formed like a spring or snail shell, as of cymose inflorescences.

Heliotropic

The movement of leaves and shoots under the influence of light: *cf*. Geotropic, Phototropic.

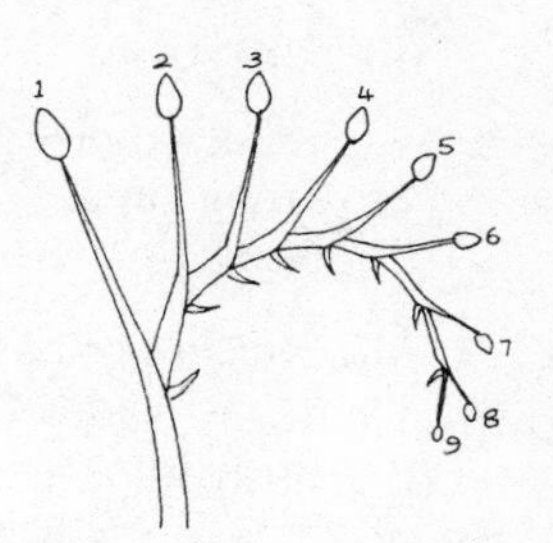

HELICOID MONOCHASIAL CYME (THE NUMBERS SHOW THE ORDER OF FLOWERS DEVELOPING)

Helmet

Arched upper part of the corolla in some flowers, especially orchids. *Same as* Hood.

Helve

Old name for the handle or shaft of a tool such as a trowel.

Hen-and-Chickens

A flower which has a ring of small florets around the central one, as with some daisies. Also a clump containing one large central rosette plant surrounded by a ring of small offsets, as in the illustration.

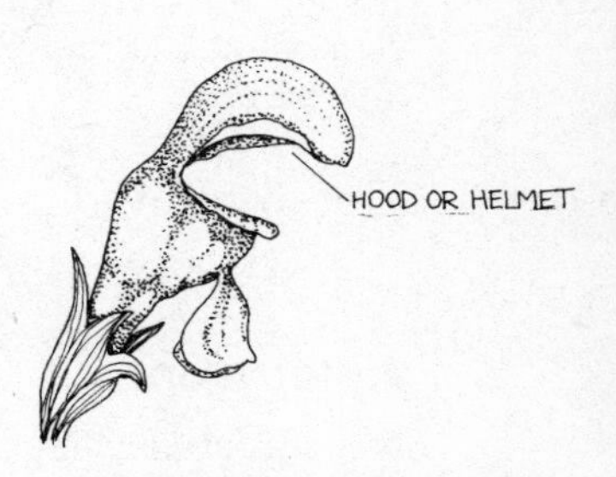

WHITE DEADNETTLE

Herb

(1) Any non-woody vascular plant.
(2) A plant valued for its aromatic, savoury or medicinal properties.

Herbaceous

Of plants with soft top growth rather than woody growth; they can be annual, biennial or perennial, but in most cases the top growth dies back in the winter. Often applied to the taller, non-woody perennial varieties such as golden rod (*Solidago*).

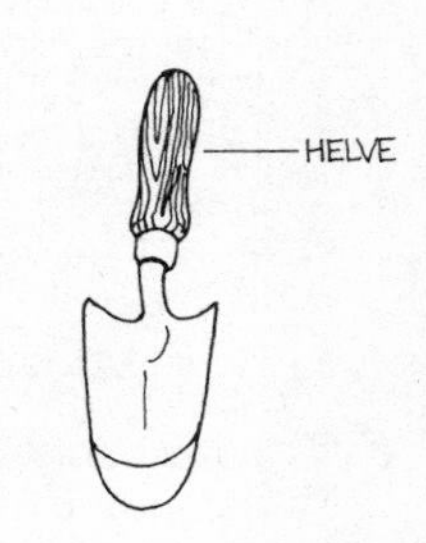

Herbaceous Border

Garden border stocked usually with perennial flowering plants.

Herbal

Book containing lists and descriptions of herbs used for medicinal or culinary purposes.

HEN-AND-CHICKENS FORMATION (SEMPERVIVUMS)

Herbarium

(1) A plant museum, or any place or building used for housing a classified collection of preserved plants.
(2) A collection of plants, usually dried, contained together in a showcase or room.

Herbary

A garden of herbs; sometimes called a physic garden.

Herbicide

A chemical product used to control weeds.

Hermaphrodite

Bisexual flowers which have both male and female organs, so do not require a separate pollinator unless the plant is self-sterile.

Hessian

Strong, coarse material much used in the past for covering frames and plants for frost protection, also fixed to stakes as a windbreak; now almost totally superseded by plastic alternatives.

Heterochlamydeous

Having distinct whorls of sepals and petals.

Heterogeneous

Not uniform; composed of parts of different kinds: *cf.* Homogeneous.

Heteromorphic

Having several forms.

Heterophyllous

Of a plant with more than one type of leaf.

Heterosporous

Producing two kinds of spores representing two sexes: *cf.* Homosporous.

Heterostylous

Having styles, and often stamens, of varying lengths in different flowers within a species: *cf.* Homostylous.

Heterotrophic

Dependent upon green plants for carbon (as in some fungi, etc.).

Heterozygous

Where both genes (alleles) of a gene-pair are different. Thus the gametes produced will not all be the same and the plant will not breed true. F_1 hybrids are heterozygous for many gene-pairs: *cf.* Homozygous.

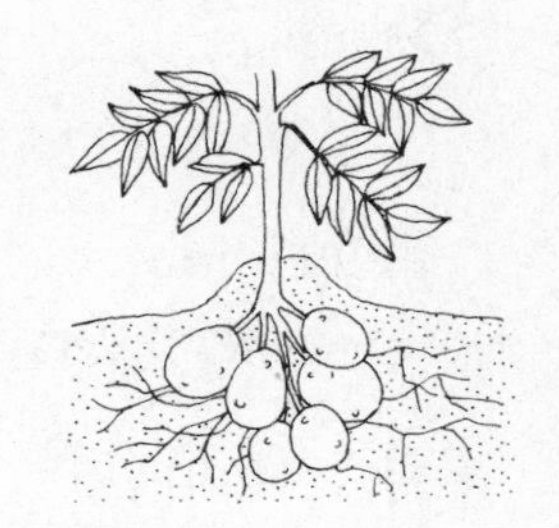

POTATOES HILLED-UP

Hill-up

Almost obsolete term: to bank the soil around the stems, as in the ridging of potatoes.

Hillock

Small mound or heap of soil, e.g. a molehill.

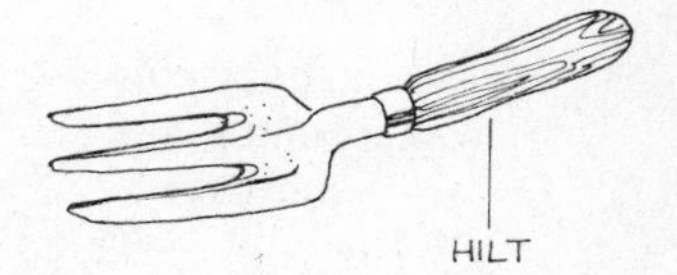

Hilt

Handle of a tool, especially a small hand-tool such as a trowel. *Same as* Haft and Helve.

Hilum

The scar or mark on a seed where it was attached.

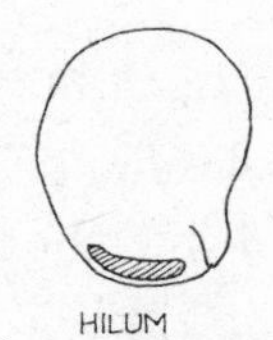

HILUM OF
BROAD BEAN

Hip

The fruit of the rose, in particular the 'dog rose' (*Rosa canina*). The botanical name for a hip is *cynarrhodion.*

Hippocrepiform, Hippocrepian

Horseshoe-shaped.

Hirsute

Covered with coarse, dense hairs.

Hirsutulous

Diminutive of hirsute; slightly hirsute.

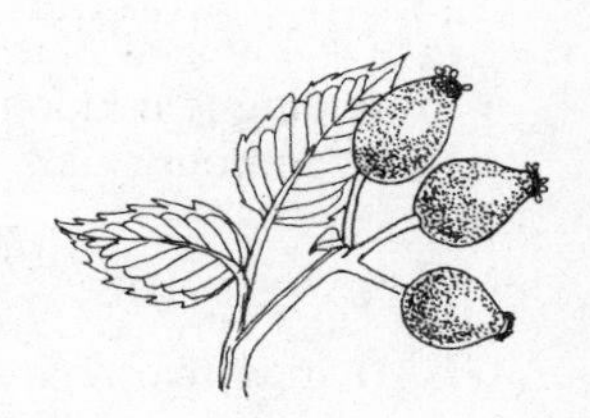

ROSE HIPS

Hirtellous

Slightly and softly hirsute.

Hispid

Covered with stiff, dense hairs.

Hispidulous

Diminutive of hispid; slightly hispid.

Hoary

Having grey, greyish or white hairs; covered with short whitish hairs, usually applied to foliage.

Hoe

Long-handled tool with metal blade for loosening soil, weeding, etc. The Dutch hoe is pushed away from the user, the draw hoe pulled toward the user, but there are many different shapes and designs.

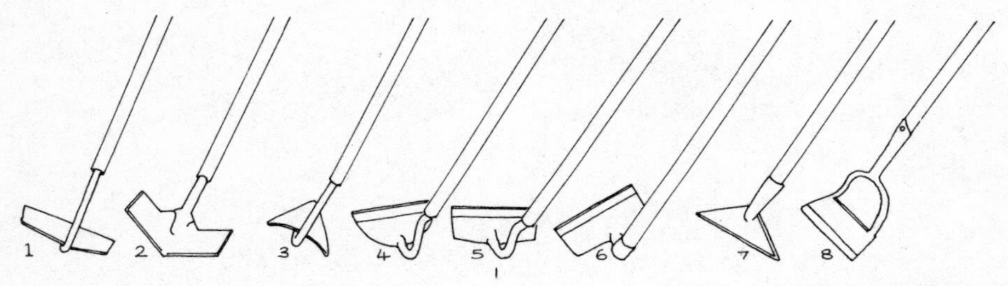

SOME TYPES OF HOE
1. DRAW HOE
2. SCUFFLE HOE
3. DRILL HOE
4. HALF MOON SWAN NECK HOE
5. SWAN NECK HOE
6. SHORT NECK HOE
7. TRIANGULAR HOE
8. DUTCH HOE

Hollow

A depression or basin, often in woodland; a small valley, often circular or nearly so.

Hollow-tine Fork

A fork with hollow, round tines with a slit on one surface, used for aerating lawns, etc.

Holm

(1) An area of flat ground beside or close to a river and subject to flooding.
(2) A small island, usually in a river or estuary.

Holt

(1) Small wood or copse.
(2) A wooded hill.
(3) An orchard.
(4) A badger's den or refuge.

Homochlamydeous

Flowers with calyx and corolla similar; having two identical whorls of perianth segments (sometimes called tepals).

Homoeomerous

Lichens having the algal cells distributed throughout the thallus.

Homogeneous

(1) Of uniform structure.
(2) Of similar kind.
(3) Of the same sex: *cf.* Heterogeneous.

Homologous

Of common evolutionary descent and similar basic structure.

Homonym

Any of two or more identical names applied to different types of plant, only one of which can be legitimate; a name rejected as already in use by another genus or species.

Homosporous

Producing spores of just one kind: *cf.* Heterosporous.

Homostylous

Having styles of the same length: *cf.* Heterostylous.

Homozygous

Where both genes (alleles) of a gene-pair are identical. Plants will breed true if most of their genes are homozygous. A self-pollinated plant will not produce entirely homozygous offspring unless it is entirely homozygous itself: *cf.* Heterozygous.

HONEY FUNGUS GROWING AT THE BASE OF A TREE

Honey Fungus

Armillaria mellea is a fungus which attacks a wide variety of trees, travelling underground to infect the roots of one tree from another and eventually impregnating the area under the bark of the trunk. When honey-coloured toadstools appear in clumps around the base of the tree, it is fatally infected and should be burned together with all its roots and branches. There is little chance of a cure once the toadstools appear, but in the early stages some chemical control may be possible. Potatoes grown in soil where previously an infected tree grew may develop blemishes or necrotic patches above and below the skin. Bootlace fungus (*Armillaria tabescens*) is very similar to honey fungus but grows only on the trunk of deciduous trees.

HONEY FUNGUS MYCELIUM UNDER THE BARK OF A TREE

Honey-guide

Markings on the perianth, usually in the form of spots or lines, said to direct insects to the nectar in a flower. *Same as* Nectar-guide.

Honeydew

The sticky substance excreted by sap-sucking insects such as aphids and scale insects, creating a tacky coating on leaves and stems. Aphids are 'farmed' by ants to yield a supply of honeydew as a diet supplement.

HONEY-GUIDES ON PETALS

Hood

See Helmet.

Hoof and Horn

A valuable organic fertilizer made from the ground-up hooves and horns of animals from the abattoir. The coarser grades release the nutrients more slowly than the fine grades. It is normally dug into the ground at the rate of 2 oz (56 g) per square yard, and is also a useful addition to potting compost.

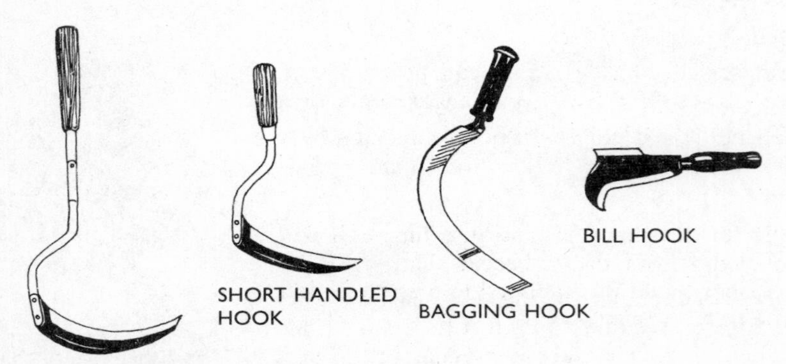

Hook

Short- or long-handled cutting implement in a variety of shapes and sizes for particular purposes; a sickle.

Hop Manure

Spent hops from breweries, useful for improving the texture of soils and adding humus. On its own has no great food value but various chemicals are normally added to improve its nutritional functions.

Hormones

Organic products of living cells. Natural hormones are found in live tissue, both animal and vegetable, and induce or control growth, etc. Synthetic hormones are usually based upon organic chemicals used to assist in rooting, fruit setting, growth retarding, weed control, etc.

Horticulture

The art of garden cultivation, distinct from botany, which is the science or study of plants.

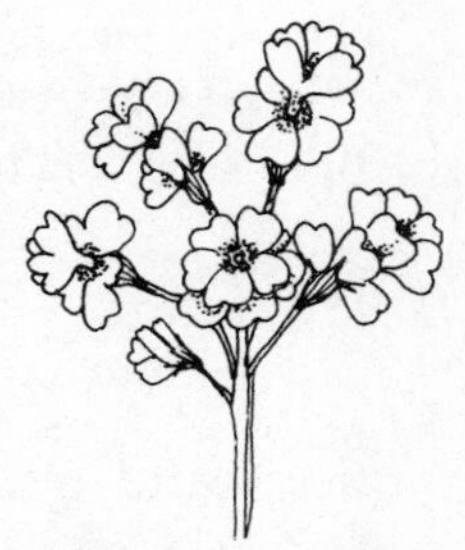

HOSE-IN-HOSE (PRIMULA)

Hose

Flexible tubing for conveying fluids to plants; previously made of fabric or rubber, now more often of plastic.

Hose-in-Hose

A floral abnormality where one flower is carried within another; often seen in polyanthus, etc. Also called 'cup-and-saucer' formation.

Host

Plant that supports a parasite which can be an insect, virus, another plant, etc. Mistletoe is an example of a parasitic plant living upon a host (tree).

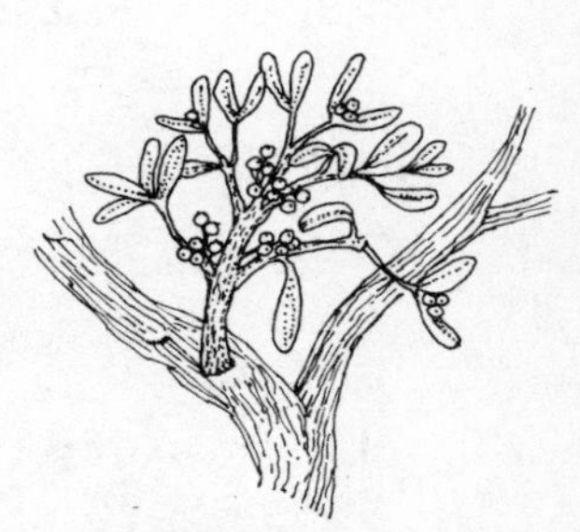

FRUITING PLANT ON HOST (MISTLETOE)

Hostile Soil

Soil unsuitable for the plants set into it; a limy soil will be hostile to members of the Ericaceae family, for instance, and seashore sand incorporated into any soil will make it hostile to most plants.

Hot-water Treatment

A method of destroying certain pests by immersing infected plants or parts of plants in water at precisely controlled temperatures. If the water is not hot enough the pest will not be killed, if too hot the plant may be destroyed. The treatment is widely used commercially against eelworms in chrysanthemum stools, narcissus bulbs, etc. but it has also been tried against various rusts, with limited success.

Hotbed

The Victorian method of providing bottom-heat for tender subjects. A bed of fresh manure was prepared and the natural heat generated as it decomposed was sufficient to provide ideal conditions for the delicate plants set upon it. Mainly superseded today by electric soil-warming cables.

HOVERFLY
(*Sericomyia borealis*)

Hoverfly

Beneficial creature, the larvae of which feed upon aphids.

Hull

The pod of peas and beans; the outer covering of some fruits.

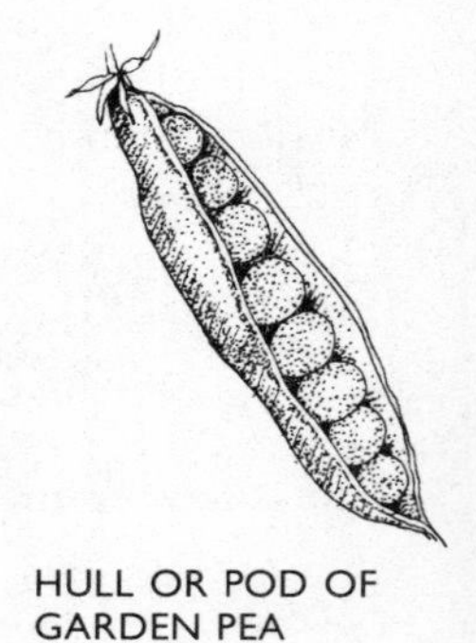

HULL OR POD OF GARDEN PEA

Humidity

The degree of moisture in the atmosphere. Relative humidity is the ratio between the amount of moisture actually present and the amount which would be present if the air were saturated at the same temperature.

Hummock

Hillock or knoll, usually in otherwise flat ground. A small hump.

Hump

Small rounded hillock or boss of earth.

Humus

The dark brown or black substance produced by the slow decomposition of organic matter in which the bacteria multiply, breaking down the various components into the simpler chemicals which plants absorb as food.

Hurdle

A type of fence panel made with flexible strips of willow or hazel woven horizontally through upright poles in a basketwork manner. The strips and poles are widely spaced; if close-packed they are wattle panels.

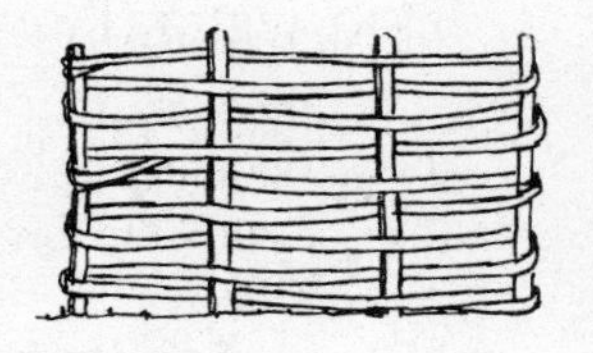

HURDLE

Hurst

Wooded eminence or small hill; an area of woodland usually on rising ground.

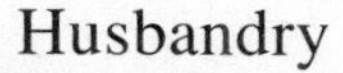

Husbandry

The tending of plants or cultivation of the ground. Husbandry may be good, bad, indifferent, etc.

Husk

The outer covering of some fruits; the shell; the dry outer case; as of corn and many other cereals.

Hyaline

Colourless, translucent. Usually applied when the subject (leaf or petal) is so thin as to be almost transparent.

Hybrid

The result of cross-fertilizing two or more species or genera of plant (or animal); technically a mongrel. Hybrids seldom breed true; some are sterile and do not breed at all.

Hybrid Vigour

The improvement, compared to the parents, that the offspring of two different varieties of some but not all plants may display when crossed.

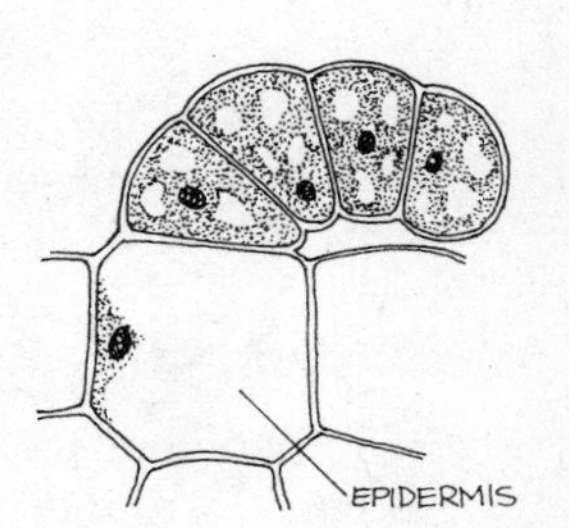

HYDATHODE OF SCARLET RUNNER

Hydathode

An epidermal water-excreting gland usually found in leaves, enabling the plant to exude water even at night when the stoma are normally closed. This excretion can often be seen in early morning on grass when it is frequently mistaken for dew.

Hydro-cooling

A method of prolonging the store life of fresh vegetables by immersion in very cold water. Widely used before refrigeration became widespread.

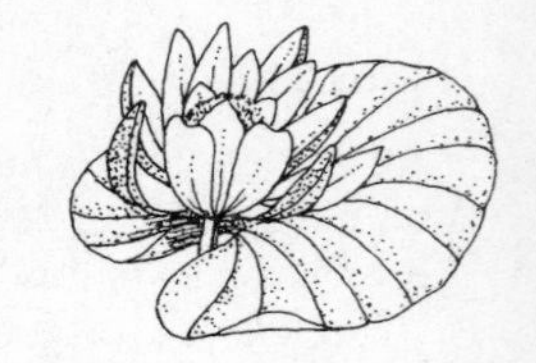

WATERLILIES ARE HYDROPHYTES

Hydro Seeding

A method of sowing grass seed in a stream of water directed at the area to be seeded. Used mainly for very large-scale works, especially where access by normal methods is difficult, such as steep hillsides.

Hydroculture

Growing plants in food-enriched water or waterlogged sterile aggregates.

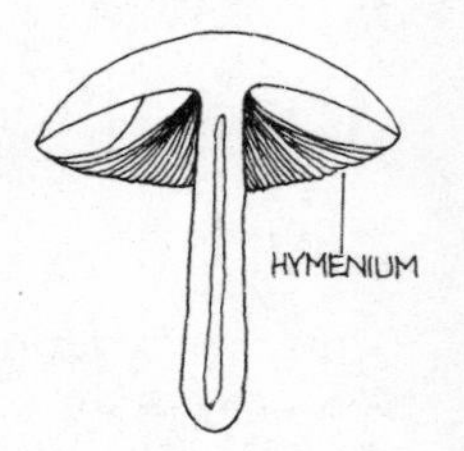

Hydroleca

Lightweight Expanded Clay Aggregate abbreviated to 'Leca'. Very lightweight granules with a hard outer surface and honeycomb interior, they absorb water and release it slowly and are often used instead of normal aggregates on greenhouse benching.

Hydrophilous

Of a plant which is pollinated through the agency of water.

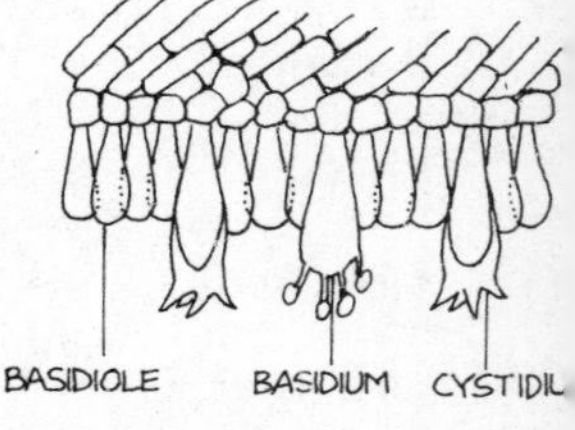

DETAILS OF A HYMENIUM

Hydrophyte

An aquatic plant such as a waterlily: *cf*. Mesophyte, Xerophyte.

Hydroponics

The study and practice of growing plants in a nutrient solution instead of soil: *see also* Soilless cultivation.

HYMENOPTERA QUEEN BEE (*Apis mellifera*)

Hygrometer

An instrument for measuring humidity.

Hygroscopic

(1) Attraction of moisture, usually from the atmosphere.
(2) Plant movements caused by absorption or loss of water.

Hymenium, *pl.* Hymenia

Spore-bearing surface on fungi.

HYPANTHIUM

ROSE HIP
(CROSS SECTION)

Hymenoptera

Order of insects with four membranous wings which includes bees, ants, etc.

Hypanthium

The flat or concave receptacle of a perigynous flower which often develops and encloses the fruits, e.g. the fleshy tissue of the rose hip.

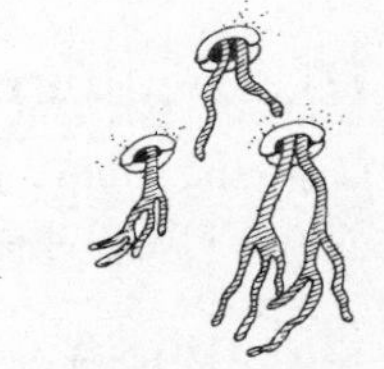

HYPHAE OF POTATO
BLIGHT EMERGING
THROUGH STOMATA

Hypertufa

An artificial medium with absorbent properties similar to tufa rock. It is made by mixing 1 part cement with 2 parts sand and 2 parts granulated peat, plus enough water to make a workable mixture. It can be shaped and moulded to imitate rocks, with pockets scooped out to hold rock plants; or old sinks and similar containers can be coated with it to give a more natural appearance.

Hyphae

Fine tubular threads of fungus mycelium.

h
m
e
h – HYPOCHIL
m – MESOCHIL
e – EPICHIL

ORCHID FLOWER

Hypochil, Hypochile

The basal part of the labellum of an orchid, or the lower lip of a flower.

Hypocotyl

The axis of an embryo below the cotyledons which will develop into the radicle as the seed germinates.

Hypogeal

Producing seed leaves below ground in germination, as in peas: *cf.* Epigeal.

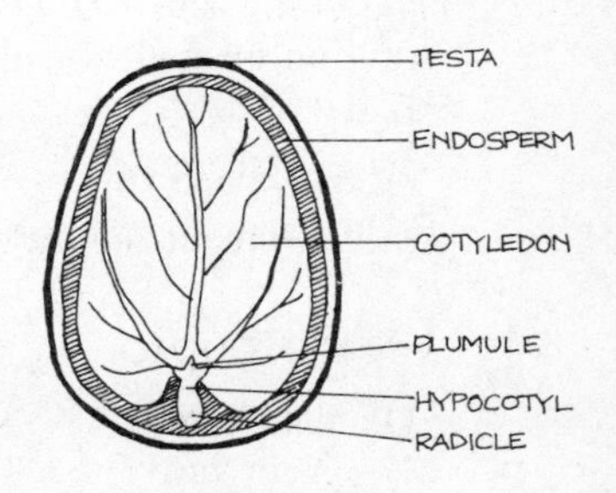

CASTOR OIL SEED
CUT IN HALF

Hypogynous

Of flowers that have the sepals, petals and stamens connected to the receptacle below the ovary, as in brambles.

I

Ichneumon Flies

Loosely used term covering a large number of insects whose larvae prey upon other insects, and are therefore beneficial to gardeners. The female of *Ichneumon suspiciosus*, for instance, lays her eggs inside the bodies of various caterpillars in springtime, and the grubs develop within the host, devouring its body as they mature.

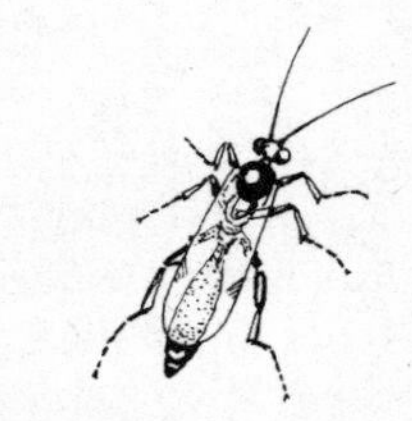

Ichneumon suspiciosus

Imago, *pl.* Imagos, Imagines

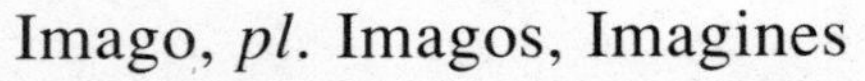

The final stage of growth of an insect after completing all its metamorphoses. The butterfly is an example.

PAINTED LADY (*Cynthia cardui*)

Imbricate

Closely overlapping, as in roof tiles and fish scales. Many tree buds have imbricate scales to protect them, such as the winter buds of horse-chestnut.

Immersed

Below the surface of water, e.g. many oxygenators such as *Potamogeton crispus* (frog's lettuce) and *Callitriche palustris* (starwort): *cf.* Emersed.

Immortelle

Another name for 'everlasting' flowers, derived from the French word for immortal.

IMBRICATED LEAVES OF MONKEY-PUZZLE TREE (*Araucaria imbricata*)

Immune

(1) Plants which are unaffected by a particular disease or pest are described as being immune to it, but immunity to one disease or pest does not imply immunity to all. Some plants are especially bred for

immunity to a specific disease or pest and these are termed immune varieties. Resistant varieties are not necessarily immune, and although they may show none of the particular characteristics of the problem themselves, may still be carrying the infection, especially where virus diseases are involved.

(2) Some insect pests may become immune to a particular pesticide if it is used too frequently; this can be avoided by alternating two or three different products.

Immunology

The study of immunity from disease.

Immutable

Unchangeable; not subject to variation.

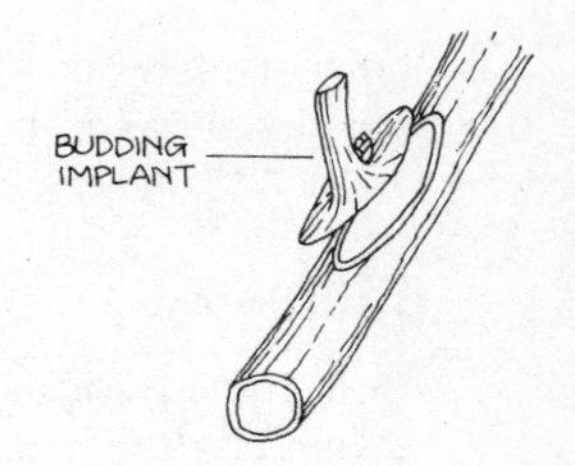

Imparipinnate

See Odd-pinnate.

Impermeable

Not permitting the passage of fluids.

Implant

An insert, e.g. a scion graft.

Inaperturate

Having no aperture, as applied to pollen grains; lacking pores or furrows.

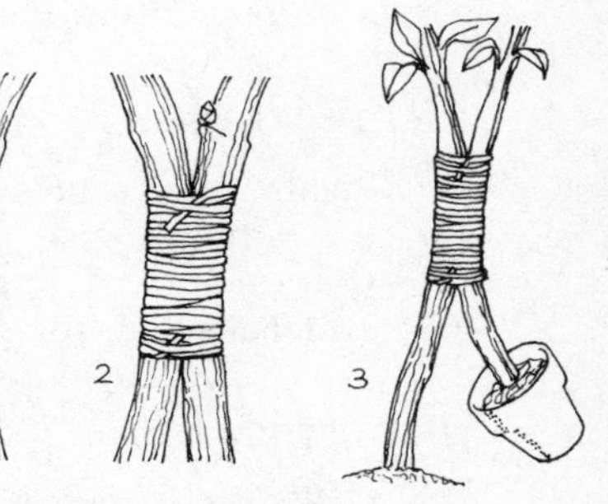

INARCHING

Inarching

A method of grafting whereby the scion and stock both continue to grow on their own roots until the union is completed, after which the scion is detached just below the joint and the stock just above. Also termed approach grafting. *See* Grafting.

Inbred

Produced by inbreeding. When a bisexual (hermaphrodite) flower pollinates itself the resulting seed is said to be inbred.

Incinerate

To burn or consume by fire.

Incinerator

A type of brazier used for burning rubbish and garden waste outdoors.

Incise

To make a cut or slit, usually in a leaf, or in a stem prior to grafting.

SLASHED OR INCISED

Incised

Of leaf margins, stipules or bracts that are cut or slashed irregularly and quite sharply; between toothed and lobed. *Same as* Slashed.

Inclined

Bent forward or backward, as of a flowerhead. Dahlias are considered unfit for the show bench if inclined.

INCLINED FLOWERHEAD

Included

Not protruding or projecting.

Incompatible

(1) Plants which cannot be crossed to produce hybrids.
(2) In grafting, some scions are incompatible with certain stocks and if grafted thereon the union will eventually fail, sometimes after several years.

Inconspicuous

Not readily noticeable, as of small, pale or green flowers, e.g. those of *Pelargonium gibbosum*.

Incumbent

Leaning or resting upon another for support, as of a plant or part of a plant.

Incurved, Incurving

Bent inward, as applied to a type of florist's chrysanthemum in which the petals turn loosely inward and upward to produce a compact spherical flower: *cf.* Recurved.

INCURVED

Indefinite

Of flower parts, such as petals, which are present in such quantity as to make a precise count difficult, usually implying, too, that the number is variable.

INDEHISCENT FRUIT (HAZELNUT)

Indehiscent

Applied to fruits which do not open or split to release seeds: *cf.* Dehiscent.

Indentation

Cut or notch in a leaf edge or petal.

Indeterminate

Of plants whose stems can continue to grow indefinitely such as the tomato; conversely, if the stem terminates in a flowerbud, no further extension is possible.

THE RUNNER BEAN PLANT IS INDETERMINATE

Indigenous

Native to a country or soil; belonging where it is found; not imported: *cf.* Exotic.

Indore Process

A method of making compost developed at the Indore Research Station in India, by which vegetable refuse and animal remains are composted together with dung, urine and lime.

Indumentum

A coating or covering of fine felt-like hairs as found, for instance, on the underside of rhododendron leaves; woolly pubescence.

Induplicate

Folded or rolled inward.

Indurate

Hardened, toughened, weathered.

Indusiate

Having an indusium.

Indusium

(1) The almost transparent membrane which covers and protects the spores in the sorus of most ferns, eventually rupturing to release the spores.
(2) The pupa case of some insects.

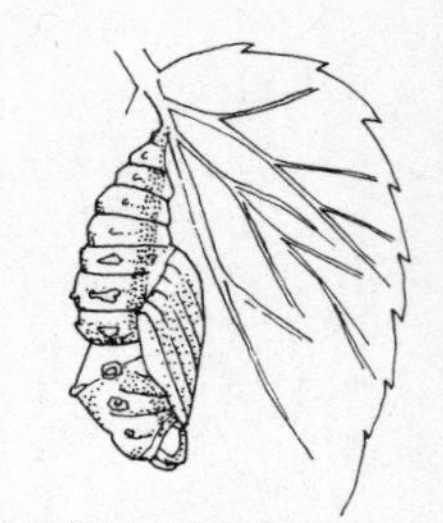

THE PUPA CASE OF RED ADMIRAL BUTTERFLY (INDUSIUM)

Infect

To contaminate; implant disease; produce micro-organisms in a plant or environment.

Inferior

Lower; below.
(1) Of a calyx situated below the ovary.
(2) Of an ovary with sepals, petals and stamens attached at its apex: *cf.* Superior.

Infertile

Not fertile; applicable to soils deficient in nutrients or to fruit trees and bushes which fail to crop.

Infest

To swarm or congregate in large numbers; as blackfly will infest untreated broad bean plants or greenfly the young shoots of roses, etc.

Infield

The area of farmland around or close to the farmhouse or dwellings; the fields nearest to the farm buildings.

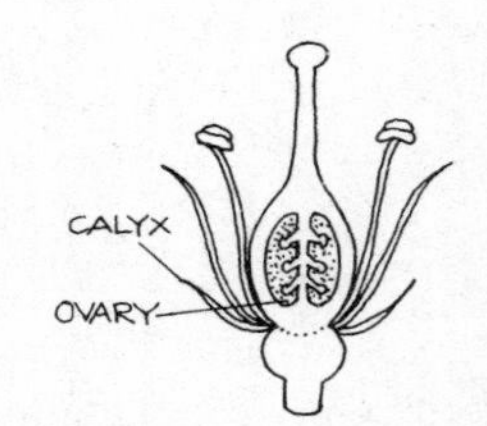

(1) INFERIOR CALYX

Inflorescence

The flowering part of a plant; individual florets or a group comprising a flowerhead. A general imprecise term used to describe the arrangement of the flowers on a stem or branch; the mode of flowering. Inflorescences can be of several types including:

Spike – a raceme (i.e. a main axis bearing stalked flowers, as in lupin or foxglove) with numerous, nearly stalkless flowers on a vertical stem, with the youngest flowers nearest the apex, e.g. lavender.

Panicle – a raceme with a complex branching system, each branch representing an individual raceme, with the younger flowers at the apex of each branch; e.g. gypsophila.

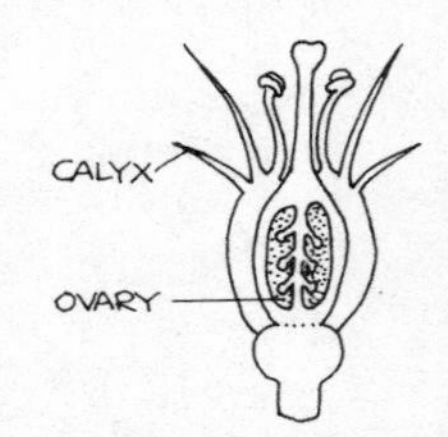

(2) INFERIOR OVARY

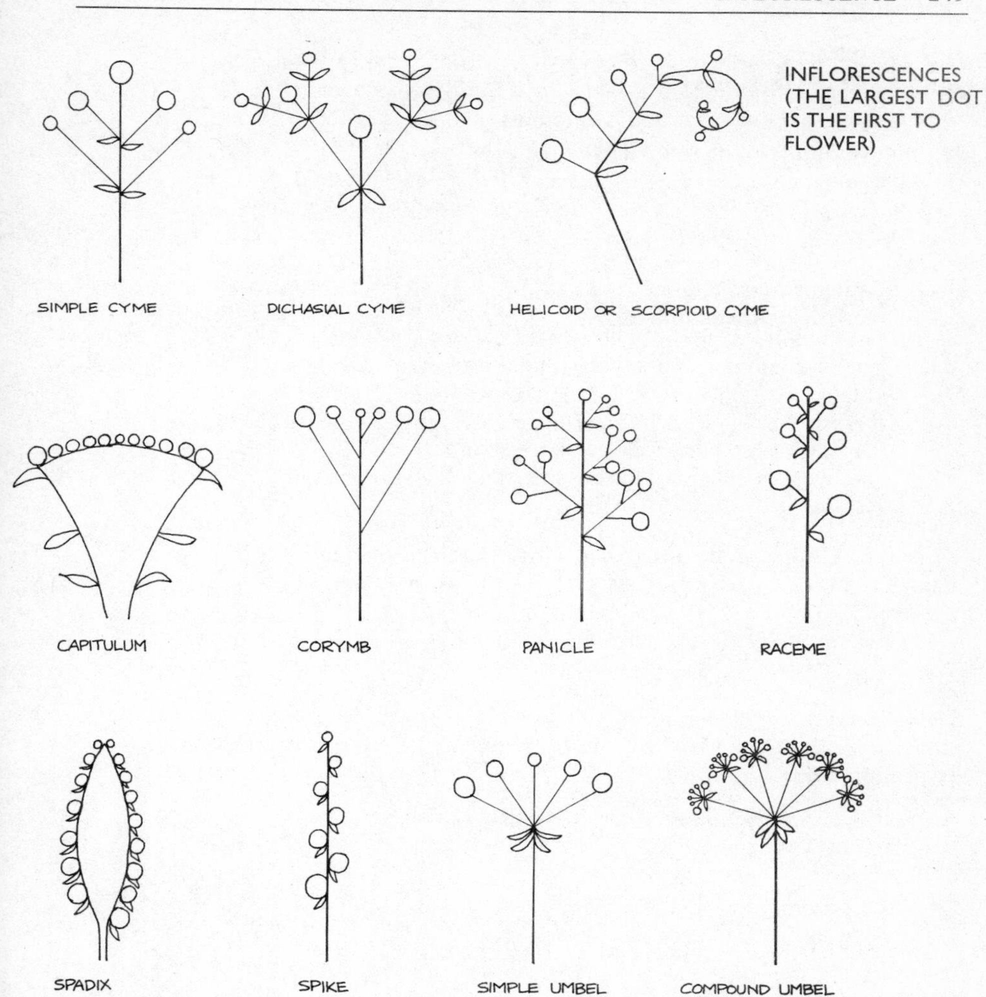

Umbel – a raceme with individual flower stalks arising from a central point at the top of the main stem, as in an umbrella, the inner stalks being shorter than the outer, thus presenting a flat or nearly flat top to the flowerhead. Umbels can be simple (e.g. *Prunus cerasus*) or compound as in fennel and cow-parsley.

Capitulum – a raceme consisting of a tightly packed head of almost stalkless flowers, as in composites such as the daisy family (*Bellis*).

Corymb – a fairly flat-topped flower cluster of small flowerheads on stalks which arise from different points on the main stem, as distinct from an umbel where they radiate from a single point, e.g. *Prunus mahaleb*.

Cyme – a portmanteau word covering a variety of inflorescences, all of which terminate in a flower, successive flowers being borne on new flowerstems. Cymes may be simple or compound, dome-shaped or flattish, or one-sided (helicoid).

Spadix – a spike with a fleshy or swollen central axis, e.g. cuckoo-pint (*Arum maculatum*).

Informal

(1) The opposite of formal when applied to garden design – not laid out in geometrical patterns. Woodland gardens, wild and rock gardens are examples.
(2) Flower varieties in which the head is irregular, such as in some chrysanthemums, dahlias, etc.

Infrafoliar

Beneath the foliage; the flowers of African violet are often naturally produced infrafoliarly and have to be manually exposed.

Infrapetiolar

Below the petiole or leaf-stalk.

Infraspecific

Any unit of classification below the rank of species, e.g. cultivar.

Infructescence

A cluster of fruits resulting from an inflorescence.

Infructuous

Unfruitful, barren.

Infundibular

Funnel-shaped, as the flowers of convolvulus.

INFUNDIBULAR

Innes

See John Innes composts.

Inorganic

Of substances which contain no carbon, being neither animal nor vegetable, including a vast range of chemical products which can be added to the soil as fertilizers.

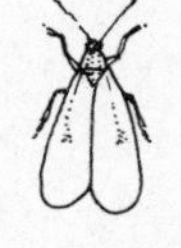

INSECTS

Insect

A small animal with body divided into segments. Over a million have been listed so far worldwide and new ones are being discovered every year. Some are harmful, others beneficial; some have wings, others are wingless.

Insecticide

A chemical used to control insects, applied in the form of a liquid, powder or smoke.

Insectivorous

Insect-eating, as of plants that are adapted for trapping insects to supplement their nutrient intake. In nature, most live in boggy areas and have a very restricted root system, deriving most of the nitrogen they need from their victims.

VENUS FLY-TRAP IS AN INSECTIVOROUS PLANT

Inserted

Attached to or growing out of another organ, as of leaflets growing out of cladodes.

Instant Gardening

Producing an instant effect by planting mature, often flowering subjects purchased from nurseries or garden centres in containers. In this way the seasonal planting restrictions which apply to many plants can be largely overcome.

Integument

The outer covering of an ovule which eventually becomes the seed coat, husk or skin.

Intercalary

Inserted between or among others.

Intercropping

(1) Growing two crops in succession on the same piece of ground.

(2) Raising a quick-growing crop between rows of slower-growing plants: *cf.* Catch crop.
(3) Planting bush fruits in an orchard between top fruit trees, e.g. gooseberries between apples.

Interfoliar

Among the leaves; as of a plant with flowers which are produced within the foliage.

Intergeneric

Of hybrids that result from crosses between plants of different genera: *cf.* Interspecific.

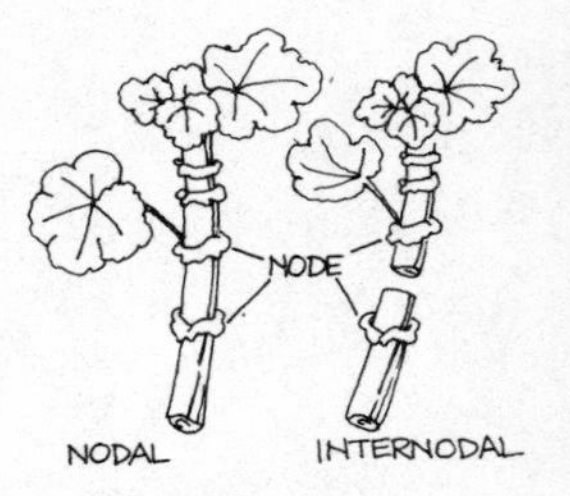

CUTTINGS

Intermediate

Chrysanthemums, having flowers between incurved and reflexed.

Internodal Cuttings

Cuttings which have not been trimmed back to a node (leaf joint) but have a length of internodal stem retained. Some plants, including ivy-leaf pelargoniums and chrysanthemums, will root internodally, others only produce roots at the nodes.

Internode

That portion of the stem of a plant lying between two nodes (leaf joints).

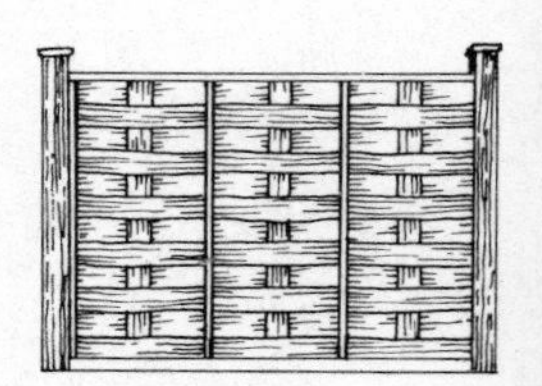
INTERWOVEN FENCING

Interspecific

Of hybrids that result from crosses between species of one genus: *cf.* Intergeneric.

Interwoven

Applied to a type of fencing panel in which the slats are interlaced both horizontally and vertically.

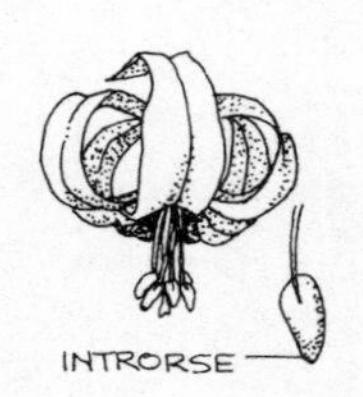

Introrse

Applicable to anthers which are inward-bending, attached at the back to face toward the centre of a flower: *cf.* Extrorse.

Inverted

Upside down.

ASTRANTIA

Involucre

The whorl of bracts enclosing a number of flowers. A characteristic of the daisy family (*Bellis*), it is also a feature of the teasel and many other plants.

Involute

Upward- or inward-rolled margins, mainly applied to leaf margins but also to petals.

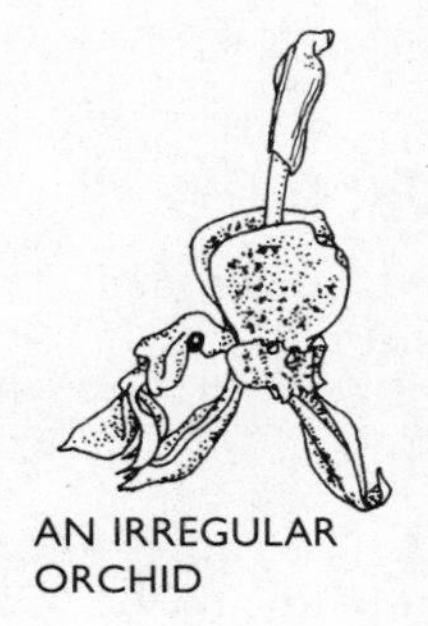
AN IRREGULAR ORCHID

Irregular

Not regular, as applicable to a flower that is a compound of different types of petal, e.g. salvias and some orchids.

Island Bed

An isolated flowerbed surrounded by lawn or paths and visible from all sides.

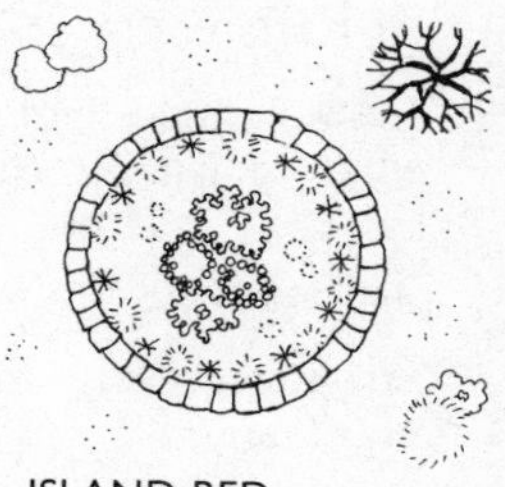
ISLAND BED

Isobilateral

With corresponding opposite parts alike, applicable to leaves where both surfaces are identical, as in iris.

Isomerous

Made up of corresponding parts or segments, as of a flower which has the same number of parts in each whorl.

J

Jacket

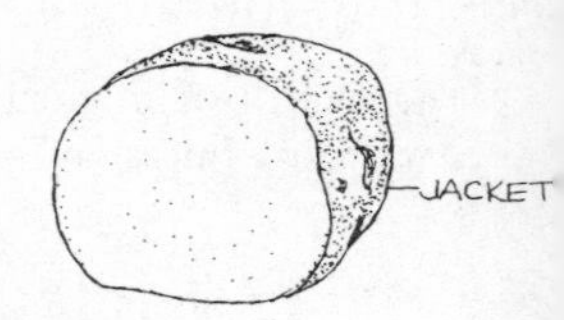

POTATO

The outer coat or covering of a potato; the skin or epidermis.

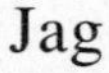

Jag

A sharp division or cleft, as the leaf margins of the meadow thistle (*Cirsium dissectum*).

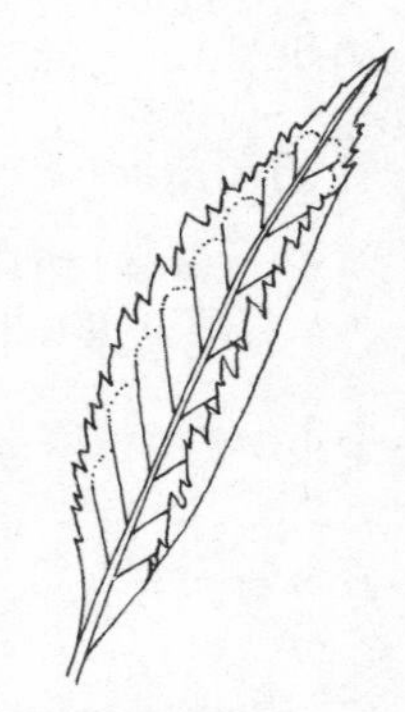
JAG MEADOW THISTLE (*Cirsium dissectum*)

Jardinière

Ornamental stand or container for displaying a growing plant, normally used indoors.

Jenneting

A type of early apple, now seldom grown, certainly not commercially.

John Innes Composts

The John Innes Horticultural Institute was founded in 1904 with a bequest from a London businessman of Scottish descent for '. . . the improvement of horticulture by experiment and research'.

Between the wars experimental work at the Institute resulted in the introduction of the famous John Innes Composts – an attempt to produce a high-quality product to a standard range of formulae as a replacement for the great variety of growing media then marketed by almost every nurseryman as his own product – many derived from ordinary soil and 'sterilized', if at all, in an old copper behind the potting shed.

The basic ingredient of such composts is sterilized, well-rotted loam; and since loams vary from area to area, it follows there is no such thing as a standard J.I. Compost. Although most are sold today in sealed plastic bags, the compost from a Fenland area will bear little resemblance to one, say, from Devon or Bedfordshire. All may be made exactly to the J.I. formulae, all are good, yet they may perform differently with various plant subjects.

The earlier 'John Innes Formulae' consisted of only two recipes, a Seed Sowing Compost and a Potting Compost, but the range has now been extended as follows:

J.I. SEED COMPOST (for seed sowing or rooting cuttings). 2 parts well rotted and sterilized loam, 1 part granulated peat, 1 part sharp washed river sand; to which must be added 2 lb (0.9 kg) superphosphate and 1 lb (0.45 kg) ground chalk or ground limestone per cubic yard (0.45 kg per 0.76 cubic metre).

J.I. POTTING COMPOSTS (1, 2 & 3) require the inclusion of J.I. Base fertilizer. Formula as follows: (parts by weight) 2 hoof and horn, $\frac{1}{8}$ in. (3.1 mm) grist (13 per cent nitrogen); 2 superphosphate of lime (18 per cent phosphoric acid); 1 sulphate of potash (48 per cent pure potash) giving an analysis of N:5.1, P:7.2, K:9.7 per cent.

J.I. POTTING COMPOST No. 1 (for the initial potting). Parts by bulk: 7 sterilized loam, 3 peat, 2 sand, to which add 5 lb (2.25 kg) J.I. Base and 1 lb (0.45 kg) ground chalk per cubic yard (cubic metre).

J.I. POTTING COMPOST No. 2 (for potting on). As for No. 1 but double the amount of J.I. Base and ground chalk.

J.I. POTTING COMPOST No. 3 (for final potting of stock plants). As for No. 1 but with three times the quantities of J.I. Base and ground chalk.

For lime-hating plants substitute flowers of sulphur for the ground chalk or limestone. All these soil-based composts are relatively unstable and should be used within a month or so of manufacture and stored under dry conditions.

JARDINIÈRE

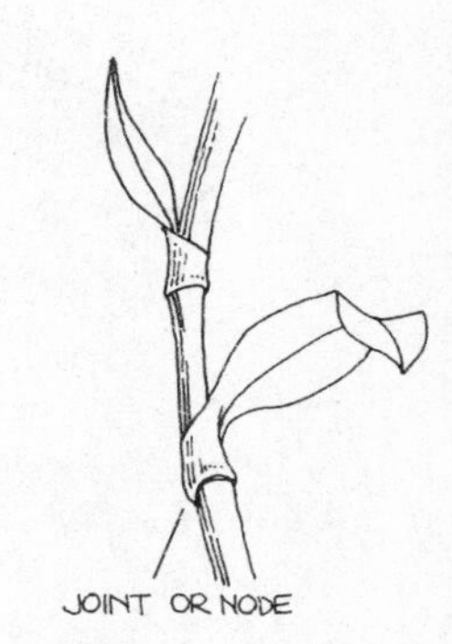

JOINT OR NODE

Joint

The point on a plant stem from which a leaf or leaf-bud grows; more usually termed node.

Jugate

Paired; having leaflets or leaves in pairs.

JUGATE

Jugum

A pair of opposite leaves.

Juice

The liquid content of fruits or vegetables.

June Drop

The seasonal shedding of many immature fruits on apple and other top-fruit trees. It is nature's way of reducing the crop to manageable proportions relative to the tree's bearing capabilities; or it can sometimes be caused by other factors such as drought or poor pollination.

Jungle

Tangled vegetation or undergrowth; a wild tangled mass of dense forest.

Jute

Corchorus olitorius and *C. capsularis*, grown for the fibre obtained from the inner bark and, in the tropics, for the young shoots, which are eaten. The fibres are woven into mats, sacks, etc. and also used for making coarse cloth and cordage.

Juvenile

The distinctly immature stage in the development of certain plants, when either the leaf-shape or some other feature differs from the adult stage. Eucalyptus normally carries juvenile and adult leaves, and the common ivy has juvenile climbing stems and adult non-climbing stems.

K

Kainit, Kainite

A natural mineral found in salt deposits, seldom used today except occasionally as a winter dressing for orchards. It was originally the K in NPK formulations which encouraged flower and fruit development, containing sulphate of potash, sodium chloride, sulphate of magnesium, etc. Modern formulations of high-potassium fertilizers are much purer and their effects more predictable.

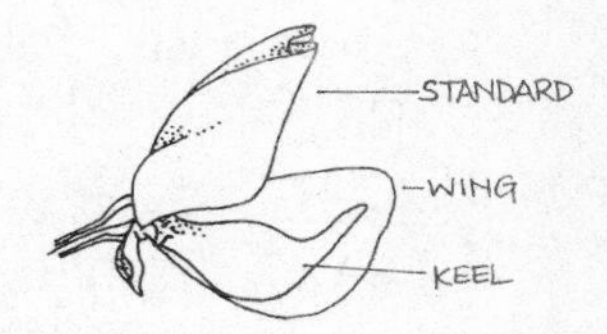

COROLLA OF SWEET PEA (ONE WING REMOVED)

Kapok

The fine cottonwool-like material surrounding the seeds of the kapok or silk-cotton tree (*Ceiba pentandra*), widely used for stuffing cushions, furniture and clothing before plastic foam was developed.

Keel

The two lower petals of most leguminous flowers, united or partially joined to form a keel similar to that of a boat. The pea is a common example, as are the vetches.

Keiki

Hawaiian term, now in common use by orchid growers, to describe an offshoot or offset on certain plants. Some of the long-caned dendrobiums produce adventitious growths (keikis) from the nodes which can be detached and grown on to flower earlier than plants raised from seed.

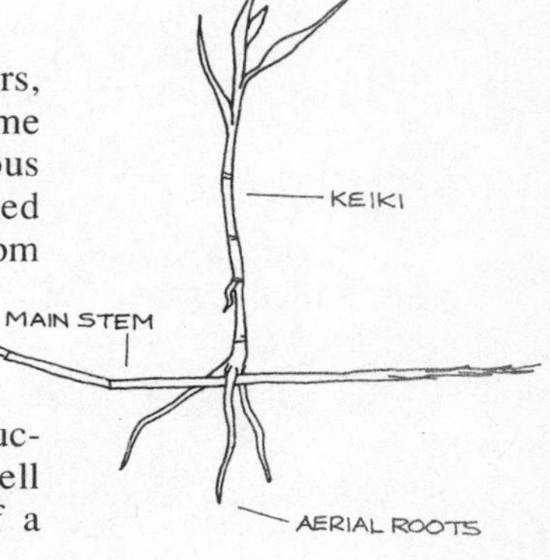

Keiki Paste

A hormone compound sometimes applied (not always successfully) to an orchid node in an effort to alter the cell structure so that a keiki will be produced instead of a flower.

Kelp

A large form of seaweed which can be burned to yield potash, iodine, etc. or rotted down and processed as a garden fertilizer.

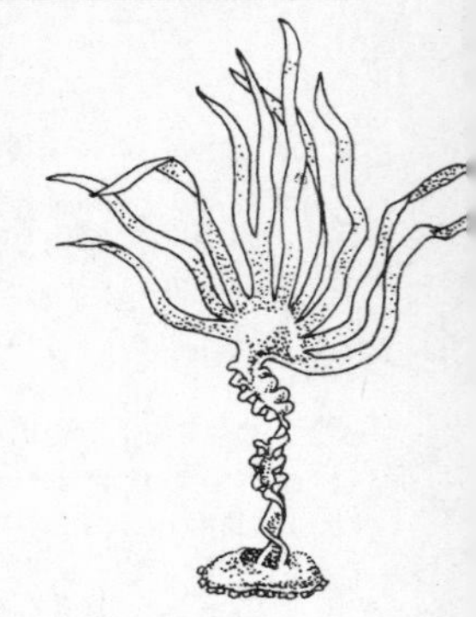

KELP

Kerf

(1) A cut or notch, the groove made by a saw.
(2) The face of a cut surface such as the cut end of a felled tree.
(3) A single layer of cut turf, hay, etc.

Kernel

A seed within a hard shell; the edible part of a nut.

KERNEL

Kex

A dry, usually hollow, herbaceous stem, or any tall umbelliferous plant.

Key

A samara; winged seed as produced by sycamores and maples.

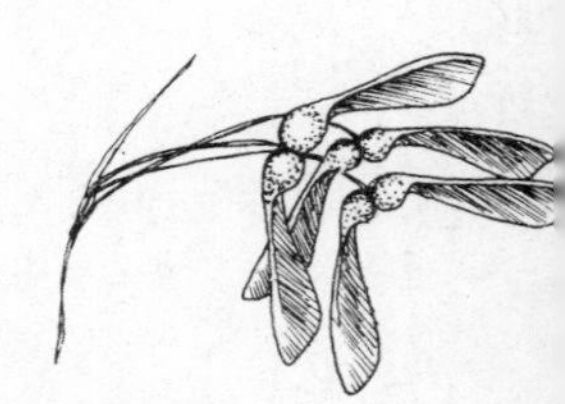

KEYS OF SILVER MAPLE

Kidney Bean

Usually, the French dwarf bean, especially its seeds.

Kidney Potato

Any long oval-shaped variety of potato tuber, e.g. the 'Jersey Royal'.

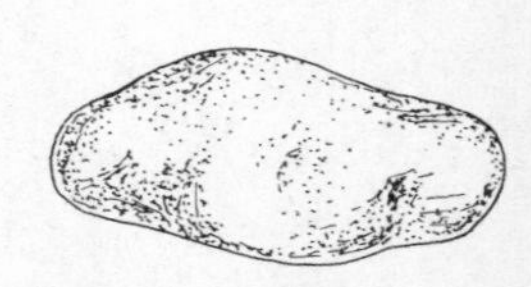

KIDNEY POTATO

Kinin

A plant hormone which promotes cell division, used commercially as a preservative for cut flowers.

Kinnikinic

A mixture of dried sumach leaves, bark of willow, etc. used by the American Indians as a substitute for tobacco or for mixing with it.

Kino

The astringent gum of various trees used in medicine and in tanning.

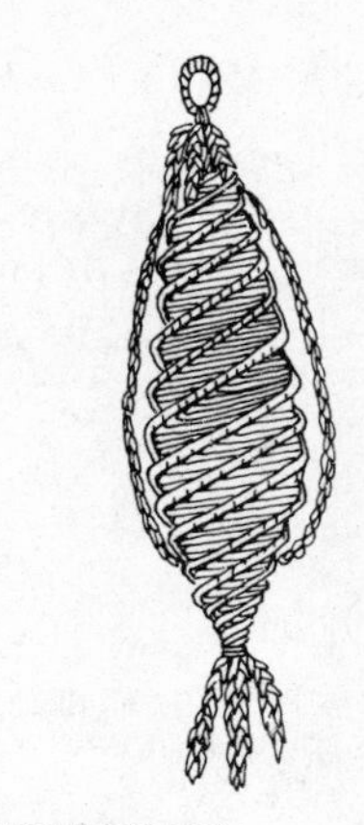

CORN-DOLLY (KIRN-DOLLIE)

Kirn

Scottish term for the last sheaf or handful of the harvest, from which derives the kirn-dollie (corn-dolly), a dressed-up figure made from the kirn.

Kitchen Garden

A functional rather than ornamental garden reserved for the production of vegetables and fruits for culinary use; often surrounded by a wall in earlier times, and including greenhouses for raising out-of-season produce.

Kittul

The strong black fibres from the leaf-stalks of the kittul or jaggery palm (*Caryota urens*).

Klendusik

Plants which are able to repel or withstand disease by means of some protective mechanism.

Klinostat

A revolving plant stand used in experiments with factors affecting the orientation of plant growth.

Knag

A knot in wood, a peg; the base of a branch where it is inserted into the trunk of a tree.

Knar

A protuberance on the trunk of a tree, usually covered with bark, or on a root.

Knee

An upward-growing root of swamp trees which enables them to breathe.

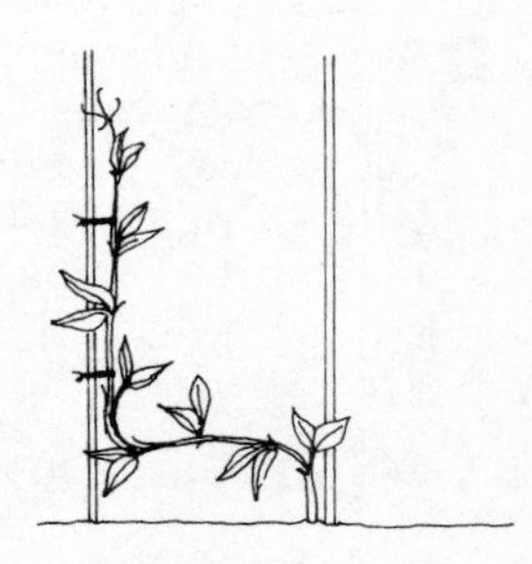

KNEEING

Kneeing

A method of training cordon sweet peas by which the plant stem is detached from its stake, bent horizontally just above ground level, then refastened to a further stake. The restriction of the sap flow thus produced is said to result in an increase in flower production. Also known as layering.

Knife

There are several types of knife specially designed for garden use such as the budding knife, grafting knife, pruning knife, dissecting knife, etc.

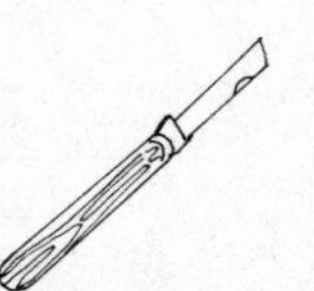

DISSECTING KNIFE WITH DETACHABLE BLADE   PRUNING KNIFE   BUDDING KNIFE   ASPARAGUS KNIFE

Knitch

The colloquial name for a faggot in some parts of the country.

Knob

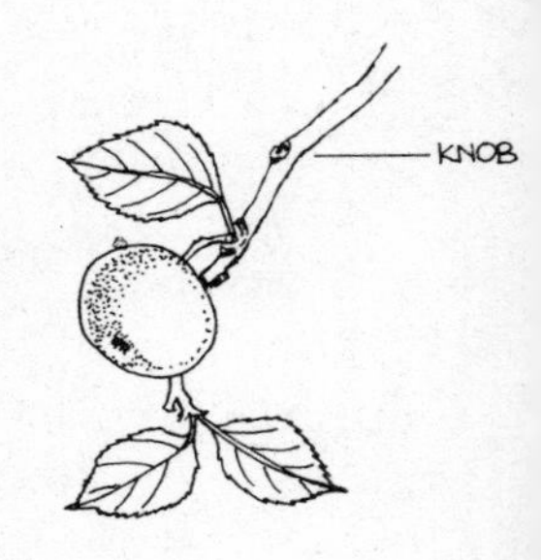

As used by fruit growers, the swollen point at which the previous year's fruit stalk was joined to the branch, and from which later fruit buds will develop.

Knock-down

A pesticide which is quick-acting but not always lethal, e.g. pyrethrum, which was originally derived from the species *Chrysanthemum cinerariifolium*; pyrethrum had little lasting effect on either plants or animals (and some insects), and many flies attacked with it and seen to collapse could recover within a short time. This natural product has now largely been abandoned in favour of more potent synthetic formulations. The term can also apply to the time taken to achieve knock-down.

Knoll

Hillock; small hill or the top of a hill.

Knop

Obsolete word for a knob, tuft or bud.

Knopper Gall

A type of abnormal ridged growth of acorns caused by a parasitic insect.

Knosp

An unopened flowerbud.

Knot

(1) Node on the stem of a plant.
(2) Hard mass formed in the trunk of a tree at the point where a branch originates.
(3) Round, cross-grained piece in a cut board which may fall out as the timber dries; the base of the branch set into the trunkwood.

Knot Garden

Flowerbeds or, more often, low hedges laid out between paths in such a manner as to form intricate patterns when seen from above. Early examples in the sixteenth century were constructed mainly of plants such as lavender and box, but later ones often included floral subjects. Some still exist.

A TYPICAL KNOT GARDEN DESIGN

Knur(r)

A hard ball or knot of wood; a hard excrescence on the trunk of a tree. *Same as* Nur(r).

Kokum Butter

An edible fat derived from the nuts of the kokum tree (*Garcinia indica*).

Kumara

Maori name for the sweet potato.

Kurrajong

Australian name for various native trees with fibrous bark.

Kyanize

The injection of a corrosive sublimate into the pores of wood to preserve from dry rot; from the name of its originator, John H. Kyan (1774–1830).

L

Labellum

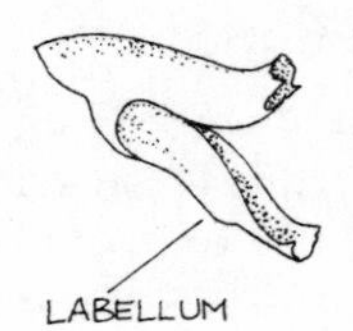

The lip-like structure of certain flowers, often applied to the lower petals of orchids.

Labiate

(1) Lipped. Usually applied to a flower with two lips, upper and lower, or divided.
(2) A member of the plant family Labiatae.

Labium

The lower lip of a labiate corolla.

Labyrinth

See Maze.

Lac

LACED GARDEN PINK

The resinous incrustation on certain trees in the East Indies, produced by an insect. Shellac is manufactured from lac.

Laced

Applied to those border pinks and carnations which have petals edged with a colour that contrasts the ground colour in a narrow scalloped band, rather like lacework in appearance.

Lacerate

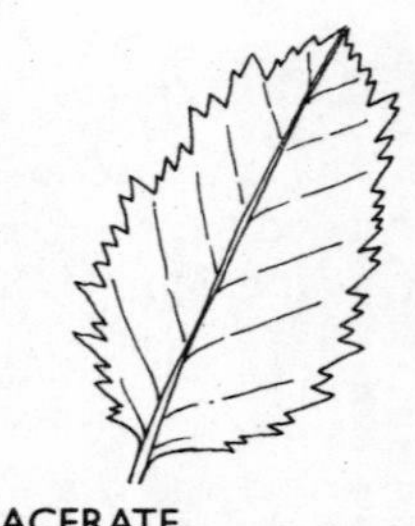
LACERATE

Of a leaf, irregularly torn or cleft, with edges cut or slit into ragged segments.

Lacewing Fly

Name applied to various insects of the order Neuroptera. The larvae eat aphids and are therefore beneficial.

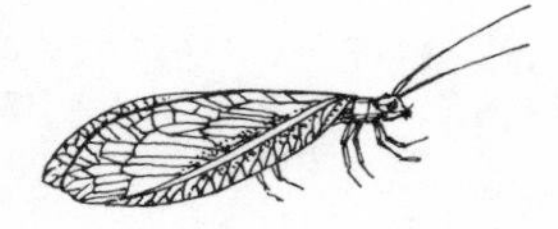

LACEWING FLY

Lachrymose

Drooping or weeping habit.

Laciniate

Jagged or fringed; cut or slashed into numerous narrow segments. Can apply to both leaves and flowers.

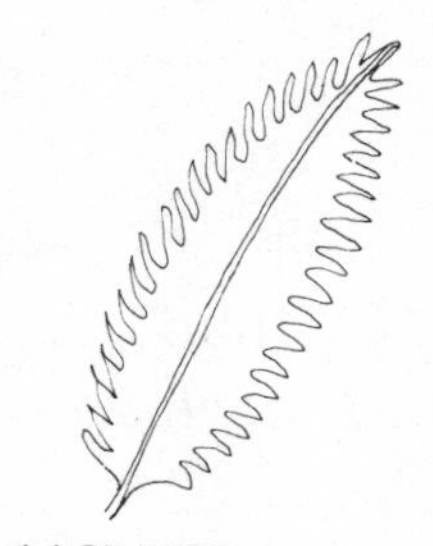

LACINIATE

Lacuna, *pl.* Lacunae

A cavity or pit; a depression.

Lacunate

Pitted.

LACUNATE APPLE

Ladanum

Gum resin exuded from plants of the genus *Cistus*, used in perfumery, cosmetics, etc.

Ladybird, LADYBUG (USA)

One of the family of beetles (Coccinellidae), many of which are brightly coloured. They feed on aphids and are very beneficial to gardeners.

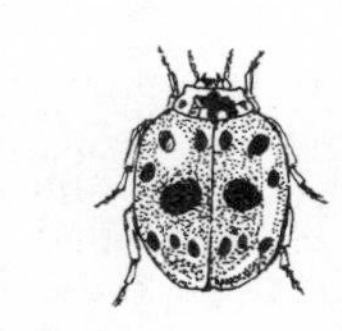

LADYBIRD
(*Anatis ocellata*)

Lamella, *pl.* Lamellae

Thin plate or layer.

Lamellate

Having, consisting of, or covered with thin plates or layers.

Lamellicorn

A member of the Lamellicornia, the former name for a group of beetles which includes the cockchafer and others which have the ends of the antennae expanded into flattened plates. Until quite recently they were a serious pest in Britain and Europe, the adults attacking and defoliating trees in 'plague' years, and the larvae feeding upon root

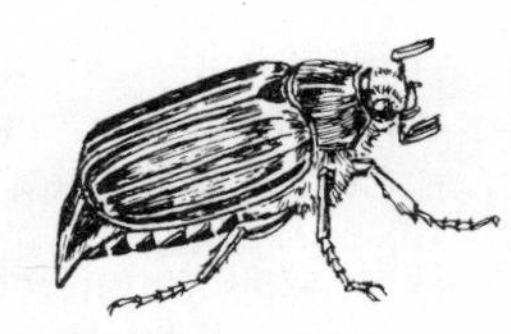

COCKCHAFER

crops, cereals and even grape vines. A concentrated effort was made to eliminate them in the 1950s and 1960s, and now they are rarely seen in this country.

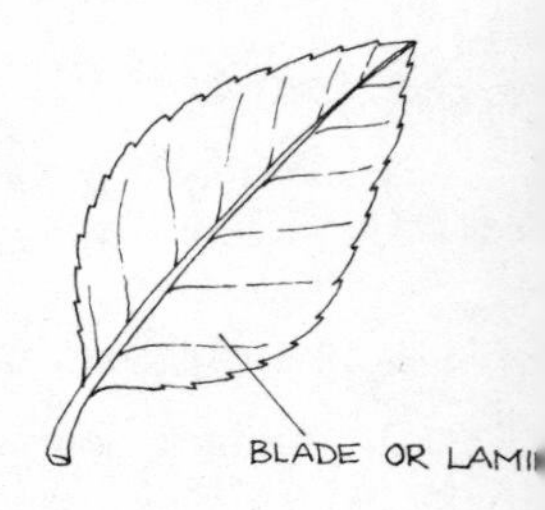

Lamina

The blade of a leaf or petal, usually thin except in the case of the seaweeds where the fronds are often large and leathery.

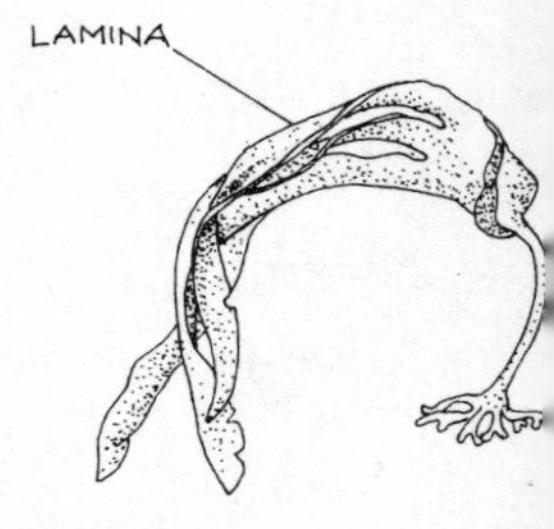

SEAWEED
(*Laminaria digitata*)

Lanate

Woolly, with long curly hairs, usually matted or intertwined.

Lanceolate

Shaped like a lance-head, as of leaves tapering at both ends, much longer than broad, and wider below the middle.

Lantern Cloche

A type of individual glass cloche popular in the nineteenth century for protecting one favoured subject. It was shaped like a lantern or bird-cage, often with a side-opening door, and a handle on top for carrying.

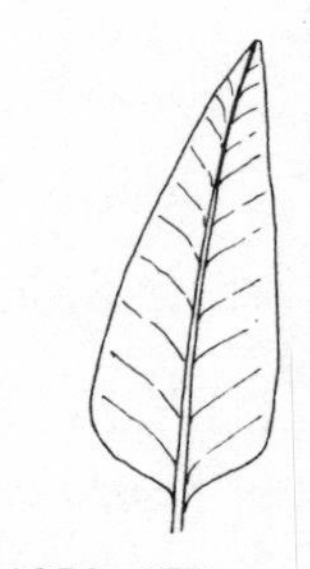

LANCEOLATE

Lanuginose

Downy or woolly but with shorter hairs than lanate.

Lanulose

Very short hairs, downy or woolly.

Larva

The stage in the development of an insect between leaving the egg and transformation into pupa; commonly called caterpillar, grub or maggot. Usually it is the most destructive period of the insect life cycle.

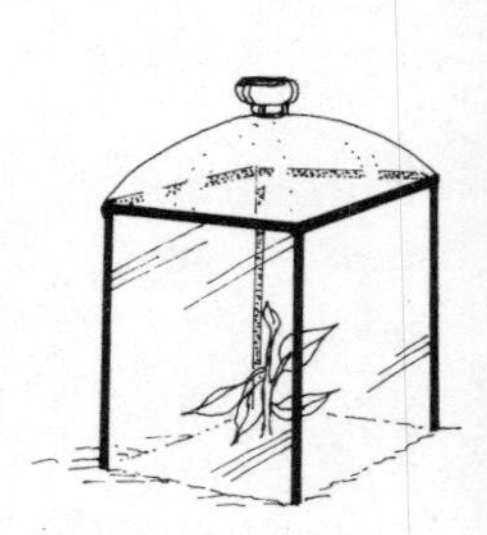

LANTERN CLOCHE

Latent

In horticultural sense, hidden, but often applied to immature buds which remain undeveloped until stimulated into growth by pruning or similar means.

LARVA OF THE GOOSEBERRY SAWFLY

Lateral

Attached to or near the side; a side-shoot or bud, especially one arising in the leaf-axil of a larger stem.

Latex

The milky sap of certain plants, e.g. rubber trees and dandelions.

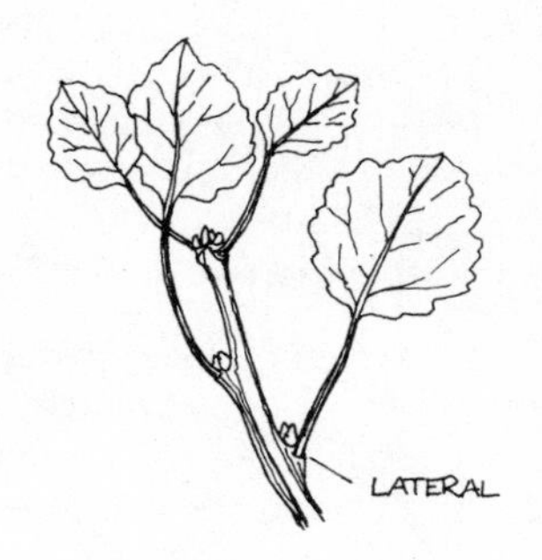

Lath House

A structure, usually in hot climates, covered with narrow wood slats or brushwood to reduce the power of the sun while affording some protection against wind and slight frost. Lath houses are very popular in the southern states of America, in Mexico, and many other sunny areas of the world.

Laticiferous

Containing or producing latex.

Lattice, Latticework

Same as trelliswork, but more usually applied to metal or plastic products.

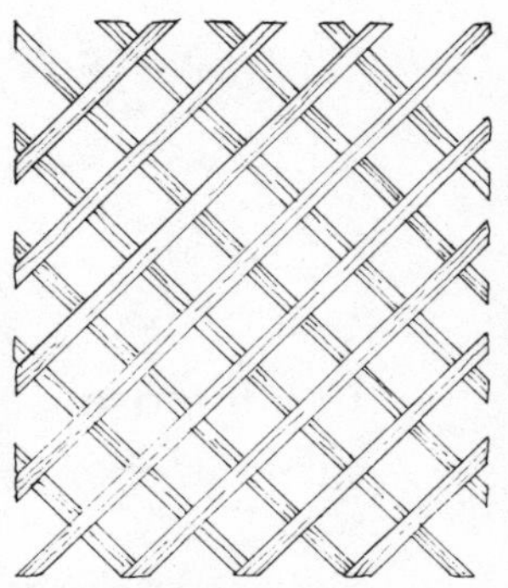
LATTICEWORK

Lawn

A more or less flat area of ground covered with grass kept short by regular mowing. Lawns are a feature of most gardens in Britain and USA but are not so widely favoured elsewhere in the world except perhaps in parts of Australasia.

Lawn Rake

A rake with radiating spring tines used for removing matted grass, moss and leaves from lawns.

LAWN RAKE

Lawn Sand

Each manufacturer has his own formula but basically they all consist of a mixture of 3 parts sulphate of ammonia and 1 part sulphate of iron to 20 parts silver sand (all by weight). Mercuric compounds are often added to control moss. Lawn sand is best applied in spring or late summer at the rate 4 oz per square yard (100 g per square metre).

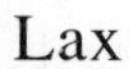

Lax

Loose, relaxed, widely spaced: *cf.* Congested.

Layering

A method of propagation by inducing shoots to root while still attached to the parent plant. It is used mainly for climbing and lax-stemmed subjects, but strawberries, carnations, jasmines and many other plants can also be layered.

LAYERING BORDER CARNATIONS

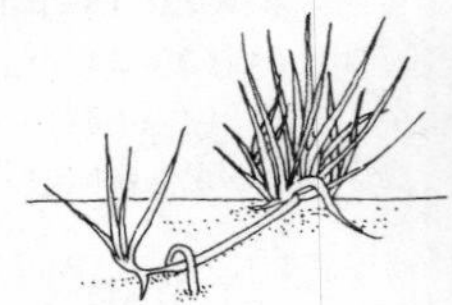

Laying

Hedging, as practised before the advent of the mechanical hedge-trimmer. After removing all unwanted growth, the remaining thick stems were cut almost through but not severed, then bent over at an angle of not less than 45 degrees and interwoven between vertical stakes driven into the centre of the hedge. Laying was a highly skilled job in earlier times and hedgers took great pride in the results of their efforts: *cf.* Hedging, Pleaching.

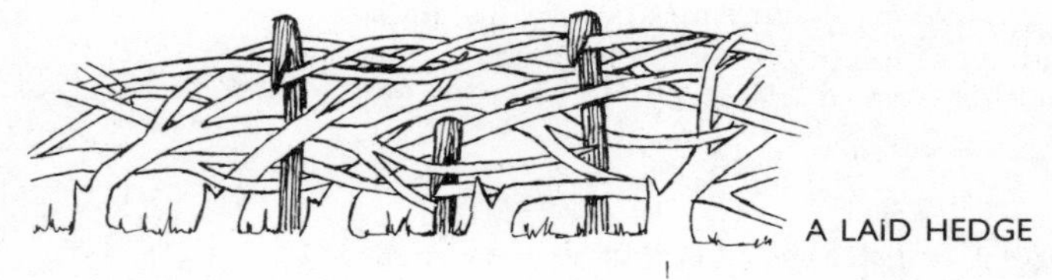
A LAiD HEDGE

Lea

An area of open ground, especially grassland.

Leaching

The loss of soluble (and some insoluble) minerals from the growing medium during the normal watering process or rain. Some minerals leach faster than others, sometimes producing an imbalance in the soil's chemical content with an adverse effect upon the plants growing therein. In general, leaching is more of a problem with pot plants than with those grown in the garden.

Lead-headed Nail

See Wall nail.

Leader

(1) A main stem of a plant from which laterals are produced.
(2) A shoot growing at the apex of a stem or main branch.

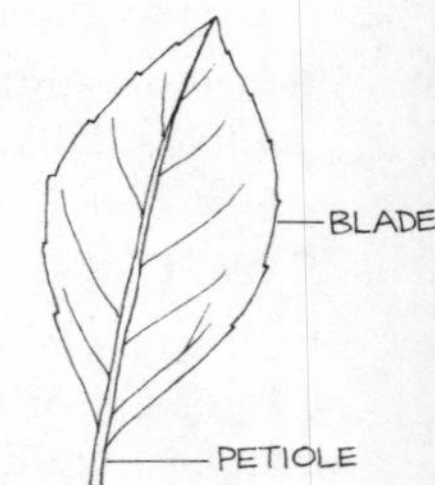

LEAF

LEAF-BUD CUTTING

Leaf

A lateral organ attached to the stem or axis of a plant below its growing point, performing the work of transpiration and carbon-assimilation, and manufacturing food by photosynthesis.

PEACH LEAF CURL

Leaf-bud Cutting

A cutting which includes a single leaf, a bud in its axil and a heel of the stem from which it was taken.

Leaf Curl

A fungal disease of peaches (peach leaf curl), nectarines and almond trees in which the leaves curl up and thicken, and change colour from green to yellow and dark red. All affected leaves should be removed and burned. A fungicide should be applied in spring, just before the buds begin to swell, and again in autumn when the leaves start to fall.

LEAF CUTTER BEES

Leaf Cutter

A type of bee, similar to the honey bee but solitary, with orange hairs on the underside and powerful jaws for shearing off pieces of leaf. In June and July the female cuts oval sections from the leaves of roses, laburnums, lilacs, etc. and rolls them into cylinders in which she lays her eggs.

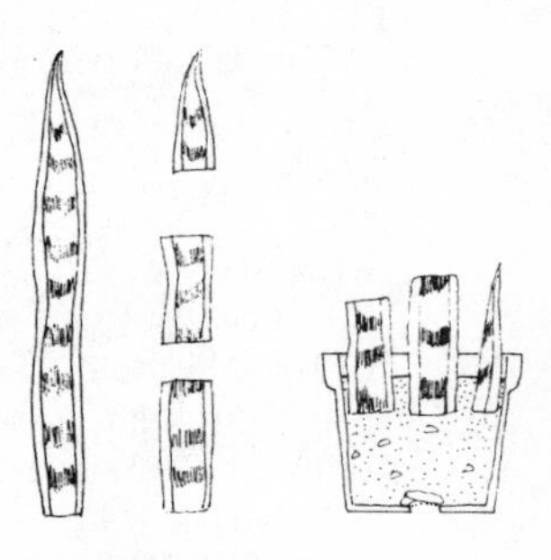
SANSEVIERIA LEAF CUTTINGS

Leaf Cutting

A method of plant propagation by using leaves or portions of them. Many plants can be propagated successfully by this method including some begonias, saintpaulias and most succulent stonecrops.

Leaf Miner

The larva of a group of small flies and some moths, a smooth-bodied grub which tunnels into the leaves of many plants, leaving a maze-like pattern of semi-transparent lines or a pale blister patch. Damaged leaves should be burned and any visible grubs despatched by crushing. Control in the greenhouse by spraying or fumigating the plants after removal of affected foliage.

LEAF MINER DAMAGE

Leaf Mould

The product of stacking and composting fallen leaves, or the residue of partially decayed leaves under trees.

Leaf Roll

A virus disease of potatoes. Leaves of other plants can also be rolled, especially when attacked by sap-sucking pests such as aphids and some types of sawfly whose larvae can often be found inside the rolled leaf.

LEAF ROLL OF POTATO

Leaf Scorch

An abnormal condition of leaves which, under glass, is often caused by splashing the foliage with water, so that the sun's rays are concentrated as with a magnifying glass and scorch the leaf. However, it can also be a symptom of potash deficiency on fruit trees and bushes, when the leaf-margins become crisp and dry, especially if an excess of nitrogen is present in the soil.

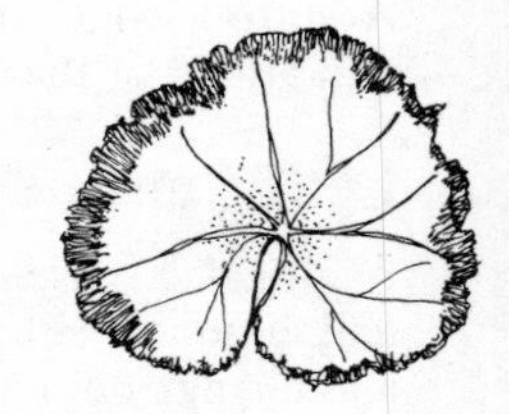

POTASH DEFICIENCY LEAF SCORCH

Leaf Spot

Visible sign of several unrelated virus diseases which can attack the leaves of many plants including roses, black-currants, celery, etc. The roundish spots may be of various colours, but in every instance the infected leaves should be removed and burned and the plant treated with a fungicide.

LEAF SPOT

Leaf Stalk

The petiole whereby the leaf is attached to the stem of the plant.

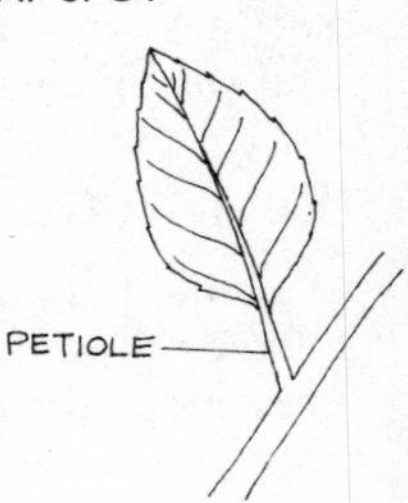

Leafhopper

A member, like the froghopper, of the suborder Homoptera, often very colourful, usually quite small, and capable of jumping and flying. Many are pests and can be found on a variety of plants including potatoes and nettles, shrubs and grasses; they may also cause considerable damage in greenhouses. Control is by the use of pesticides.

LEAFHOPPER

Leaflet

A small leaf; the subdivision of a compound leaf.

Leafy Gall

An abnormal growth, usually on the underside of a leaf, caused by bacteria. Attacked pot plants should be burned together with the soil in which they have been grown; in open ground where infected plants have grown, similar crops should be avoided for several years.

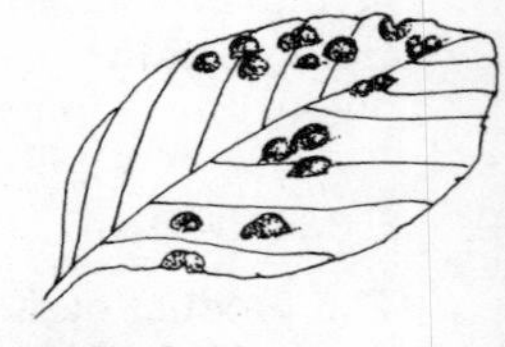

LEAFY GALL

Leather

LEATHERJACKET

Leather scraps or dust can be dug into the ground as a slow-release fertilizer rich in nitrogen.

Leatherjackets

The larvae of the daddy-longlegs or cranefly. The parents are harmless but the maggots will attack the roots of many plants including flowers, vegetables and pot plants. They can also be a pest in lawns by attacking the grass roots and causing dead patches to appear. Control is by the regular use of pesticides.

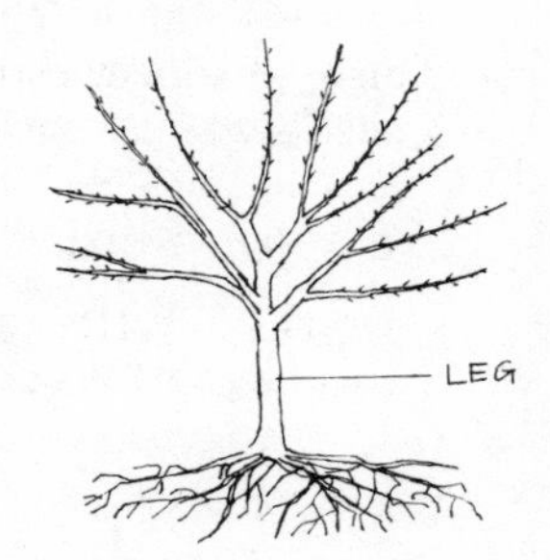

Leg

The short stem on a shrub below the point where the branches begin, e.g. gooseberry bush.

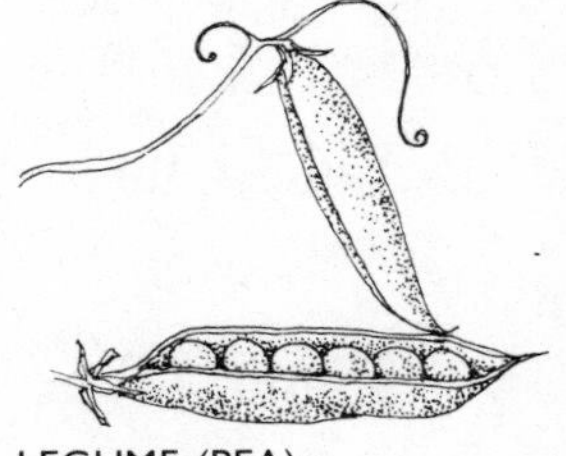
LEGUME (PEA)

Leghorn

A type of straw used for making hats, etc.; the stems of a Mediterranean wheat cut green and bleached, originally imported from Leghorn in Italy.

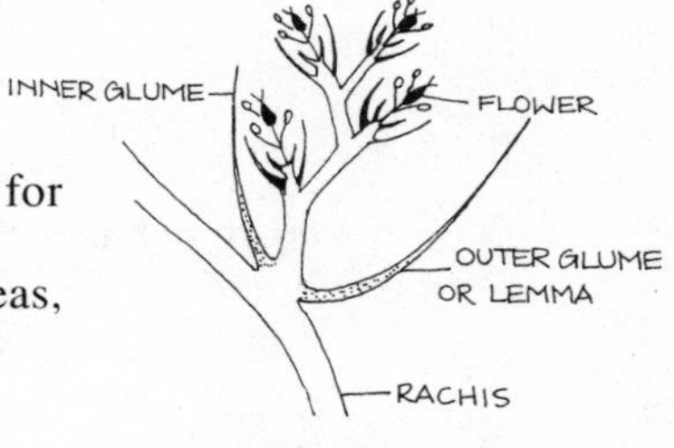

Legume

(1) A pod, as in peas and beans; a vegetable used for food.
(2) A member of the plant family Leguminosae (peas, beans and their relatives).

Leguminous

Relating to or resembling a legume; pod-producing. Any member of the pea or bean family is leguminous, as are several other plants.

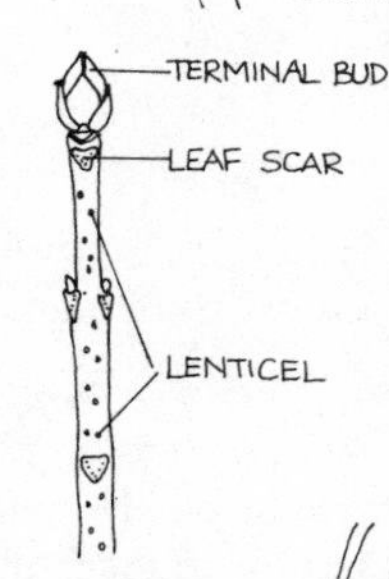

Lemma, *pl.* Lemmata

The lower of the two bracts or glumes enclosing the flower in some members of the grass family (Gramineae). *See* Valve.

Lenticel

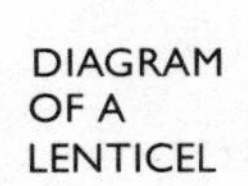

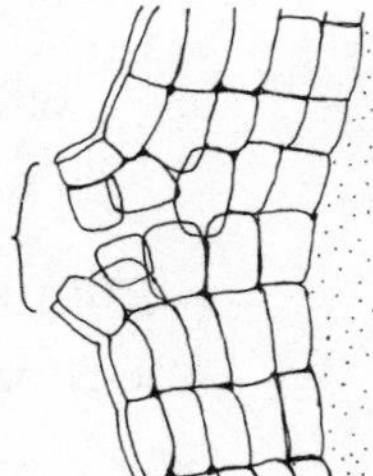

Aeration pore penetrating the young bark of stems and some roots.

Lenticular

Shaped like a lens or lentil seed; roughly circular but convex on both surfaces. Often applied to the corky spots on the bark of young trees.

LEPIDOPTERA
SWALLOWTAIL

Lepidoptera

Order of insects comprising butterflies and moths, having two large wings covered with flattened scales.

Lepidote

Scaly, scurfy, as the undersides of some rhododendron leaves.

LICHEN
(*Xanthoria parietina*)
ON A TILE

Leucoplast

A starch-forming colourless body in protoplasm.

Liana, Liane

Technically any climbing plant, but usually applied to the woody, vigorous vines often found festooning tropical trees.

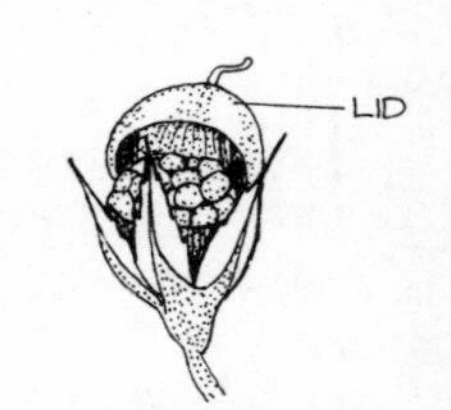

CIRCUMSCISSLE
CAPSULE (*Anagallis*)

Lichen

Plant organism composed of fungus and an alga in association, often found growing on trees, rocks, tiles, etc. and usually coloured grey, green or yellow. Some species are edible, others are used in the manufacture of cosmetics, dyes, etc.

Lid

Operculum, a cover or lid-like structure, as the top of the circumscissile capsule of the fruit of *Anagallis*.

Light

Frame or cloche. Many commercial greenhouses in the past were constructed of 'Dutch lights', basically a sheet of horticultural glass framed with timber, several of which could be bolted together to provide protection for tender crops.

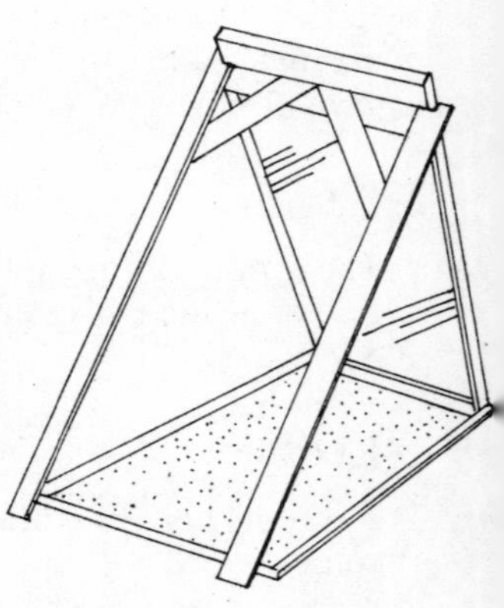

DUTCH LIGHTS IN
USE

Ligneous

Woody or wooden.

Lignin

Hardening material within the cell-walls of woody tissues.

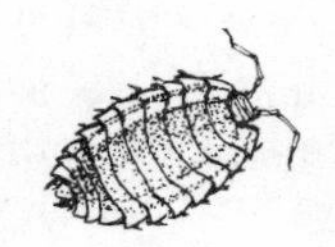

THE WOODLOUSE IS SEMI-LIGNIVOROUS BUT IT CAN ALSO EAT VEGETATION

Lignivorous

A creature that feeds on wood or woody material.

Lignotuber

The swollen base of a woody plant such as certain eucalyptus species. If top growth is killed or cut down, sucker shoots will normally arise from the tubers.

Lignum

(1) Wood.
(2) A wiry Australian shrub, one of the *Polygonum* family.

Ligulate

Strap-shaped, as applicable to a petal, leaf or corolla.

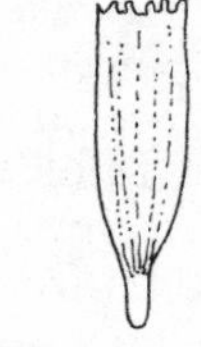

LIGULATE

Ligule

A strap-shaped projection from the top of the sheath in some grasses and palms; can also apply to a similar scale on a petal.

Liliaceous

Lily-like.

Limb

(1) A main branch of a tree or shrub.
(2) The free or expanded part of a petal, sepal or leaf.

Limbate

Having a distinct or different-coloured edge or border, as in some carnation flowers.

LIMBATE
(*Begonia crispa marginata*)

Lime

A chemical compound supplying calcium to the soil and reducing its acidity. The amount of lime present in the soil determines whether it is alkaline, neutral or acid.

Lime-hater

A calcifuge; any plant unable to thrive in alkaline soil, e.g. many ericas: *cf.* Calcicole.

HEATHER – A CALCIFUGE

Lime Sulphur, SULFUR (USA)

A favourite fungicide and acaricide with Victorian gardeners for the control of scab on apples and pears, and against red spider mites on fruit bushes, strawberries, etc. Now largely superseded by modern chemical products such as Benomyl and Captan.

Limestone

A type of calcium carbonate, similar to but harder than chalk. Limestone rock is widely used in garden construction, especially for rockeries where the rate of release of the lime is too slow to affect plants which are lime-haters.

Linear

Long and narrow, as the leaf blades of many grasses; having parallel or near-parallel sides, and being at least twelve times as long as it is wide.

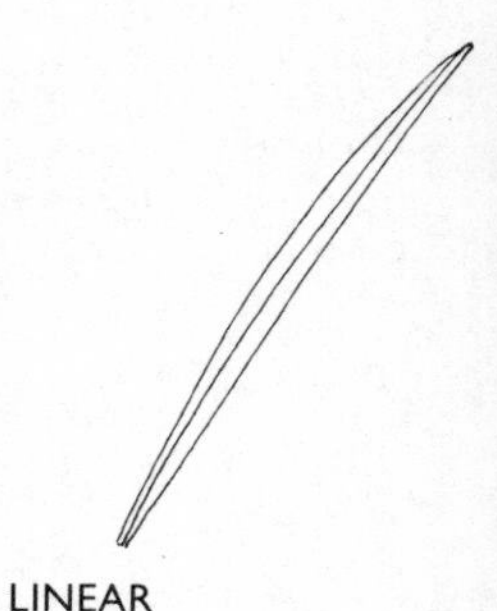
LINEAR

Lineate

Marked with lines, usually longitudinally, as in many crocus leaves.

Lineolate

Marked with fine lines, more delicately than when lineate.

Lingulate

Tongue-shaped or having a tongue-shaped appendage.

Linn

The wood of linden or lime trees.

LINEATE FOLIAGE (*Crocus asturicus*)

Linnaeus

Carl von Linné (1707–78), Swedish naturalist and founder of the artificial system of classification, after whom the Linnean Society was named. Although Linnaeus' scheme for plant classification is not used today, his system of nomenclature is the basis of all plant and animal names, and the rules used to establish them: *see also* Binominal.

Lip

See Labiate.

Liquid Manure

A mixture that can be made by soaking animal manures and using the resulting liquid, diluted to meet the needs of any particular crop. Vegetable waste can be collected in a suitable container and allowed to rot down to produce a thick brown liquid which, when diluted, also makes an excellent fertilizer.

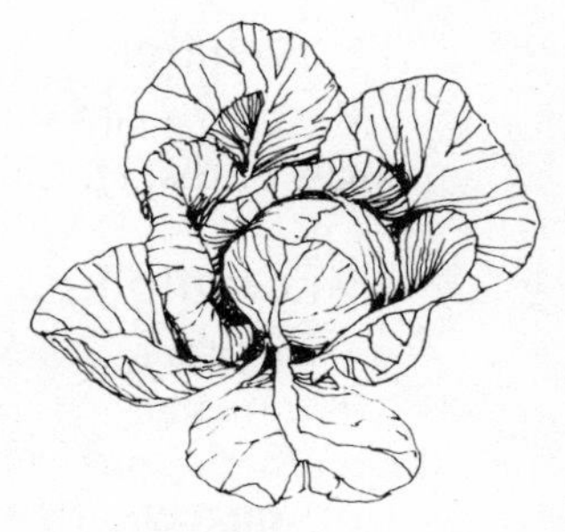

LOAF (CABBAGE)

Lithophyte

A plant that grows on rocks or stones, needing little or no soil, and obtaining most of its nourishment from the atmosphere, as in lichens and some orchid varieties.

Loaf

The solid round head of cabbage or lettuce.

Loam

(1) Fertile soil containing neither an excess of sand (too light) or of clay (too heavy) and being rich in humus.
(2) The top 6 in (15 cm) of soil below the surface of pasture, excluding the grass layer.

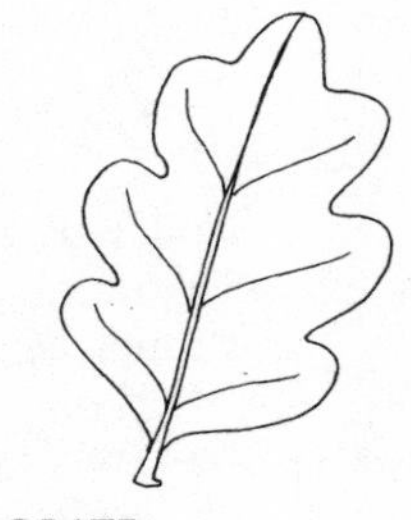

LOBATE

Lobate

Having lobes; applied mainly to leaves.

Lobe

The rounded parts of some leaves and petals which are divided, halfway or less to the centre, not sufficiently to make separate leaflets.

Lobulate

Having small lobes.

Lobule

A small lobe.

Locule, Loculus

A small compartment or cavity; a cell of an ovary containing the ovules, or of an anther containing the pollen.

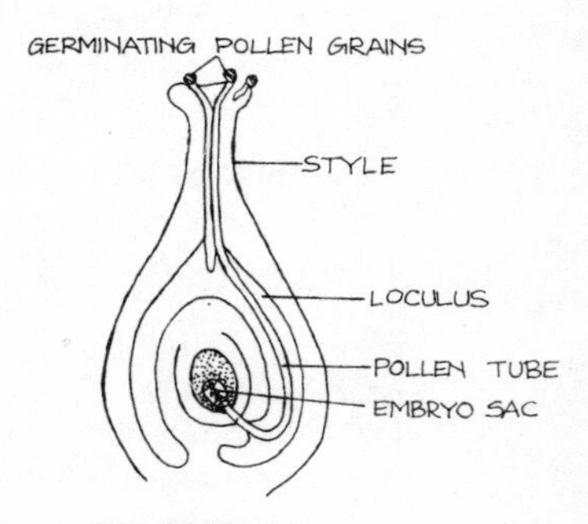

DIAGRAM OF AN OVARY

Loculicidal

Dehiscing along the back of the carpel. The opening of the seed capsule by splitting along the back seam.

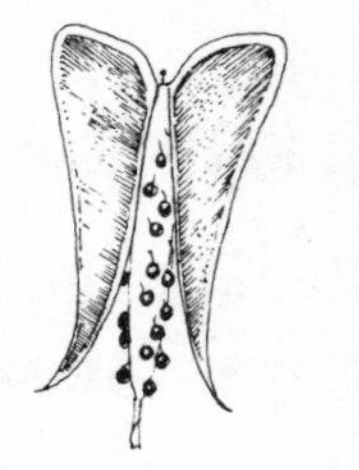

LOCULICIDAL SILICULA (SHEPHERD'S PURSE)

Lodicule

A small scale below the stamens in the flowers of most grasses.

Lomentum, *pl.* Lomenta

A pod that separates into pieces where it contracts between seeds, as in Sophoreae.

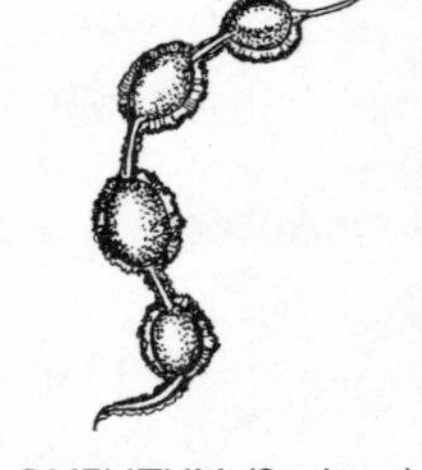

LOMENTUM (Sophora)

Long-arm Pruner

Pruning shears attached to a long handle and operated by a remotely controlled lever mechanism.

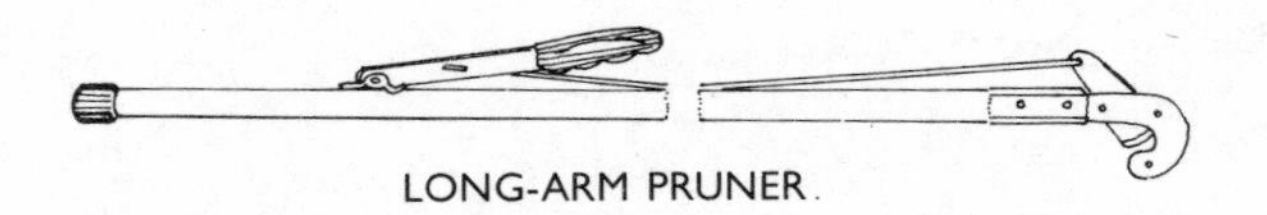

LONG-ARM PRUNER

Long-day Plant

A plant which flowers only when nights are short and there are maximum daylight hours. Day length is often controlled artificially by the use of supplementary lighting in commercial nurseries: *cf.* Short-day plant.

Long Tom

A flowerpot much deeper than usual relative to its width; seldom seen today but popular prior to the introduction of plastic pots. Most, if not all long toms, were clay pots.

LONG TOM POT

Lop

To remove smaller branches and twigs of trees, formerly with a pruning hook but now more often with long-handled shears.

Lopping Shears

Long-handled shears with short stout blades for pruning or lopping branches up to about 1 in (2.5 cm) in thickness.

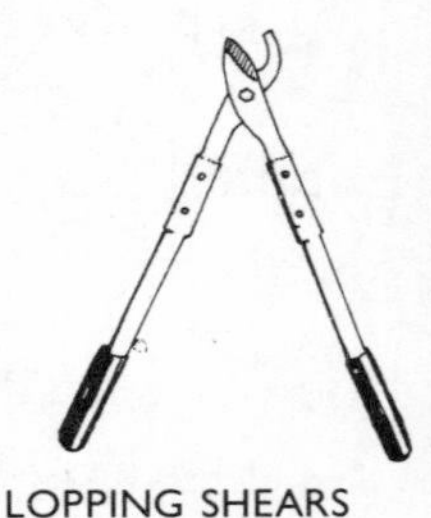

LOPPING SHEARS

Lorate

Same as ligulate – strap-shaped.

Lunate

Crescent-shaped, like a new moon.

LUNATE

Lush

Luxuriant and succulent, often applied to a well-grown grass sward.

Luxuriant

Profuse of growth, florid, lavishly ornamented, showy.

Lyrate

Of a leaf, lyre-shaped, having the terminal lobe much larger than the lateral ones.

M

Macchie

Italian form of maquis, a dense shrub or thicket: *see also* Maquis.

Macrobian

Long-lived, lasting.

Macrobiotic

Applied to seeds that remain alive in a dormant condition, sometimes for several years, until a suitable environment prevails for their germination.

Macrospore

The larger of two forms of spore; a spore producing a female gametophyte. *Same as* Megaspore.

Macrotherm

Plant which thrives in high temperatures: *cf.* Microtherm.

A MACULATE CALCEOLARIA

Maculate

Spotting or speckling on flowers or leaves, e.g. some calceolarias.

Maggot

Larva or grub of many flies, especially the blowfly; frequently used to refer to the caterpillar of the codling moth in apples. Often called a worm in USA.

Magnesium

An essential element in the production of chlorophyll in plants; a deficiency can inhibit normal development. Epsom Salts contain about 10 per cent magnesium and can be applied as a foliar feed or drench when dissolved in water and diluted. An application once a year is usually ample to meet the requirements of most subjects.

Maiden

(1) Often applied to young fruit trees, roses, etc. especially in the first year after grafting before the tree has been pruned.
(2) A plant grown from a seed or, in the case of strawberries, a newly rooted runner.

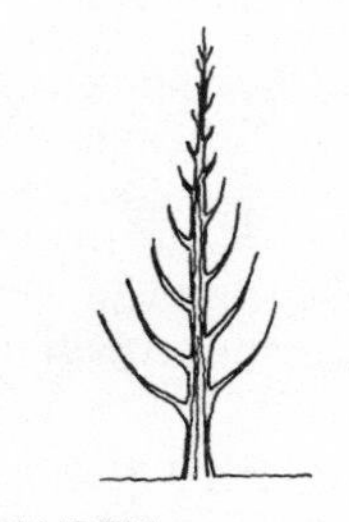

MAIDEN

Maincrop

(1) The heaviest-yielding crop of fruit or vegetables as compared with the early or late crops which are normally less fruitful.
(2) A variety (cultivar) which normally produces such a crop.

Maintenance Pruning

The gradual removal of diseased and crowded branches from elderly fruit trees to encourage new and more fruitful growth.

Male Flower

A flower having functional stamens but no carpels.

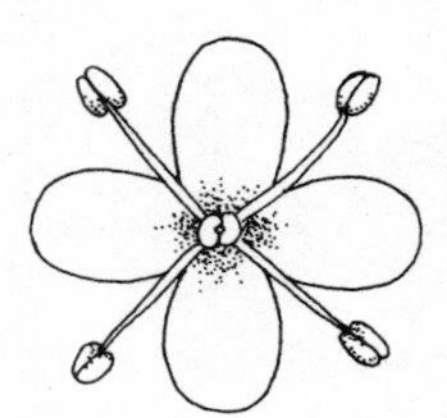

MALE FLOWER SHOWING 4 STAMENS

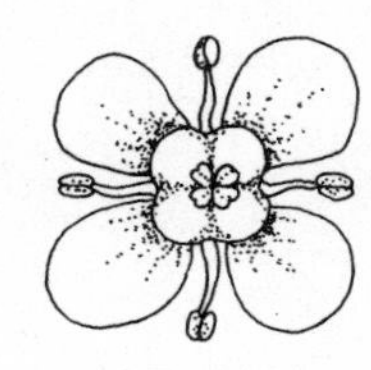

FEMALE FLOWER WITH 4 STAMINODES (FALSE STAMENS)

HOLLY (*Ilex aquifolium*)

Malic Acid

An acid present in many fruits such as apples, grapes, etc.

Malodorous

Evil-smelling, stinking, as the foliage of Herb Robert when crushed or handled.

Mamelon

Small, rounded (usually grassy) hillock or mound.

Manganese

One of the 'trace' elements essential in minute quantities to plant growth.

Manure

Formerly farmyard or stable refuse, now virtually any bulky material which supplies humus and plant nutrients to the soil, e.g. hop manure.

MANURE DRAG

Manure Drag

A long-handled type of fork or rake with several sharply bent tines, used for shifting farmyard manure from stalls, etc.

Manure Fork

A long-handled fork with slightly curved, widely spaced tines, used for stacking and moving farmyard manure. Called a muck fork in Gloucestershire and other rural areas. Similar to a hayfork but with more tines (a hayfork has only two).

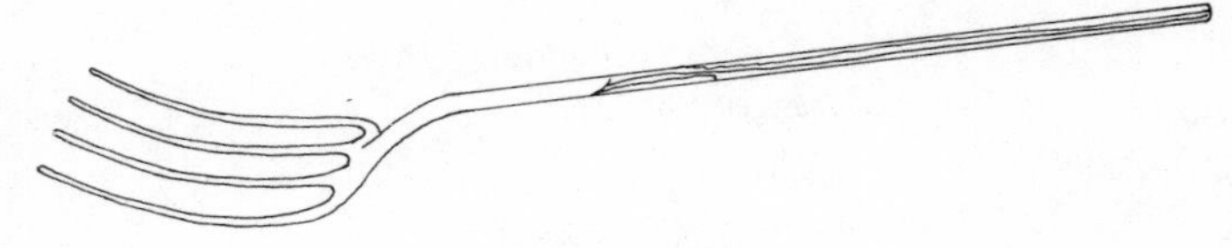

MANURE FORK

Maquis

Dense forest of trees and shrubs; the haunt of French patriots during the Second World War, from which their name is derived: *see also* Macchie.

Marbled

Flowers mottled or splashed with a colour in contrast to the base colour; can sometimes be applied to leaves.

A MARBLED IVY (*Hedera*)

Marc

The pulp remaining after the juice has been pressed from the fruit, especially in wine-making.

CRENATE LEAF MARGIN

Marcescent

Of leaves which die or wither but are retained on the plant until detached by next season's replacements, e.g. beech hedging.

Margin

Usually applied to the edges of leaves but can include the edges of other organs such as petals.

Marginal Hairs

Hairs or similar hairy outgrowths attached to the edges of leaves or other organs.

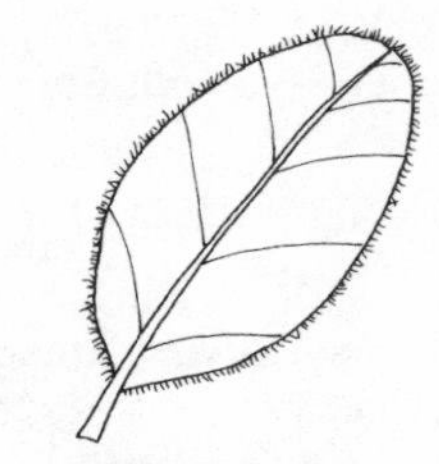

CILIATE LEAF WITH MARGINAL FINE HAIRS

Marginal Placentation

Placentation where the ovules are borne along the fused margins of a carpel, e.g. peas in a pod.

Marginal Water Plant

One that grows partially submerged in shallow water or in damp soil at the edge of a lake or pond.

Maritime

Applied to climate influenced by close proximity to the sea, often enjoying mild winters and cool summers marked by a high level of rainfall coupled with persistent winds. Plants which thrive under these conditions are mainly those with a high degree of salt-toleration.

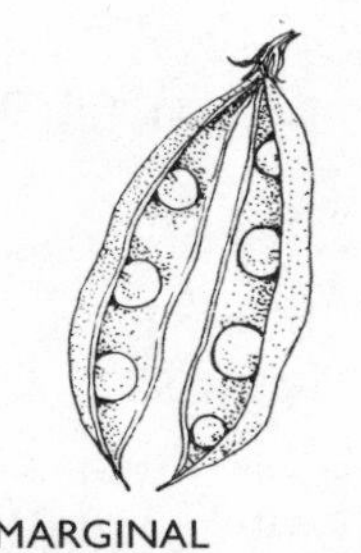

MARGINAL PLACENTATION (PEA)

Marl

A type of limy clay, often dried and granulated and then used as a fertilizer or manure.

Marlite

A type of marl which does not become pulverized by atmospheric conditions.

Marsh

Bog; morass; low-lying land often permanently water-logged even at low tide, although not always situated beside a river or the sea.

MAST OF
COMMON BEECH
(*Fagus sylvatica*)

Mast

The fruit of beech, chestnut and several other forest trees, mainly used for, or incorporated into, pig food.

Mast Year

A year in which mast is produced in larger quantities than in intervening years.

Mastic

Grafting wax or any similar substance used to seal exposed cut surfaces after pruning or grafting.

Mat

The layer of dead grass and other debris lying near the surface of a neglected lawn, which stifles the new shoots of young grass. It can be dispersed by persistent and thorough raking.

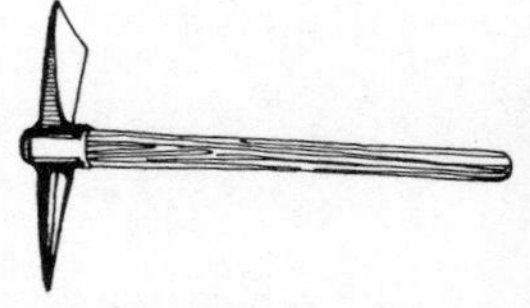

MATTOCK

Mattock

A type of pick used for heavy work. One end of the blade is pointed as with a normal pickaxe, the other is flattened like a chisel.

Maturate

To promote full development, to ripen.

Mature

Fully developed or ripe.

Maze

A complicated layout of paths and hedges with numerous dead ends and false turnings, very popular in earlier times. Some were simply paths laid in turf, allowing an overall view from above, but the more usual forms were arranged between dense high hedges. A few still exist today including that at Hampton Court in London. Also known as a labyrinth.

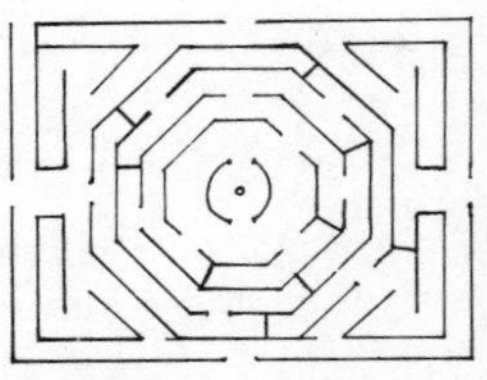

A TYPICAL
MAZE LAYOUT

Meadow

Area of grassland, especially one used for hay. A water-meadow is a low-lying area of ground, usually beside a river, which is regularly flooded.

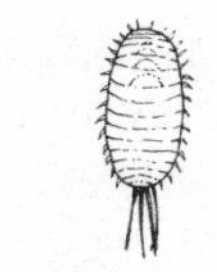

MEALY BUG

Mealy Bug

Sap-sucking insect rather like a small woodlouse in appearance but coated with a white or pale grey waxy substance resembling meal. A greenhouse pest controlled by insecticides, especially systemics.

Medium

(1) Soils of good quality having neither clay nor sand in excess – 'a good medium loam'.
(2) Flowers with heads of middling size such as dahlias and chrysanthemums which are not large- or small-flowered varieties.

Medium (Growing)

Soil or a compost mixture used for raising seeds or cuttings, pot plants, etc.

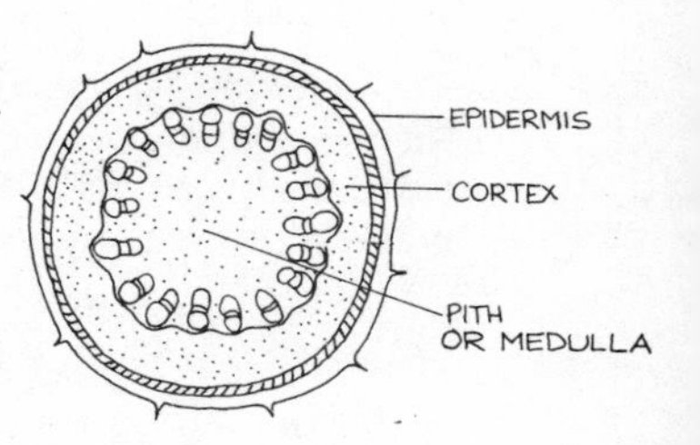

TRANVERSE SECTION OF A MAIZE STEM

Medulla

The soft internal tissue of plant stems; the spongy pith within the ring of vascular bundles.

Megaspore

See Macrospore.

Megass

The fibrous residue after the sugar cane has been processed.

Meiosis

The reduction of the number of chromosomes in cell-division during the development of a reproductive cell.

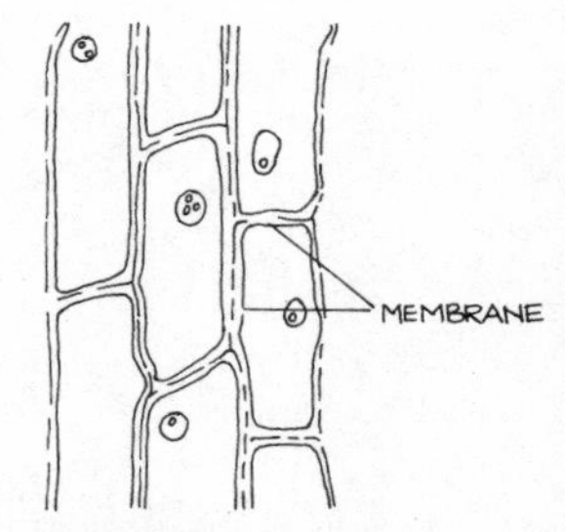

DIAGRAM OF CELLS IN ONION SKIN

Membrane

The surface layer of a cell which limits the diffusion of its contents and separates it from adjacent cells.

Membranous, Membraneous

Very thin and translucent; like a membrane; papery.

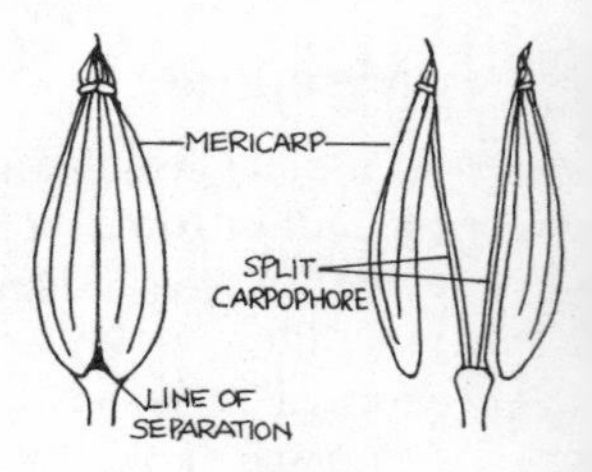

FRUIT OF FENNEL
(*Foeniculum vulgare*)

Mendel's Law, Mendelism

The theory of inheritance propounded by Gregor Mendel (1822–84) in 1866; the basis for the study of genetics which has developed into modern scientific plant breeding.

Mentum

From the Latin *mentum*, 'chin'. A forward chin-like extension of the base of the column in some orchids.

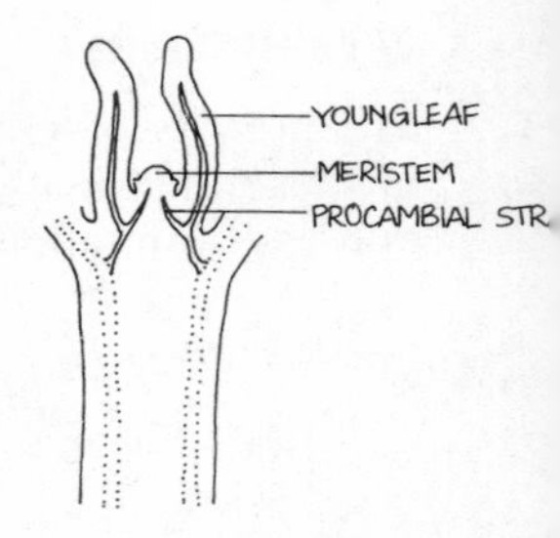

DIAGRAM OF THE MERISTEM AT THE APEX OF A SHOOT

Mericarp

A separating single-seeded part of a schizocarp. The schizocarp splits into two halves when ripe, each half being a mericarp, e.g. parsley and carrot.

Meristem

The growing tip and tissue of a plant, distinguished from the permanent and established tissues by the power of its cells to divide and form new cells; the area on a plant where growth is initiated in both root and stem.

Mesh Pot

A plastic container with large rectangular perforations. Unsuitable for holding soil without lining, it is used mainly as a decorative cover for a plant in a normal pot, or for planting aquatic subjects, such as waterlilies, in a pond.

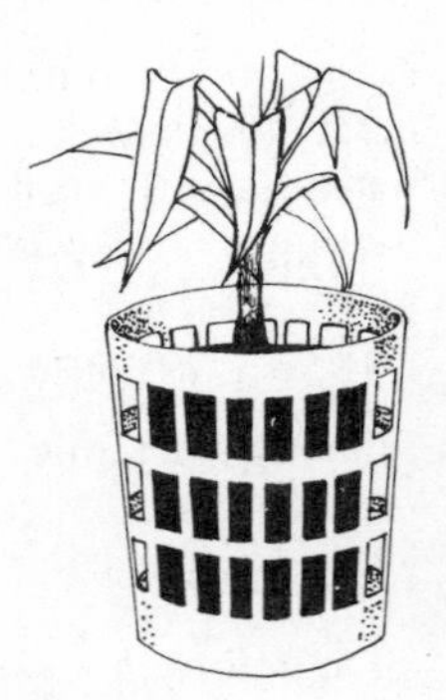
MESH POT

Mesocarp

The middle wall of a ripe ovary (fruit). The wall normally consists of three layers, the exocarp or outer layer, the mesocarp and the endocarp or inner layer.

Mesochile

The middle part of the lip in some orchids.

Mesophyll

The spongy tissue within a leaf; its inner tissues.

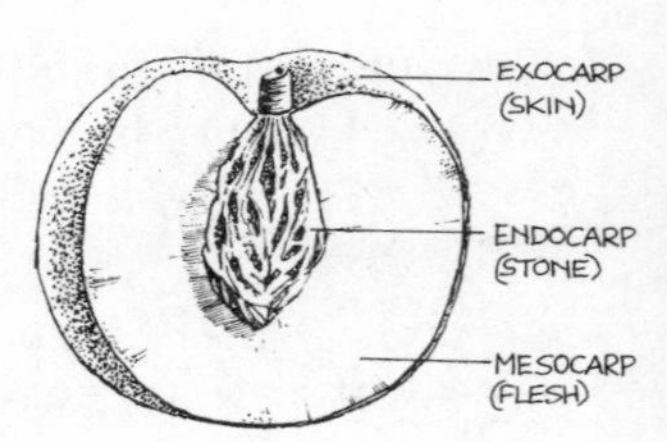

PEACH

Mesophyte

Of a plant which has moderate moisture requirement, not too wet or too dry; intermediate between a xerophyte and a hydrophyte. Mesophytes do not store water, but rely upon their roots to obtain it: *cf.* Hydrophyte, Xerophyte.

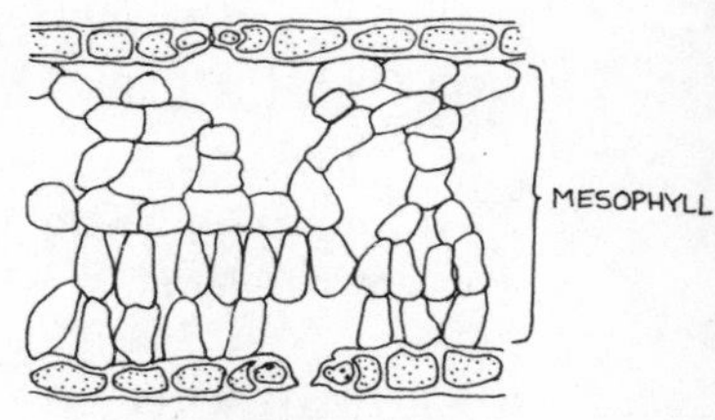

ENLARGED SECTION THROUGH A LEAF

Mesotherm

A plant which lives in temperate regions.

Metabolism

The process of growth and development relative to the intake of nutrients; including photosynthesis, transpiration and respiration.

Metaldehyde

A chemical used to control slugs and snails, usually sold in the form of pellets. It is poisonous to humans, birds and domestic pets and should therefore be used with great care.

Methylated Spirit

A volatile alcoholic fluid useful for localized control of mealy bugs and woolly aphids. Colonies can be painted with a small, stiff brush dipped in the spirit – a method particularly effective on cacti and infested leaf axils.

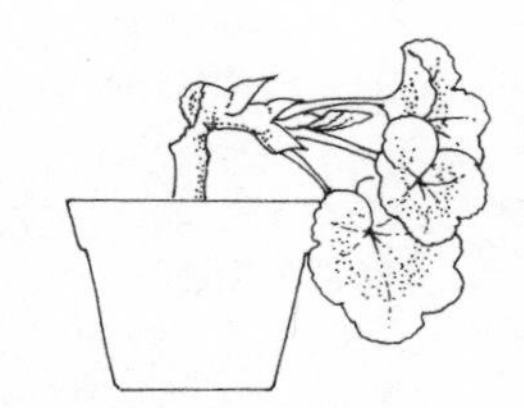
MICE DAMAGE

Mice Damage

In the garden, mice eat large seeds such as peas and beans, seedlings, bulbs, corms and tubers, and in the greenhouse will often destroy mature pot plants by gnawing through the stems just above pot level. A variety of poisons are available or the mice can be trapped with any of several baits, the best of which is probably nut-milk chocolate.

A BOTTLE GARDEN PROVIDES PLANTS WITH A MICRO-CLIMATE

Micro-climate

Usually applied to the artificial climate inside a propagator, cloche, frame or greenhouse, but more widely to describe a particular geographical location, or even a sheltered garden. At the other extreme, a refrigerator provides a micro-climate.

Micro-organism

Any organism invisible to the naked eye such as bacteria and protozoa, unicellular algae, fungus and viruses.

Microbe

A micro-organism such as one of the bacteria which causes disease or fermentation.

Microbiology

The study of micro-organisms.

Micronutrient

A nutritive substance required by a plant in minute quantities, e.g. trace elements.

Microphyllous

Having or bearing small leaves.

Microphyte

A very tiny plant.

Micropyle

The orifice in the coat or outer layer of the ovule through which the pollen tube enters at fertilization. The passage between the stigma and ovule.

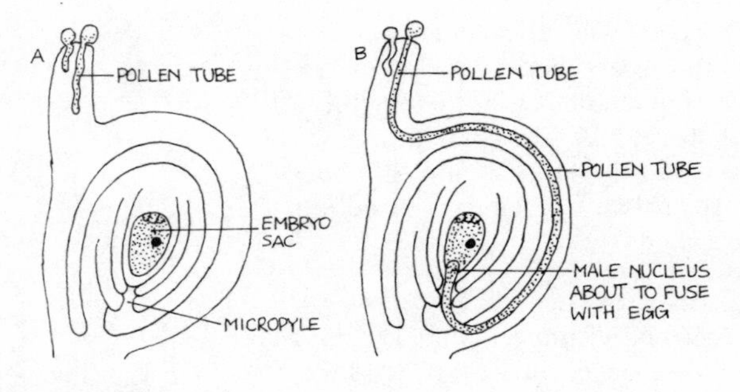

TWO STAGES OF FERTILIZATION SHOWING THE PROGRESS OF THE POLLEN TUBE THROUGH THE MICROPYLE

Microsome

A small particle within a cell, not visible with an ordinary microscope.

Microspecies

Generally only of interest to the specialist; a group of plants so closely resembling each other that they are classified under one species heading. The form *Rubus fruticosus* is often seen, meaning the group including all the microspecies.

Microspore

The opposite to megaspore; the smaller of the two forms of spore, developing into a male gametophyte.

Microtherm

A plant which thrives in low temperatures: *cf.* Macrotherm.

Microtubule

Any of the more or less rigid structures in the cytoplasm of many plants.

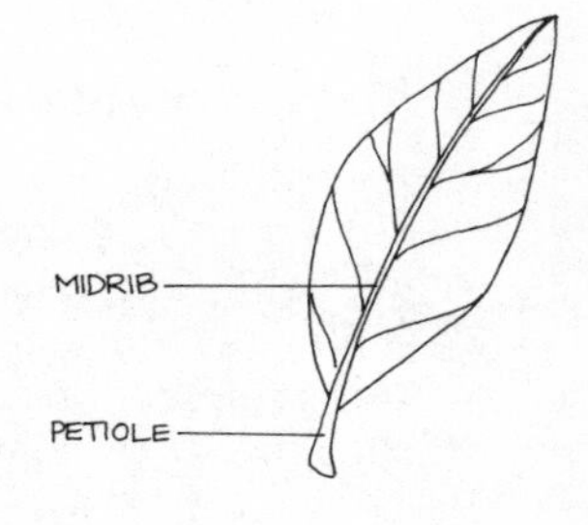

Midrib

(1) The main rib of a leaf, a continuation of the leaf-stalk or petiole.
(2) The central stem of a pinnate leaf to which the leaflets are attached.

Mildew

Minute fungi growing on plants exposed to damp conditions. Powdery mildew can be controlled by dusting with a sulphur-based fungicide, downy mildew by the use of a systemic. Prevention is easier than cure, notably by improved ventilation.

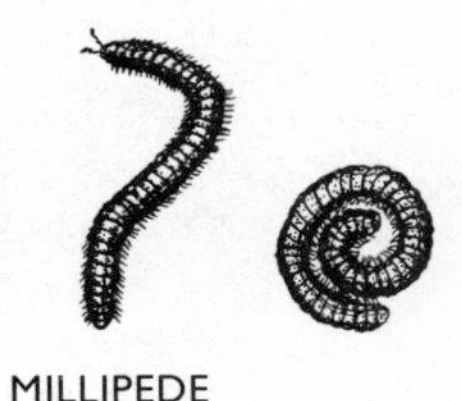
MILLIPEDE

Millipede, Millepede

A myriapod with numerous legs placed in double pairs on each body segment. Millipedes, usually black or dark grey, coil into a tight spiral when disturbed. They attack the roots of several plants including peas, beans, potatoes, etc.

Mimicry

Imitation, with reference to plants which look like other plants or objects, e.g. lithops, similar to pebbles, and the harmless white deadnettle, resembling the stinging nettle.

Lithops gracilidelineata AMONG PEBBLES

Mineral

Any substance which is neither animal nor vegetable.

Mineral Deficiency

The lack of one or more of the essential minerals normally present in the soil.

Minimus

Extremely small variety of plant; the smallest of a group.

Mire

Swampy ground; a boggy area.

Mist Propagation

A technique developed for the propagation of cuttings, especially hardwood cuttings, reducing transpiration and increasing the relative humidity by keeping them moist. A mist is produced by a series of fine jets placed at intervals of approximately 3 ft (90 cm), switched on and off either by time control apparatus or an 'electronic leaf'. The system is widely used commercially and small units are available for amateur use.

MIST PROPAGATION

Mite

Any of several kinds of arachnid; distinguished from insects especially by having eight legs and no antennae, e.g. red spider mite. Most are destructive pests controlled by chemicals called acaricides.

Eriophyes MITE WHICH ATTACKS SYCAMORE LEAVES

Mitosis

The elaborate process of cell-division in plants.

Mixture

The name for plant-growing media in USA; in Britain more often called compost. Most, if not all American mixtures, are peat-based but a British compost can be either soil- or peat-based.

Mock Trenching

Also known as double-digging or bastard trenching, by which the soil is aerated to a depth of about 2 ft (60 cm) without elevating the subsoil: *cf.* Digging, Full trenching, Ridging, Trenching.

MOCK TRENCHING

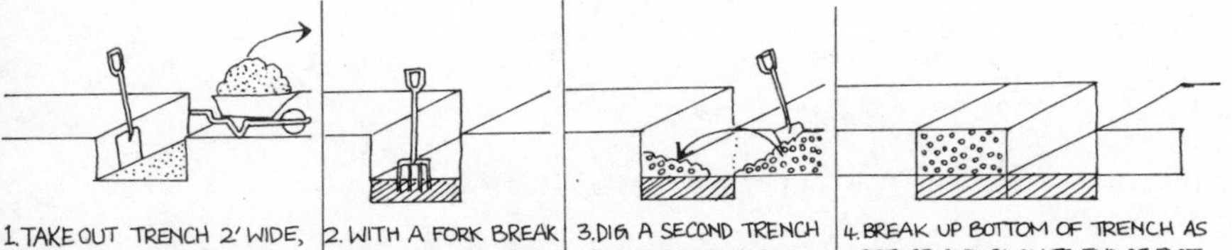

Moisture Meter

An instrument for measuring the amount of moisture in the soil, consisting of a probe attached to a calibrated dial.

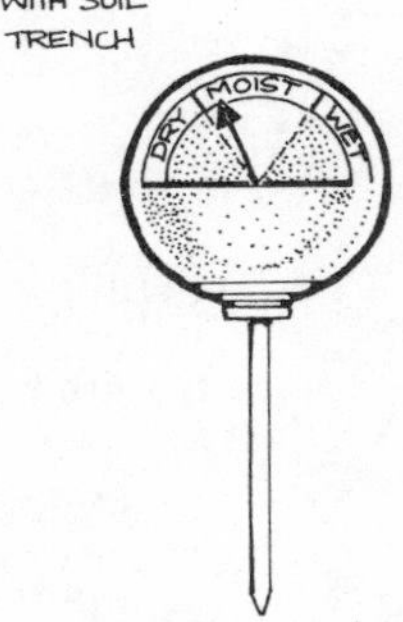

MOISTURE METER

Mole

A burrowing mammal not often welcomed in a garden because of the damage it does to the roots of plants, lawns, etc. in its constant hunt for worms, insects and numerous pests of the soil. Moles can be deterred by placing a handful of mothballs in the run; this will do them no harm but encourage them to move to another area.

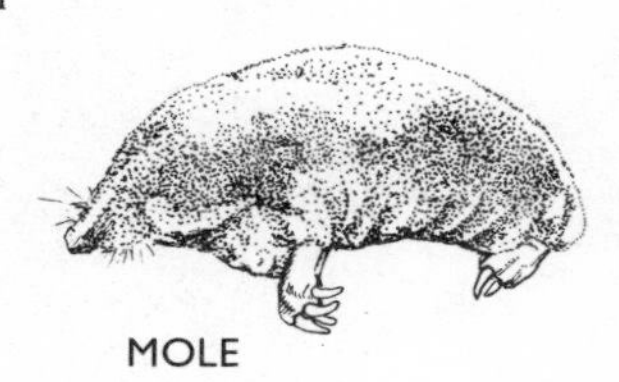

MOLE

Molluscicide

A chemical for killing slugs and snails.

Monadelphous

Having stamens with filaments united in a single group or bundle: *cf.* Diadelphous, Polyadelphous.

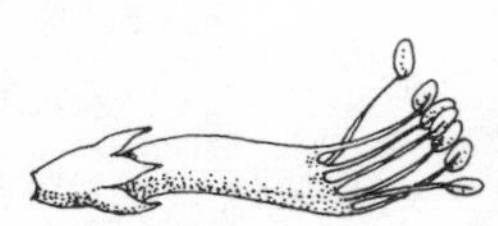

MONADELPHOUS

Moniliform

Joined together like a string of beads as in *Senecio rowleyanus.*

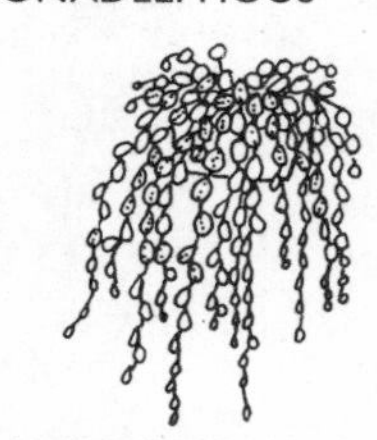

MONILIFORM
STRING OF PEARLS
(*Senecio rowleyanus*)

Monocarpic

Applicable to plants which die after fruiting; not normally applied to annuals but to those which may grow for several years before flowering, fruiting and then dying, e.g. the bromeliads and sempervivums.

Monocarpous

MONOCARPOUS

An ovary of a single carpel; having a single ovary; producing only one fruit.

Monocephalic

Bearing one flowerhead only.

Monochasium

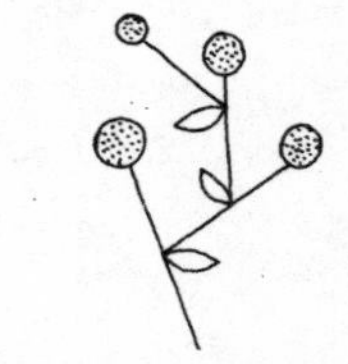

MONOCHASIUM

A cymose inflorescence in which each axis in turn produces one branch.

Monochlamydeous

Having a single floral envelope not distinguished into sepals and petals.

Monocolpate

Of pollen grains with a single groove or aperture.

Monocotyledon, Monocot

Refers to a plant bearing only one cotyledon or seed leaf, including all the grasses, lilies, bromeliads, orchids, palms, etc.

Monoecious

Hermaphrodite plants where the male and female flowers are carried separately but on the same plant, as in the hazel where the catkins are the male flowers and the small buds are the female: *cf.* Dioecious.

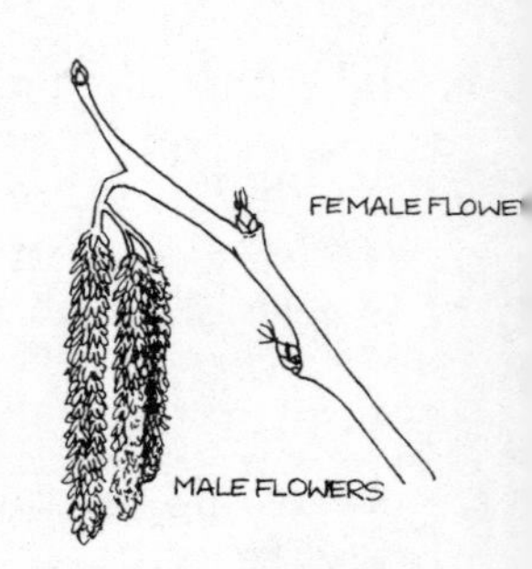

COMMON HAZEL IS MONOECIOUS (*Corylus avellana*)

Monogeneric

A family containing a single genus, e.g. *Ginkgo*.

Monopetalous

Having a single petal or with the petals united.

Monophyletic

Derived from a single ancestral line.

Monopodial

Originally a member of a fabled tribe of one-footed men with a foot large enough to be used as a sunshade. Botanically, having the growth of the stem or rhizome continuing indefinitely without branching, as in some orchids: *cf.* Sympodial.

Monotypic

Refers to any group containing a single species, such as a monotypic family.

Monstrous

A type of fasciation where a large number of tiny heads are produced. In cacti this condition is distinguished from 'cristate' (having a single growing point extended in a line) by having multiple growing points on the same branch, not merely having many small heads.

MONSTROUS
(*Celosia*)

Moor

Area of unenclosed, often heather-covered, sparsely wooded land.

Mor

Humus formed under acid conditions; the opposite to mull.

Moraine

The gravelly detritus and debris on the edges and at the foot of a glacier, often where the glacier has retreated (e.g. after the Ice Ages) leaving the moraine *in situ*, usually with fresh water flowing under and through it at certain times of the year. In many respects similar to the scree found at the foot of mountains and cliffs, but normally much drier. In nature plants grown in scree or moraine enjoy very good drainage and minimum humus. Moraine beds can be created in gardens with a supply of running water, otherwise a scree bed will be more suitable.

Morass

A swamp or swampy area; a bog.

Morphogenesis

The origin and development of an organ or organism.

Morphology

In plants, the study of the external structure of an organ.

Morphosis

The manner or order of development of an organ or organism.

Mosaic

The symptoms of one of several types of virus to be found in horticulture. It can be spread by insects, birds and humans, and once infected there is no cure for a plant. Some viruses are benign and the infected plants are grown for their unusually attractive foliage, e.g. *Abutilon megapotamicum* 'variegatum': *see also* Net veining.

MOSS
(*Sphagnum palustre*)

Moss

Small primitive herbaceous cryptogamous plants growing in bogs or damp areas, or on wood, stone, etc.

Moss Pole

A wooden or plastic pole to which sphagnum moss has been wired, used for training up a climbing or trailing plant; or a cylinder of wire netting tightly packed with sphagnum moss for a similar purpose. Sometimes called a 'totem pole' in USA.

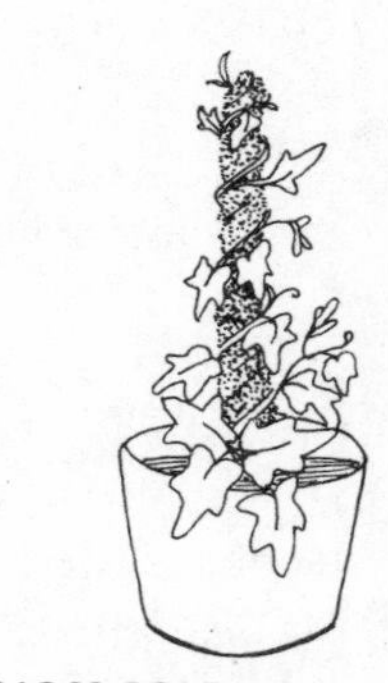

MOSS POLE

Moth

The harmless mature stage of certain caterpillars; unfortunately the grubs that hatch out from their eggs are very destructive and may cause serious damage.

Motile

Self-propelling, as in sperms and spores; capable of movement.

Mould, MOLD (USA)

(1) Term used in earlier times for soil, particularly loose, friable earth; soil rich in organic matter.
(2) Fungous growth forming on surfaces under moist conditions.

Mouldy

Covered with a furry fungous growth.

Mound

Bank of earth; heap; hillock, e.g. a molehill.

Mouth

The open end of any bell-shaped flower such as a daffodil; behind the mouth is the throat.

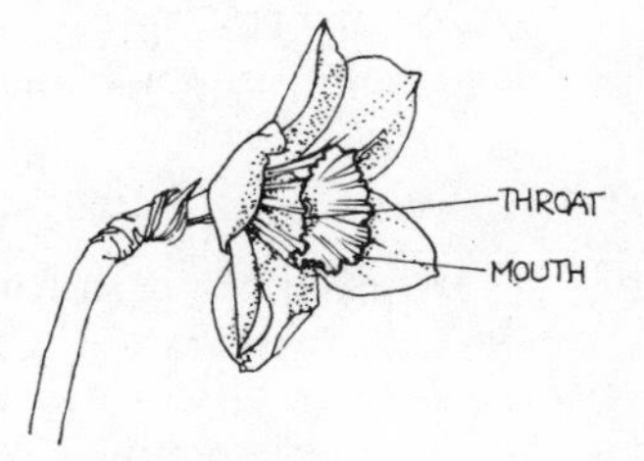

NARCISSUS

Mucilage

A gluey mixture of carbohydrates in plants; the slimy secretion of slugs and snails.

Muck

Farmyard manure.

Muck Fork

See Manure fork.

Mucro

A short sharp point or tip, as at the apex of some leaves.

MUCRONATE

Mucronate

A type of leaf terminated by a short stiff point or mucro.

Mucronulate

The diminutive of mucronate; terminating in a very small mucro.

Mud

Wet soft soil or earthy matter, not solid but not quite liquid.

Mulch

A top-dressing applied to the soil, usually consisting of a bulky organic material such as decayed leaves, grass mowings or peat, often applied to conserve moisture and suppress weeds. Black plastic sheeting can be used as a mulch but supplies no nutrients or humus to the soil.

Mull

Humus formed under alkaline conditions; the opposite to Mor.

Multiciliate

Having an abundance of cilia (hairs). Hairy or furry.

Multi-dibber

A piece of board fitted with numerous small projections, dowels or pegs, used for making several small holes in a seed tray or bed simultaneously. Can be used for seed-sowing, pricking-out seedlings, striking cuttings, etc.

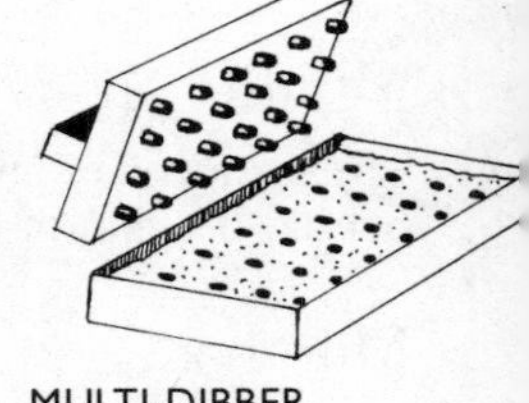

MULTI-DIBBER

Multipartite

Divided into many parts, e.g. certain leaves.

Multiply

To increase by propagation.

Multiseriate

Of flower parts carried in many whorls or series.

Muricate

Rough or warty, with short sharp points or prickles on the epidermis.

MURICATE GOOSEBERRIES

Mushroom Compost

The material in which mushrooms are grown, consisting mainly of peat, composted straw and nitrogenous fertilizer; animal manures are seldom nowadays included in the formulae. Spent mushroom compost is widely available and is a useful addition to garden soils since it increases the supply of organic matter and water retention, but it has virtually no nutritional properties.

Mushroom Flies

Sciarid and phorid flies which unfortunately do not restrict their attentions to mushrooms. They lay eggs that develop into tiny maggots which burrow into and up the stems of many plants and destroy them. Control is difficult but possible with modern insecticides. Sometimes called fungus gnats, particularly in USA.

MUSHROOM FLY (*Sciara thomae*)

Mutant

The correct terminology for what gardeners usually call a 'sport' – the accidental variation from an original; not to be confused with variations deliberately produced by breeding or hybridizing seed. Mutants may be either undesirable or beneficial, and in nature only the best will survive – the basic process of evolution. Mutation is the actual event by which a gene in an individual plant becomes different from that inherited from its parents; a mutant is an individual carrying such a gene.

Mycelium

The microscopic thread-like growth from which the fruiting bodies of fungi develop; correctly applied to the spawn used in mushroom cultivation.

Mycology

The science and study of fungi as practised by mycologists.

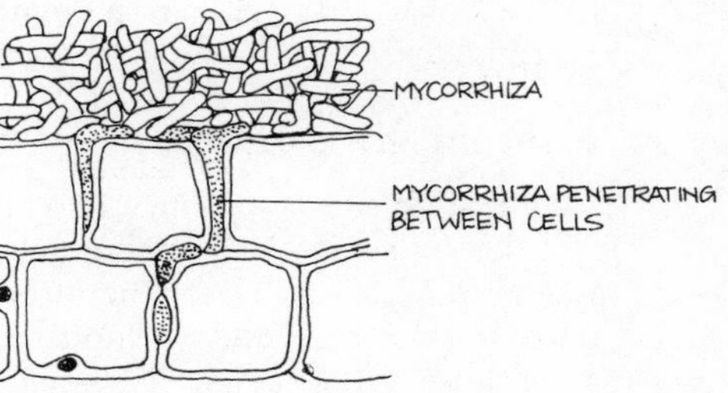

Mycorrhiza, Mycorhiza

Mycelium which penetrates the root system of a higher plant and supplies it with material from humus, a function usually performed by root hairs.

Mycotrophy

The condition of a plant living in symbiosis with a fungus.

Myriapod

A centipede or millipede – animals with many legs. Centipedes are carnivorous, feeding upon other insects, especially those living underground, and they are beneficial to the gardener; millipedes will eat both insects and vegetable matter, so can be a pest.

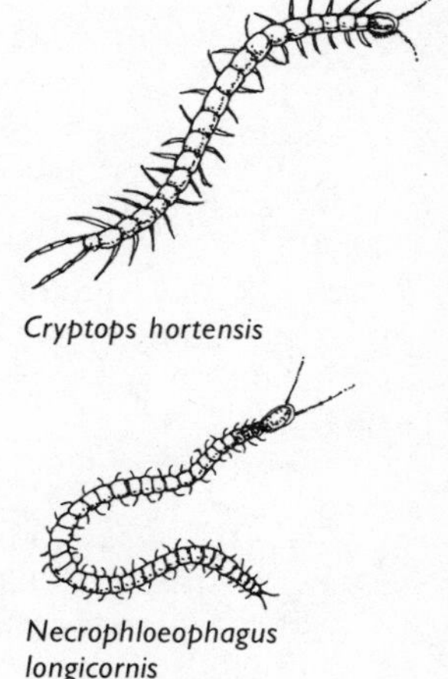
Cryptops hortensis

Necrophloeophagus longicornis

TWO CENTIPEDES OR MYRIAPODS

Myxomycete

Slime fungus.

N

Nana

Small; dwarf. Usually applied to the diminutive form of a plant.

Naphthalene

Chemical obtained from coal-tar, etc. One of those used in the past for soil sterilization but now almost superseded by pesticides.

National Growmore

A general fertilizer produced to more or less standard formulae by several manufacturers. It first appeared during World War Two under the direction of the then Ministry of Agriculture and Fisheries but is still widely used by gardeners. It provides an equal balance of nitrogen, phosphate and potash in various concentrations.

Native

An indigenous species: *cf.* Exotic.

Natural Break

A side shoot produced naturally on a plant, without pinching, pruning or any chemical treatment designed to induce breaking.

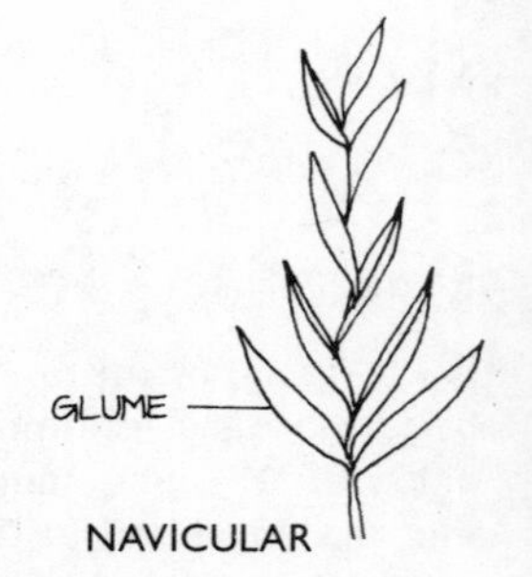

NAVICULAR

Naturalize

(1) To cause a plant to adapt to different conditions from those in its natural environment.
(2) To plant permanently, usually in a random manner, after which subjects are allowed to increase their numbers by natural means. Particularly applied to bulbs scattered haphazardly to give a 'natural' effect.

Naturalized

A plant of foreign origin which establishes itself in its new habitat, e.g. rhododendrons which came originally from China and Tibet but are now a feature of many areas of England, Scotland and Wales where they grow profusely in the wild.

Navicular

Boat-shaped, as the glumes of many grasses.

Neb

Part of the handle of a scythe or the spout of a watering can.

Neck

The upper part of a bulb or corm where the top growth (flower stem and foliage) joins the base.

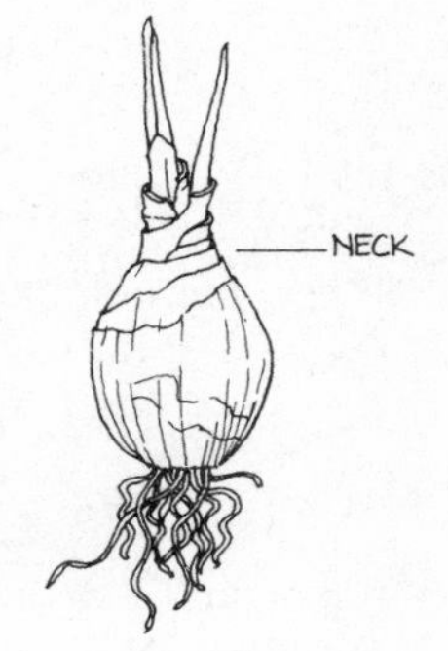

Neck Rot

Decay appearing at the neck of a bulb or corm, caused usually by a fungus which attacks a variety of subjects including onions, garlic and shallots.

Necrosis

Death of plant tissue; can be small dead areas or spots on leaves, fruit, etc., or pedicel necrosis which sometimes withers the stems of peonies and roses.

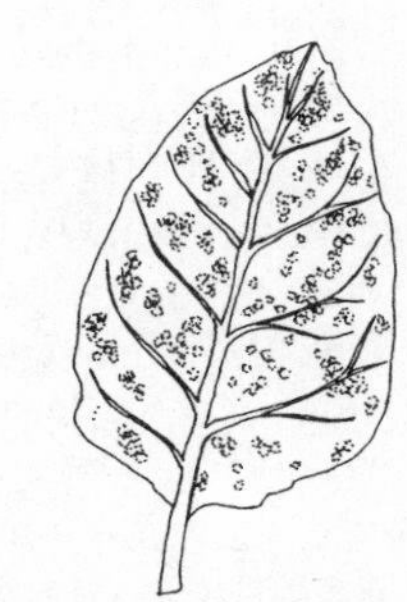

NECROTIC SPOTS ON A LEAF

Nectar

The honey secreted by the glands of plants which attracts insects to pollinate the flowers. From the Greek *nektar*, Homer's 'beverage of the gods'.

Nectar-guide

See Honey-guide.

Nectarine

A variety of smooth-skinned peach; said to be 'as sweet as nectar'.

Nectary

The glandular organ in flowers that secretes the nectar.

Needle

The stiff, narrow leaf of a conifer, e.g. pine, spruce, larch.

NEEDLES OF THE STONE PINE (*Pinus pinea*)

Nematicide

A chemical product for controlling or destroying nematodes.

Nematode

Small insegmented worm of slender cylindrical shape, resembling a piece of fine white cotton thread. Usually found in badly drained or poorly cultivated soils. An eelworm is a typical nematode.

Nervate

Of leaves having ribs, or the arrangement of the foliar veins.

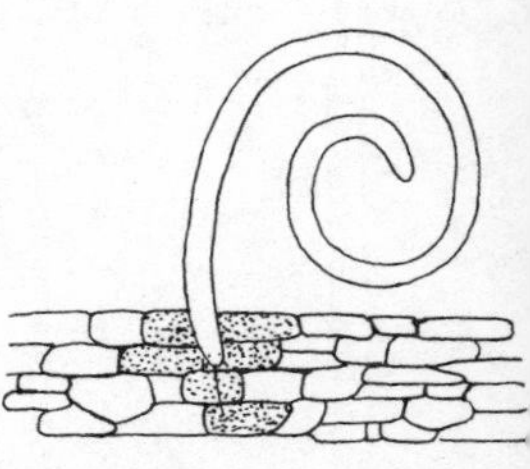

NEMATODE FEEDING ON ROOT TISSUE; NOTE DAMAGED AREA AROUND THE HEAD

Nerve

The midrib of a leaf or other single unbranched prominent rib.

Nervure

One of the principal veins of a leaf: *cf.* Vein.

Net Veining, Net Virus

A virus, similar to lace netting which produces symptoms in the leaves of some plants, e.g. *Pelargonium peltatum* 'White Mesh'. In many cases the virus is not harmful and the effect is considered attractive: *see also* Mosaic.

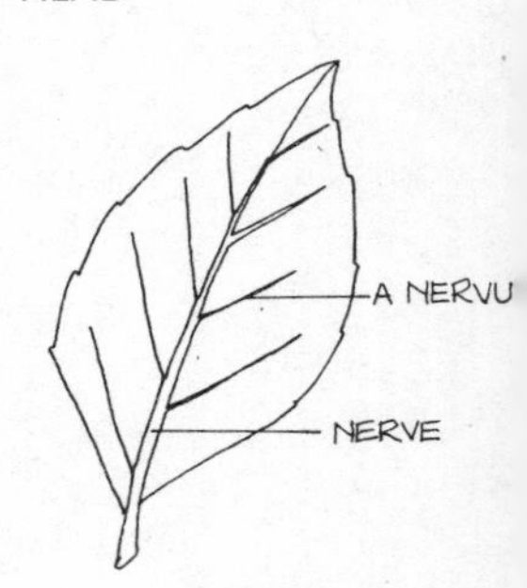

Netting

Garden netting can be made of natural materials, wire, or a variety of synthetics. It is used for many purposes such as support for climbing plants (peas, beans, etc.), protecting crops from bird and animal damage, or as a windbreak. Galvanized wire netting is used mainly for making pens for animals or for barriers such as open fencing.

'WHITE MESH' (*Pelargonium peltatum*)

Neuration

The distribution of the nervures in a leaf.

Neuter

A flower lacking sex organs and therefore sterile; having neither pistils nor stamens.

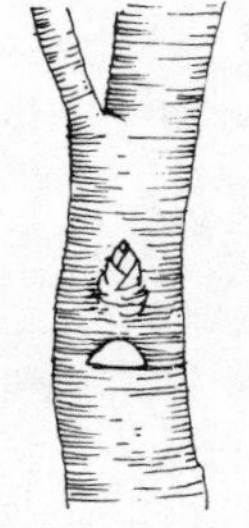

NICKING

Neutral

Soil which is neither acid nor alkaline; pH7.

Nicking

A method of inhibiting the growth of a dormant bud (usually on a fruit tree) by removing a small crescent-shaped piece of bark immediately below the bud: *cf.* Notching.

Nidus

The place in which spores or seeds develop, and insects lay their eggs. A nest or breeding place.

Nip

(1) To pinch out the growing point of a plant to promote the development of side-shoots.
(2) A slight touch of frost.

NIP

Nitro-chalk

Mixture of calcium carbonate and ammonium nitrate used as a fertilizer.

Nitrogen

The 'N' symbol in NPK formulae. An essential constituent of every organized body; in plants, mainly responsible for the promotion and wellbeing of the foliage.

No-digging

A method of cultivation which involves the application of large quantities of organic matter to the surface of the soil. No digging is involved at any time, and some remarkable results have been recorded.

No-soil Composts

Peat-based products with a variety of chemical additives which can be adjusted to suit the plant being grown. First developed at the University of California, Los Angeles in the early 1940s to provide a uniform, dependable and largely disease-free alternative to natural soils. No-soil or soilless composts are now widely used commercially in Britain and overseas. Aeration is particularly good but such composts are very light, lacking in support for the plant, and if allowed to dry out are very difficult to bring back to optimum water content. They all have a short pot-life and plants in any peat-based compost will need regular feeding after a month or rapid deterioration can result. On the plus side, they are easy to handle, being much lighter than soil, and most plants respond exceptionally well to them until they reach maturity. The rapid depletion of the world's peat stock has resulted in the increased use of coconut fibre (also known as coir) as a substitute. *See* Coconut fibre.

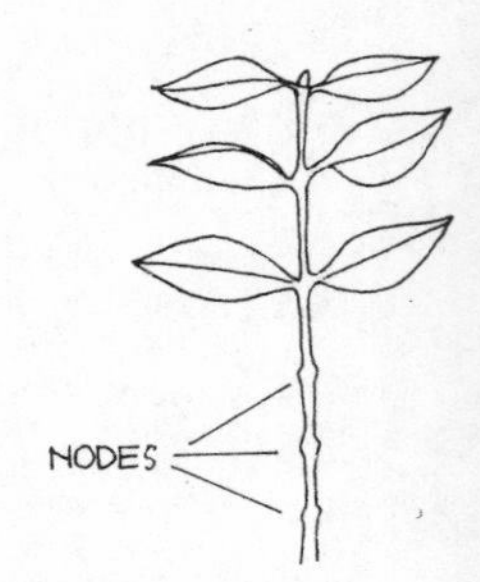

Nocturnal

Active by night, e.g. earthworms, slugs, snails, etc. Horticulturally, flowers that open at night, e.g. *Nicotiana alata*, *N. suaveolens*.

Nodation

Knottiness; a knotty area.

Node

The place, often swollen, where a leaf-stalk joins a stem, or where buds and side-shoots emerge from the stem: *cf.* Joint.

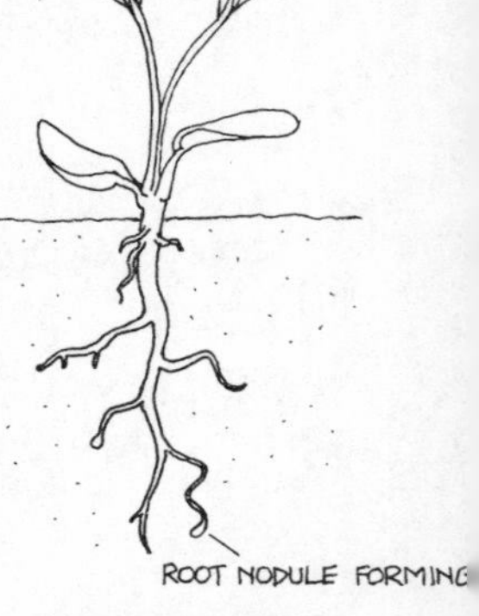

LUPIN SEEDLING

Nodose

Having many nodes situated closely together; knotty.

Nodular

Having nodules or small nodes; of or like a nodule; in the form of nodules or knots.

Nodule

Small knot-like swelling on the root of members of the Leguminoseae family (peas, beans, lupins, etc.) containing large numbers of a species of bacteria (*B. radicicola*) which attracts nitrogen from the air circulating around the roots, making it available to the plant and incidentally to other types of crop grown in the same ground in succeeding seasons.

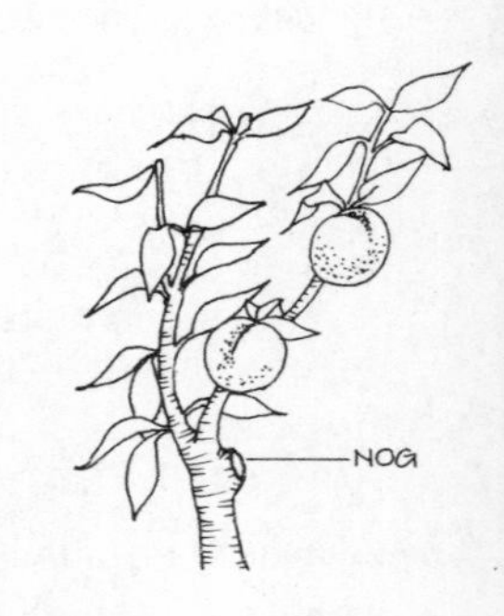

Nog

The stump or snag remaining on a tree after a branch has been removed.

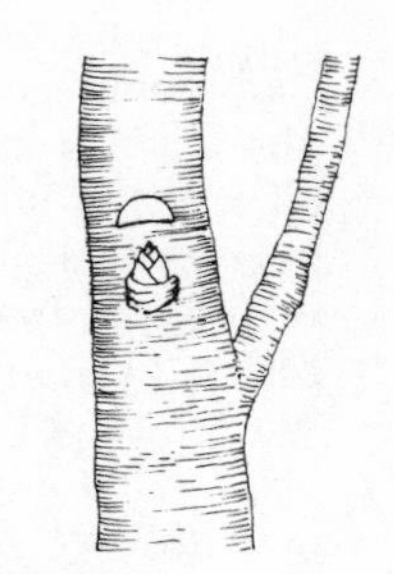

NOTCHING

Nomenclature

The systematic naming and classification of plants. *See* Classification.

Nonvascular

Plants that lack vascular tissue, such as fungi, mosses, lichens, algae, etc.

Notch

An indentation or similar V-shaped incision.

Notching

The method of removing a small crescent of bark above a dormant bud, usually on a fruit tree, to stimulate growth: *cf.* Nicking.

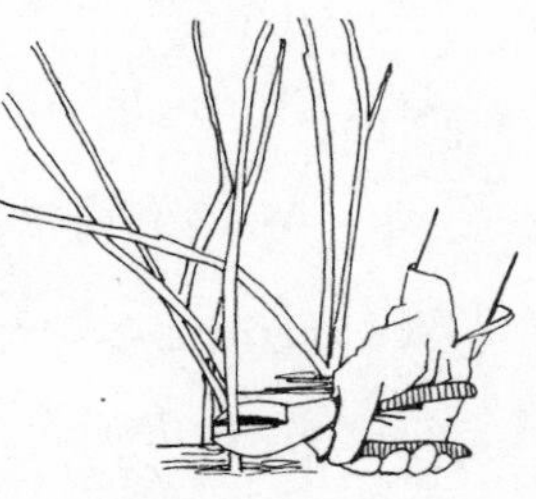

CUTTING OUT THE PRIMOCANES OF LOGANBERRIES IN AUTUMN TO ENCOURAGE THE PRODUCTION OF NOVIRAMES THE FOLLOWING YEAR

Novirame

The fruiting or flowering shoot developing from a primocane in its second year, as in brambles (*Rubus*).

NPK

Chemical symbols for the three main plant foods; N – nitrogen for foliage development, P – phosphorus for roots, K – potassium for flowers and fruit.

Nucellus

The central mass of tissue within the ovule which contains the embryo-sac.

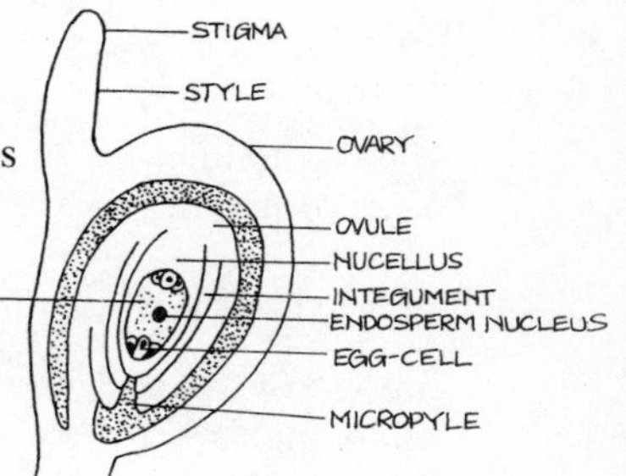

NUCELLUS

Nuciferous

Bearing or producing nuts.

Nuciform

Nut-shaped or looking like a nut.

Nuclear Stock

Plants produced vegetatively from virus-free stock, often from a single mother plant.

VACUOLE
NUCLEUS
PLASTID
CELL WALL

DIAGRAM OF A VEGETABLE CELL

Nucleus

(1) The microscopic body within a living cell which controls its division, multiplication and metabolism. The nucleus contains the chromosomes which carry the genes that regulate these processes.
(2) A very young clove of garlic; a bulblet.

Numerous

Usually taken to mean having more than ten floral parts and varying in number.

Nur(r)

See Knur(r).

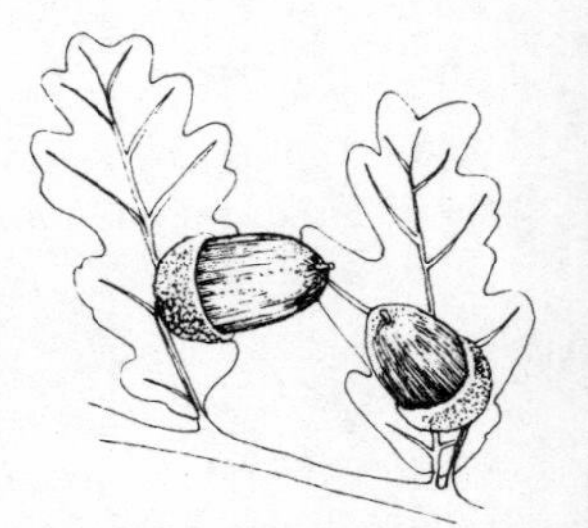

THE ACORN (*Quercus*) IS A TRUE NUT

Nursery Bed

A piece of ground set aside for raising young plants until they are mature enough to be transferred to their permanent quarters.

Nut

Popularly, any fruit whose seed is contained in a hard shell or pericarp such as hazel, beech and acorn. Walnuts, brazils and horse-chestnuts are not nuts in the botanical sense, but fruits.

Nutlet

Small nut, often one of a number within a woody pericarp.

Nymph

The immature stage of certain insects including aphids, leafhoppers, etc. Nymphs normally resemble adults but are wingless and often have different colouring. A nymph is distinguished from a pupa, which has its legs immobilized, by being active.

NYMPH OF A DRAGONFLY NOTE THE WING BUDS DEVELOPING

O

Obcompressed

Flattened from front to back.

Obconical

Of a leaf, roughly conical but with the stalk at the narrow end.

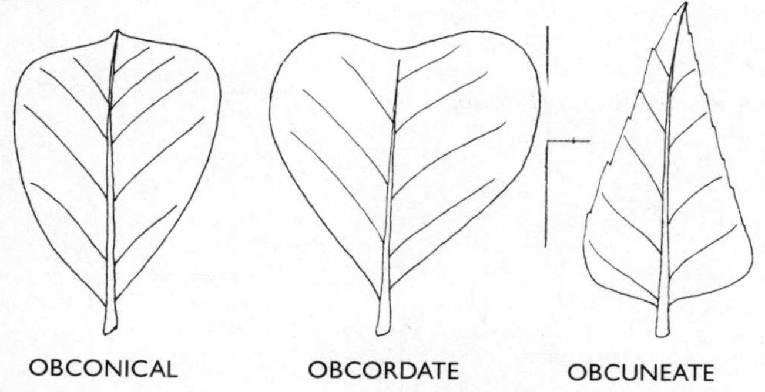

Obcordate

Of a leaf, inversely heart-shaped; the reverse of cordate.

Obcuneate

Of a leaf, inversely cuneate and wedge-shaped, attached at the broad end.

Obdeltoid

Of a leaf, inversely deltoid, narrowest at the stalk end.

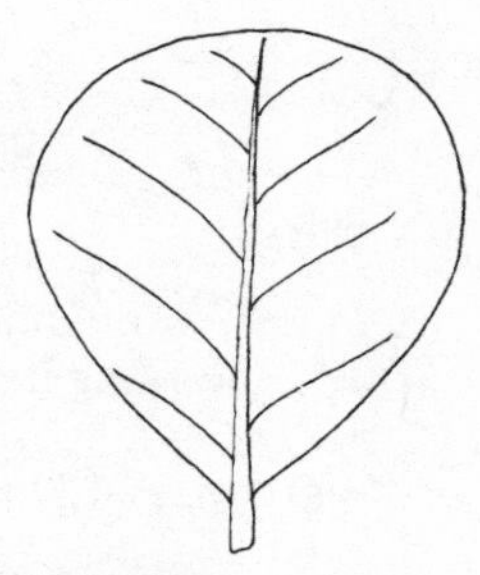

Obdiplostemonous

Having two whorls of stamens, the outer whorl opposite the petals, the inner whorl opposite the sepals.

Obelisk

An erect, usually square-sectioned masonry pillar, occasionally seen in landscape gardens and other public places, often with plants such as ivy growing up it.

OBELISK

Oblanceolate

Inversely lanceolate with the broadest part of the leaf above the centre, tapering toward the base.

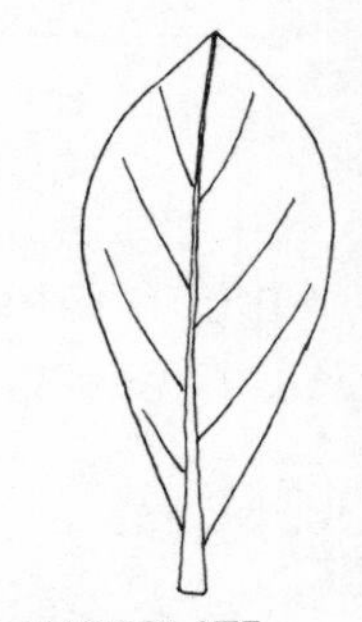

OBLANCEOLATE

Oblate

Roughly orange-shaped; almost spherical but slightly flattened top and bottom as in grapefruit.

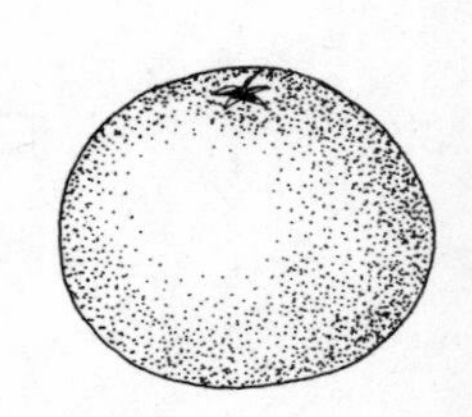

OBLATE

Obligate Parasite

A parasite entirely dependent upon its host for nutrition and therefore survival. Some host plants require a particular parasite for their wellbeing.

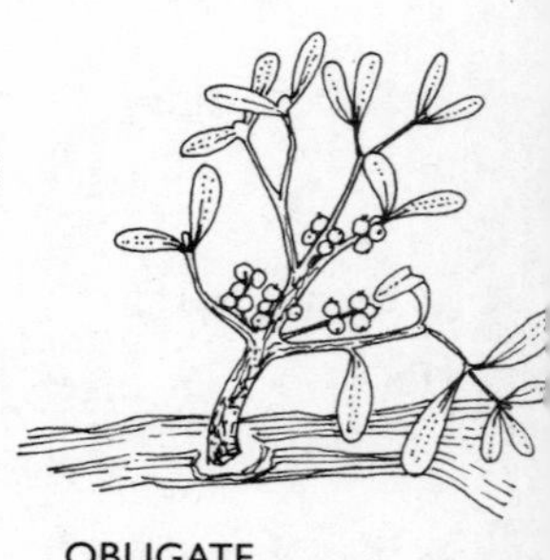

OBLIGATE MISTLETOE (*Viscum album*)

Oblique

Having unequal sides, as the leaves of *Begonia rex*.

OBLIQUE LEAF OF *Begonia rex*

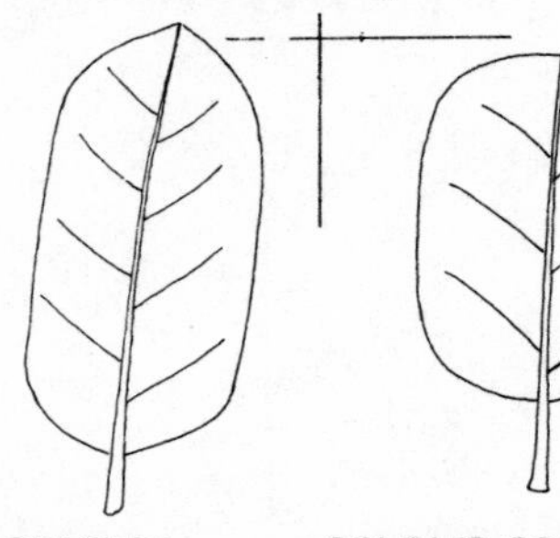

OBLONG

OBLONG-ORBICULATE

Oblong

Of a leaf with almost parallel sides, and between two and four times longer than broad.

Oblong-orbiculate

Of a leaf, basically oblong, almost as long as it is wide and appearing nearly round.

Oblong-ovate

Of a leaf, oblong with sides that curve in at either end.

Obovate

Of a leaf, basically egg-shaped in outline but with the maximum width away from the stalk: *cf.* Ovate.

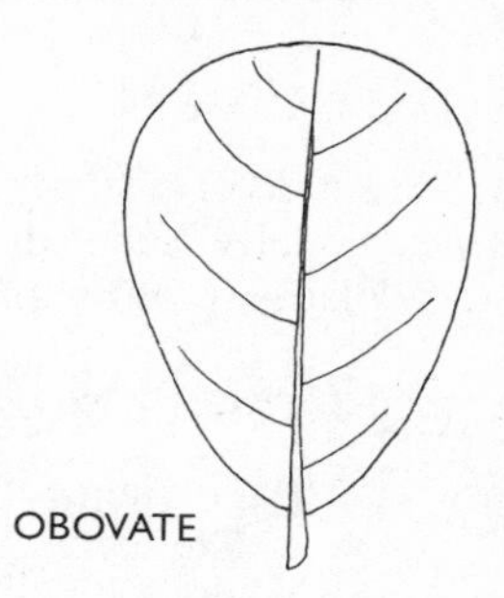

OBOVATE

Obovoid

Of leaves and petals, roughly the shape of a hen's egg with the stem at the narrow end; several fruits are obovoid, including some rose hips: *cf.* Ovoidal.

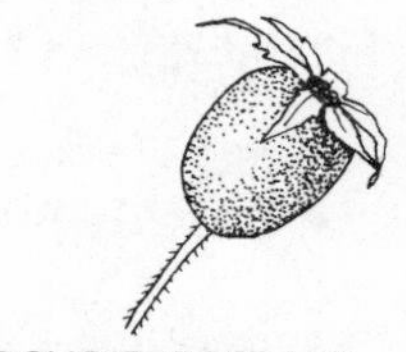

OBOVOID ROSE HIP (*Rosa villosa*)

Obpyriform

Roughly pear-shaped with the stem at the narrow end.

OBPYRIFORM FRUIT OF FIG (*Ficus carica*)

Obsolescent

Almost obsolete, in the process of disappearing. Usually applied to an organ still evident but non-functional.

Obsolete

No longer functional or fully developed; gone out of use.

Obtruncated

With the head removed; decapitated; being mutilated.

Obtuse

Of a leaf or petal that is blunt or rounded at the tip; obtusifolious.

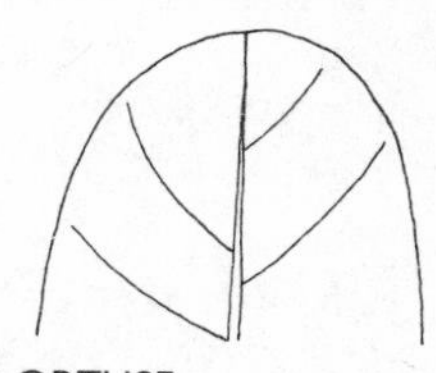

OBTUSE

Obverse

Of a leaf, narrower at the base or point of attachment than at the apex.

AN OBTRUNCATED LAUREL LEAF – RESULT OF POOR HEDGE-TRIMMING

Occluded

A wound on a tree that is completely covered by callus; can also apply to other openings that have become closed off.

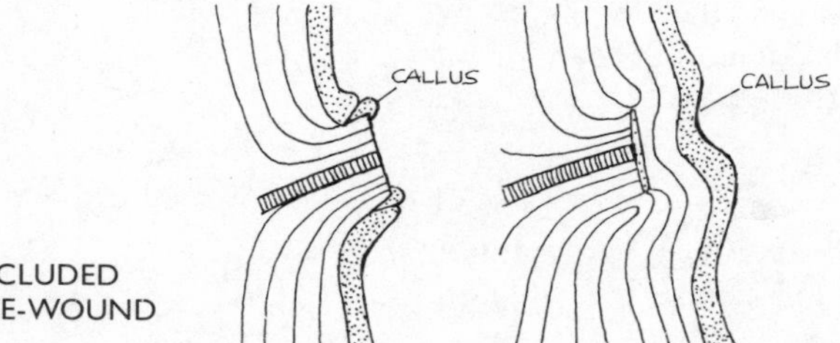

OCCLUDED TREE-WOUND

Occlusion

The process by which tree wounds callus over.

Ocrea, Ochrea

A sheath formed of two stipules united around a stem.

Odd-pinnate

Of a leaf, pinnately compound with the pairs of leaves surmounted by a single terminal leaflet: sometimes called imparipinnate.

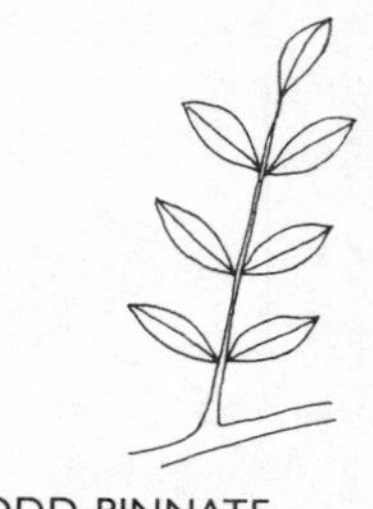

ODD-PINNATE

Odoriferous, Odorous

Sweetly scented; fragrant; emitting a pleasant smell.

Oedema, EDEMA (USA)

Non-parasitic condition characterized by the appearance, mainly on the underside of leaves, of tiny pimple-like blisters which later enlarge and turn brown and corky. Oedema is favoured by damp, warm soil when the air is moist and cool, resulting in rapid water uptake from the roots, coupled with slow release by transpiration. The plant then becomes over-turgid and some of the cells on the underside of the leaves burst. The corky areas are formed when the blisters callus over. Oedema is not a disease and can be overcome by increased ventilation and more care with watering. Leaves that are damaged will not recover but new growth will be clean if growing conditions have improved.

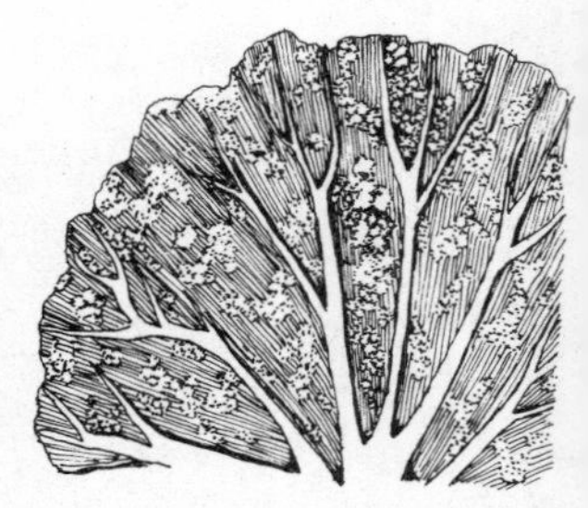

OEDEMA ON A LEAF OF *Pelargonium zonale*

Officinal

Plants that have medicinal properties; many herbs carry it as the specific name, e.g. *Rosmarinus officinalis*.

Offset

A side-shoot which develops naturally from the parent plant, usually at ground level and developing its own roots while still attached. Bulbils and cormlets are offsets, as is any young plant growing from the main rootstock.

BORAGE (*Borago officinalis*)

Offshoot

Side-shoot or branch; same as offset but mainly applied to the shoots on branches and stems rather than roots.

Oleiferous

Producing oil, e.g. some seeds and leaves.

GLADIOLUS CORM WITH OFFSETS (CORMLETS)

Olibanum

Gum-resin obtained from a species of *Boswellia*; Arabian frankincense.

Oligocarpous

Having few fruits or fruiting bodies.

Olitory

Old word for a kitchen garden or a kitchen gardener; from the Latin *olitor* – kitchen gardener.

Ombrophil(e)

A plant tolerant of much rain.

Ombrophobe

A plant intolerant of much rain.

Onion Fly

A fly which lays its eggs on onions, the larvae of which ruin the young plants; there are three generations a year. Onion sets should be soaked in an insecticide before planting, and seedlings dressed while their leaves are still looped together.

Onion Hoe

A small hand-tool with arched handle designed for weeding large onion beds in the days when this operation was performed mainly by women known as 'crawlers', from the fact that they worked on hands and knees.

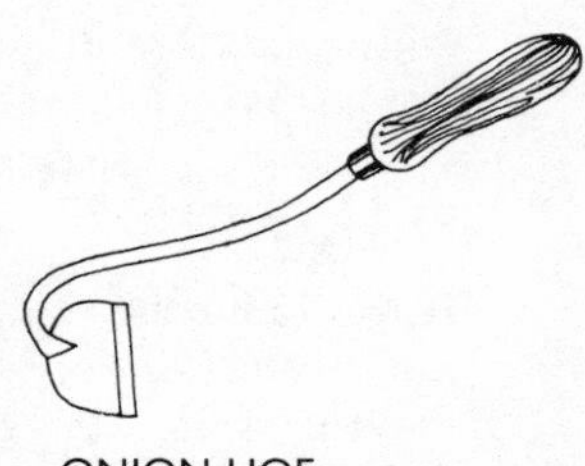

ONION HOE

Open-centre

A method of pruning trees, particularly fruit trees, which leaves the centre clear of growth, and the remaining outer branches in the shape of a cup or vase.

Operculate

Having a cap or lid.

Operculum

A cover, cap or lid, e.g. a circumscissile capsule or the protective cap at the tip of a root.

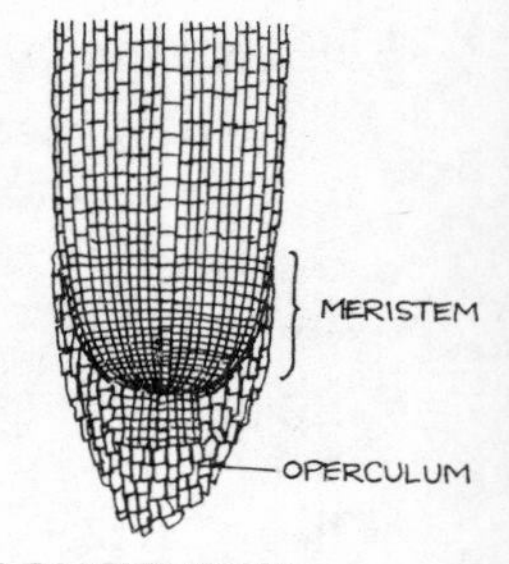

LONGITUDINAL SECTION THROUGH THE TIP OF A ROOT

Oporice

A medicine prepared mainly from quinces and pomegranates.

Opposite

(1) A pair of leaves at a node on opposite sides of the stem: *cf.* Alternate.
(2) Floral parts on the same radius.

Orangery

A heated building originally for the raising of citrus trees, especially in large eighteenth and nineteenth century country estates, usually having a solid north wall and arched windows on the south side. A few still exist today in some of the remaining stately homes.

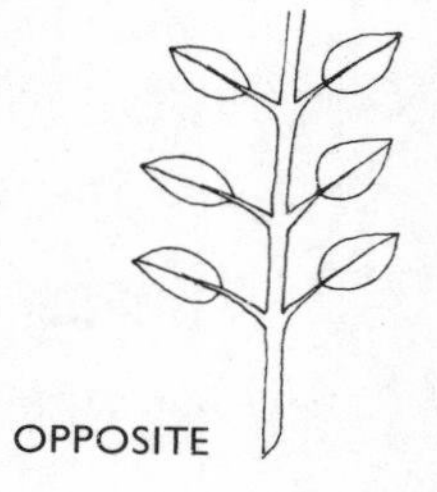
OPPOSITE

Orbicular

Of leaves and petals that are circular or nearly so; disc-shaped.

Orchard

An enclosed fruit garden or an area of land set out with fruit trees.

Orchard House

A type of unheated greenhouse (usually a lean-to structure) built against a south-facing wall. Much favoured by the gentry in the eighteenth and nineteenth centuries for supplying figs, grapes, peaches, etc. out of season.

ORBICULAR

Ordure

Dung, manure; anything unclean.

Organ

Part of a vegetable (or animal) body adapted for a special vital function.

Organic

Substances derived from the decay of leaves and other vegetation, animal waste and animal tissue.

Organic Gardening

Cultivation without the aid of chemicals of artificial origin.

Organic Surface Cultivation

See No-digging.

Ornamental Garden

A loose general term for a garden designed for pleasure rather than utility, usually having formal flowerbeds separated by paths or grass verges.

Orthotropism

The ability for the growth response of a plant to be orientated in line with a particular stimulus, such as the vertical growth of main stems under the influence of gravity.

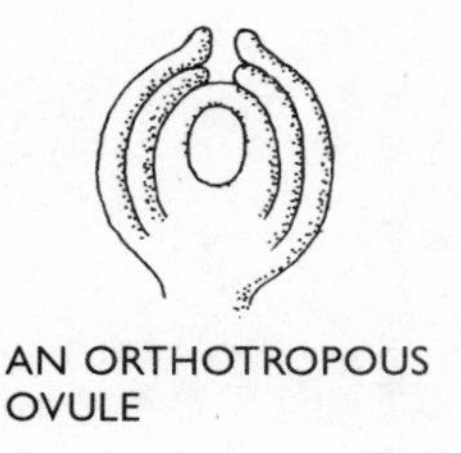

AN ORTHOTROPOUS OVULE

Orthotropous

Ovules carried on a straight funicle (stalk), not curved or bent over.

Osier

Any type of willow whose twigs are used for basket-making, furniture, etc.

Osier Bed

An area of land where osiers grow and are harvested.

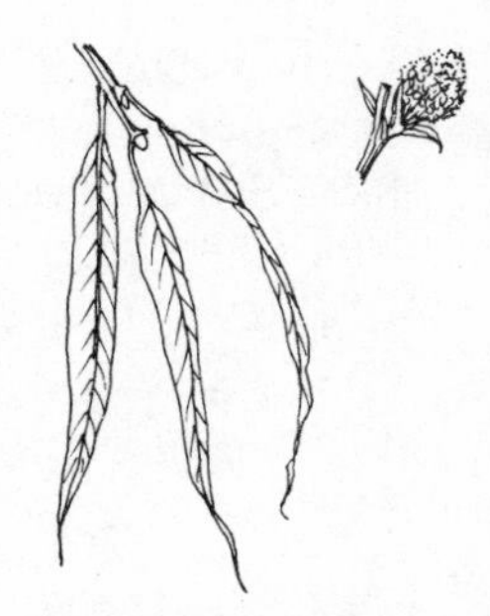

COMMON OSIER
(*Salix viminalis*)

Osmosis

The diffusion of liquids through a porous septum. The method by which nutrients are distributed throughout a plant's tissues, and by which several other functions are performed, such as the opening and closing of the stomata.

Osmunda Fibre

A material obtained by chopping up the dried roots of the royal fern (*O. regalis*), much prized by orchid growers in the past for adding to the potting compost; now rarely available and largely superseded by granulated bark or coarse peat.

OUTGROWTH ON CACTUS (*Opuntia*)

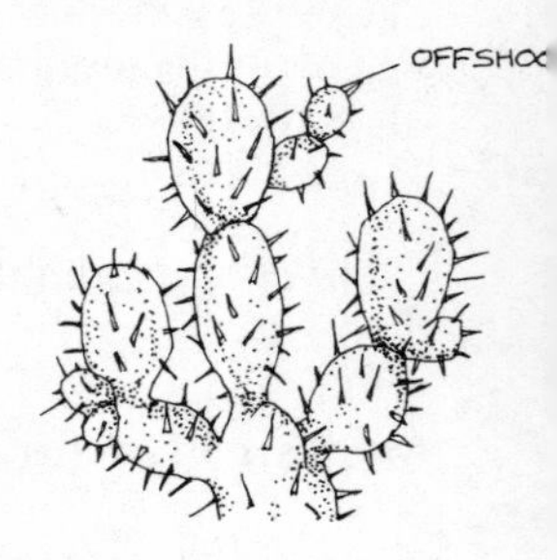

Outgrowth

Offshoot; side-shoot; excrescence. Sometimes applied to an unusual growth.

Oval

Of a leaf, petal or fruit roughly elliptic in shape and broadest at the centre, rounded at both ends, about twice as long as broad.

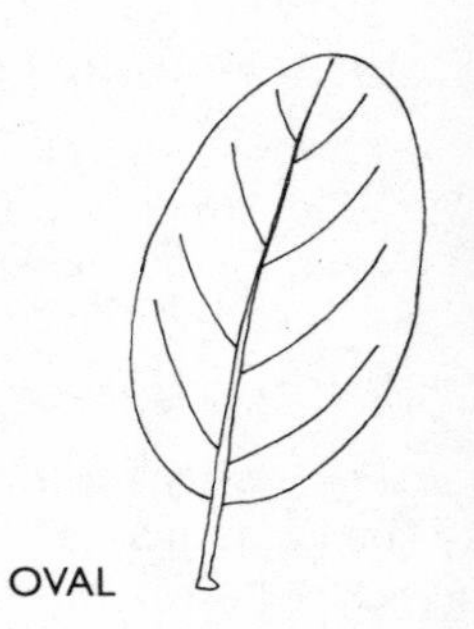

OVAL

Ovary

The female part (gynaeceum) of a flower containing the ovules which will develop into seeds after fertilization.

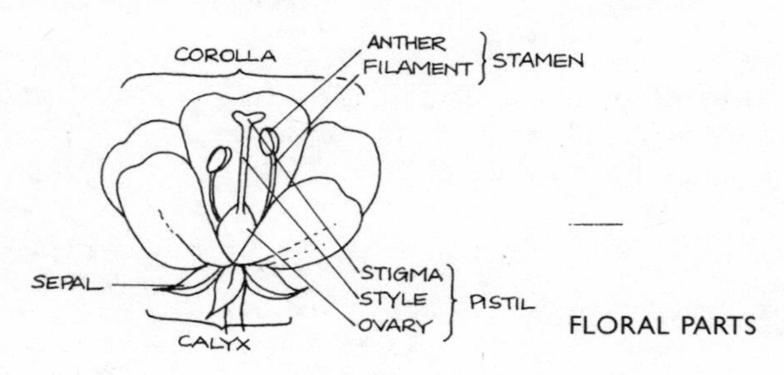

FLORAL PARTS

Ovate

Of a leaf that is roughly oval in outline but widest toward the stalk: *cf.* Obovate.

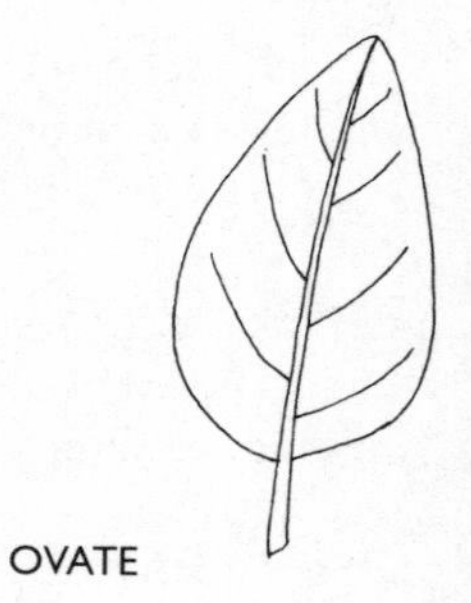

OVATE

Ovate-acuminate

Of an ovate leaf or other organ terminating in a point, e.g. the siliculas of honesty (*Lunaria annua*).

Ovate-cordate

Of a leaf, ovate with a heart-shaped base.

Over-blown

Of a flower or vegetable past its prime; too fully open and beginning to fade.

OVATE-ACUMINATE

Over-potting

The transferring of plants from small pots into others too large for the size of plant, e.g. a plant rooted in a 3 in (7.5 cm) pot being potted-on into a 10 in (25 cm) container without any intermediate stages. In such cases the soil nearest the wall of the pot will have most of its nutrients leached out before the roots have extended into it; and in those cases where the chemical content is badly out of balance – bearing in mind that the basic chemicals (NPK) leach at different rates – the roots may never penetrate into the hostile environment and the plant will not achieve its optimum development.

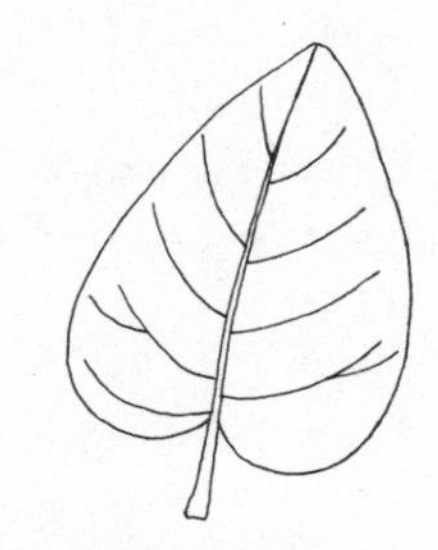
OVATE-CORDATE

Overgrow

(1) To cover with growth, e.g. ground cover plants.
(2) To grow too large or too tall for its location.

Ovicide

A chemical product for destroying the eggs of insects before they hatch.

Oviform

Egg-shaped.

Ovoid

Resembling the shape of an egg, as applicable to three-dimensional objects. Flat organs such as leaves with an oval outline are ovoidal or obovoid.

OVOID
(VICTORIA PLUM)

Ovoidal

Of leaves and petals roughly the shape of a hen's egg in outline with the stem at the broadest end: *cf.* Obovoid.

Ovulate

Having ovules; often applied when the ovules are not enclosed in a pistil.

Ovule

In flowering plants, the body which will contain the seed after fertilization.

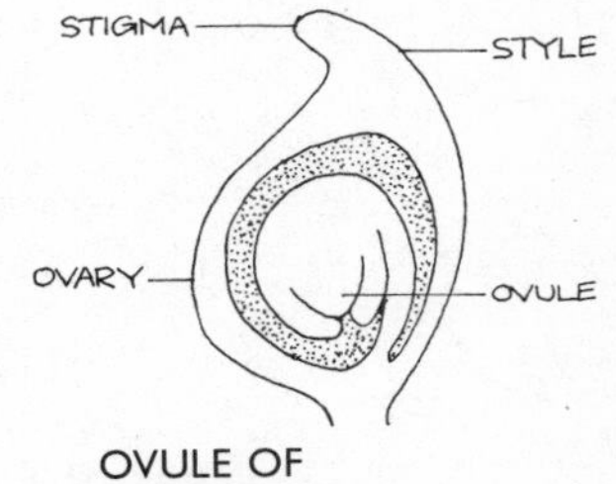

OVULE OF BUTTERCUP
(*Ranunculus acris*)

Ovuliferous Scales

A group of woody scales which form the female cones of conifers and similar trees; they carry the ovules which develop into seeds.

Ovum, *pl.* Ova

Female germ cell which may develop into a seed when fertilized by male pollen; an insect egg.

Oxalic Acid

A very powerful and poisonous acid which is present in certain plants such as sorrel and the leaves of rhubarb.

Oxidase

Any of a group of enzymes that encourage or promote oxidation in plant cells.

Oxygen

Tasteless, colourless, odourless gas essential to plant (and animal) life, and present in combination with other elements in water and most organic substances.

Oxygenator

An aquatic plant that releases oxygen into the water while submerged. Most need fairly deep water, e.g. *Ranunculus aquatilis*.

Oyster Shell

Crushed oyster shell was once considered to be an essential ingredient of bulb fibre. It contains a high proportion of calcium which is released slowly over a long period and, being porous, is able to absorb water and gradually release it as required. It has been almost completely superseded by sharp sand, perlite, etc.

P

Paired

Of leaves or other plant organs which occur in pairs at the same level on a stem.

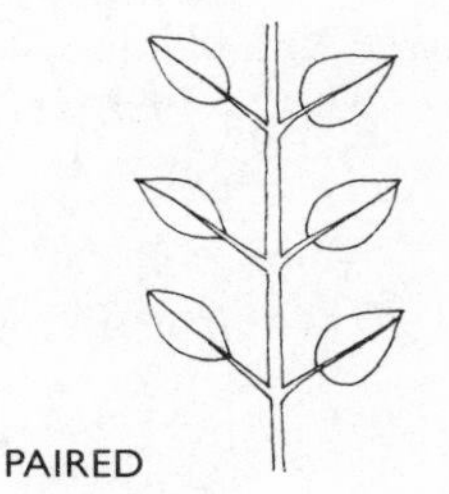
PAIRED

Palaeontology

The study of extinct plants and animals.

Pale

(1) Pointed length of wood used as a fencing stake.
(2) A bract protecting the flower in a spikelet of grasses (Gramineae).

PALE

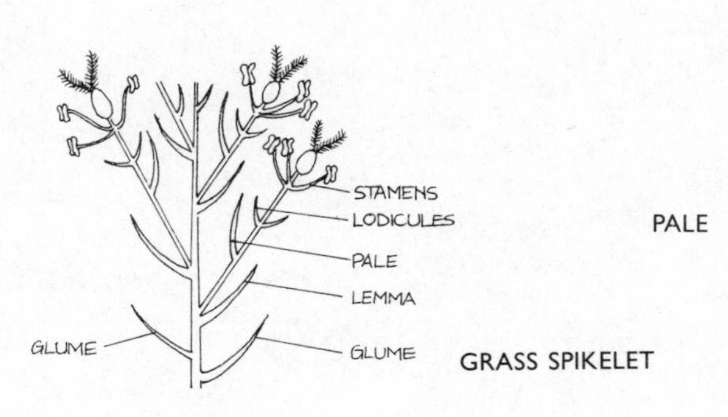

GRASS SPIKELET

Palea

(1) The upper, small, chaffy inner bract of an individual grass-flower, known also as the valvule or the upper-pale.
(2) A scale on a fern-leaf or stem.

Paleaceous

Chaff-like or chaffy in texture.

Paling

Fencing made of pales or split pales such as chestnut paling.

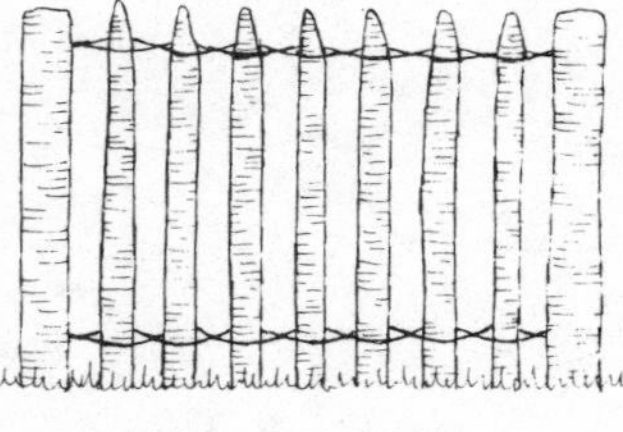

CHESTNUT PALING

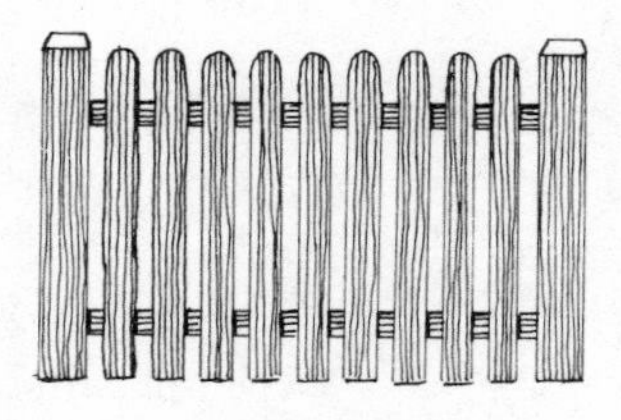

REGULAR PALING

Palisade Cells

Cells of the internal tissue of a leaf. *See* Palisade tissue.

Palisade Tissue

The main photosynthetic leaf-tissue which contains the most chlorophyll, and composed of cells set closely together with their long axes perpendicular to the surface – i.e. arranged like a palisade or protective fence.

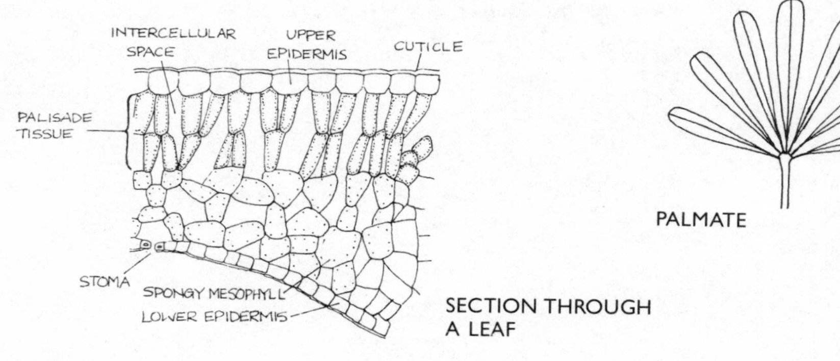

SECTION THROUGH A LEAF

PALMATE

Palmate

Of a leaf in the form of a hand, consisting of three or more leaflets.

Palmately Lobed

Like an open hand; usually applied to leaves with lobes or leaflets radiating fanwise from a common point of attachment, as in the horse-chestnut.

PALMATIFID

Palmatifid

Of a leaf cut in palmate manner more than halfway to the leaf-stalk.

PALMATIPARTITE
SILVER MAPLE
(*Acer saccharinum*)

Palmatipartite

Of a leaf palmately divided but separated nearer than halfway to the leaf-stalk.

Palmatisect

Of a leaf deeply cut in palmate manner, almost but not quite to the leaf-stalk.

Palynology

The study of fossil pollen and spores.

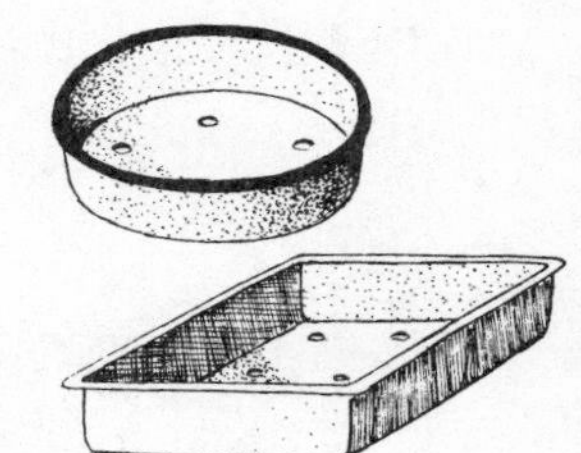

PLANT PANS

Pan

(1) A shallow dish used for raising seedlings and other shallow-rooted subjects such as alpines.
(2) A solid compacted horizontal layer of subsoil, produced by digging or ploughing to the same depth over a long period, through which roots will not penetrate. It should be broken up by double digging or the use of a chisel plough.
(3) A surface pan is caused by the topsoil compacting after heavy rain, a condition aggravated by walking on wet soil or by rolling. The remedy is through hoeing.

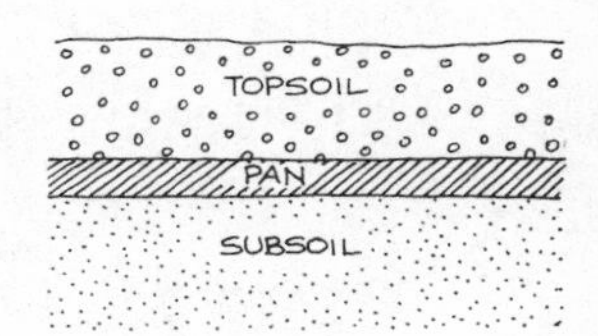

Pandurate, Panduriform

Of a leaf that is fiddle-shaped, rounded at both ends and narrower at the middle.

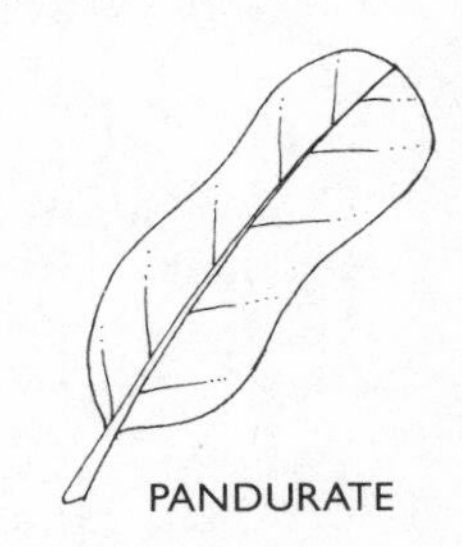

PANDURATE

Panicle

A flowerhead with several branches, either opposite or alternate; a branched raceme.

Paniculate

Arranged in a panicle.

Pannose

Like felt, soft and flexible; covered with very short, smooth and dense fur-like hairs.

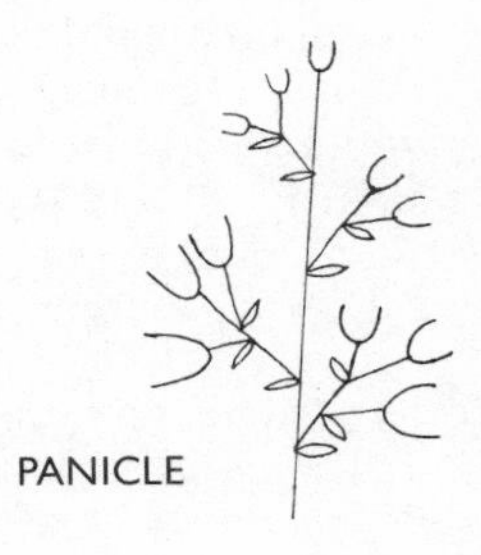

PANICLE

Papilionaceous

Reminiscent of a butterfly. Applied to some leguminous subjects such as sweet peas, where the uppermost of the five petals (the standard) is the largest and erect, with a pair of laterals or wings flanking it. Also refers to the name of the family Papilionaceae (= Leguminosae).

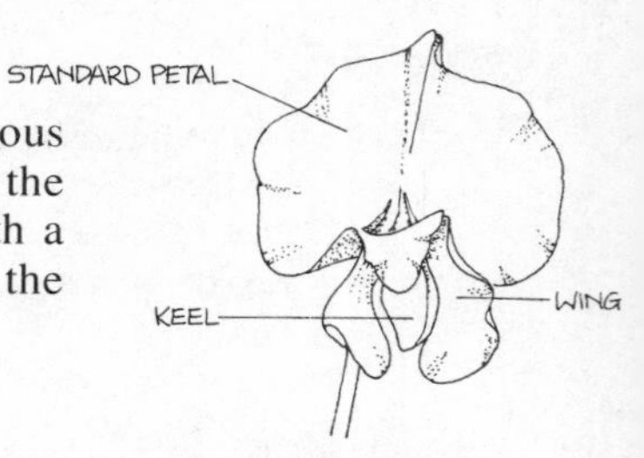

PAPILIONACEOUS SWEET PEA (*Lathyrus odoratus*)

Papilla

A tiny pipe-like or conical protuberance found on some petals and other plant surfaces.

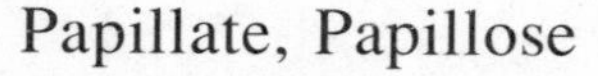

Papillate, Papillose

Having papillae; warty.

Pappus

A circle or parachute of fine hair or down which grows above the seed and assists wind-dissemination in certain plants, including composites such as dandelion, thistle, etc.

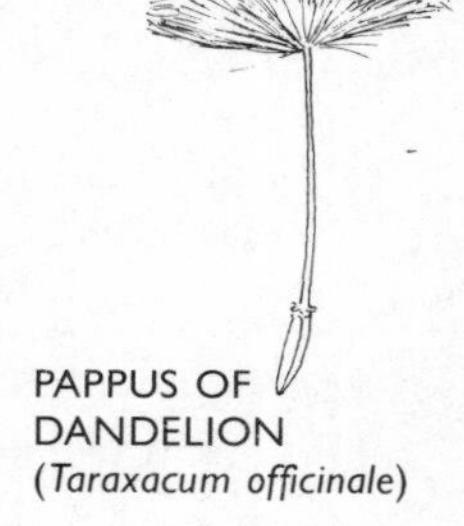

PAPPUS OF DANDELION (*Taraxacum officinale*)

Papula, Papule

Small pimple-like projection on a plant, usually on the stem.

EPIDERMIS
PARENCHYMA
MEDULLA

CROSS SECTION OF A YOUNG LIME TWIG

Papyraceous

Paper-thin or paper-like.

Paradichlorobenzene

A strong-smelling chemical used as an insecticide. It can be spread on the garden to repel a variety of pests, and a few crystals placed in the bottom of a pot will deter vine-weevil and many other soil-borne pests.

Paraffin Heater

Common appliance, fuelled by paraffin, used for heating. Fumes from such heaters are damaging to a large number of plants, as also is neat paraffin. Soil contaminated with paraffin oil will grow nothing for several seasons.

Parallel Venation

The arrangement of the veins in the leaves of grasses, etc: *cf.* Reticulate venation.

Parasite

(1) An organism which lives upon another living organism. A complete parasite has no leaves or chlorophyll, and relies entirely upon its host for nourishment; but there are also many semi-parasitic plants such as mistletoe which, because it has leaves, is able to manufacture some of its own food requirements. An efficient parasite may weaken but not kill its host – most of the rusts fall into this category.

(2) A term used loosely but incorrectly to describe a plant that climbs up another plant, fence, etc.

(3) A creature such as a nematode, bacterium, etc.

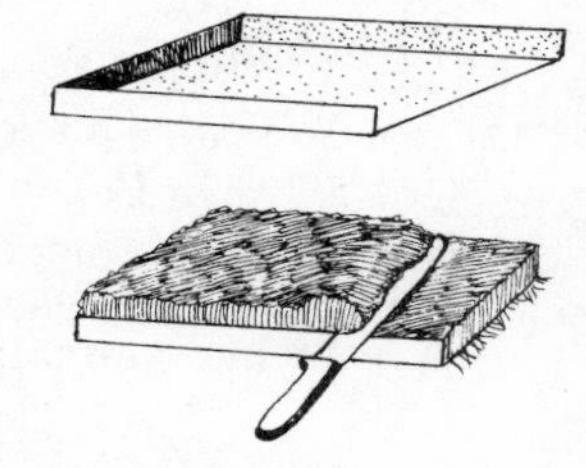

PARING BOX

Parcel

Plot or area of land with recognizable natural or artificial boundaries.

Parenchyma

The ordinary soft thin-walled tissue of plants, not differentiated into conducting or mechanical tissue. The soft part of leaves and the pulp of fruits.

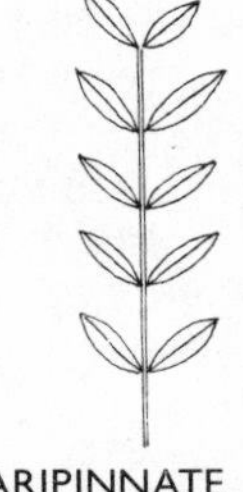

PARIPINNATE

Paring Box

A simple device used to trim turves to a standard thickness. It comprises a three-sided frame mounted on a flat base, with wooden or metal vertical sides of the height of the desired thickness. Turves are placed in it grass side down and surplus soil skimmed off with a cutting blade or double-handled knife.

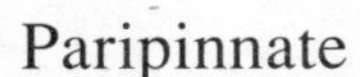

Paripinnate

Of a leaf that is evenly pinnate; having leaflets in pairs along the axis but lacking a terminal leaflet.

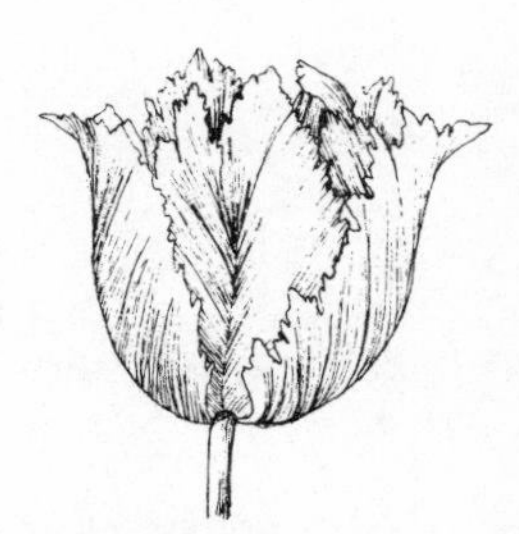

PARROT TULIP

Parrot

A class of tulip with vivid colours and slashed petals.

Parrot-beak

A type of secateur in which the cutting blades are curved and cross each other.

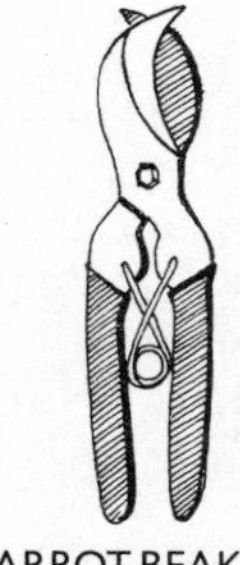

PARROT-BEAK SECATEUR

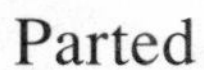

Parted

Cut or divided not quite to the base.

Parterre

A type of formal garden layout which originated in France and came to England in the seventeenth century. It consists of a level garden, usually rectangular, divided into a series of ornamental flowerbeds often separated by dwarf box edging or similar low-growing shrubs.

Parthenocarpic

Producing fruits without fertilization, as with the greenhouse cucumber and the cultivated banana.

Parthenogenesis

The development of seeds as in normal sexual reproduction but without fertilization. This occurs sporadically in many plants, e.g. dandelion, and in many animals. In some species of aphid the male is absent or very rare; the eggs produced by the female contain the full number of chromosomes and need no fertilization.

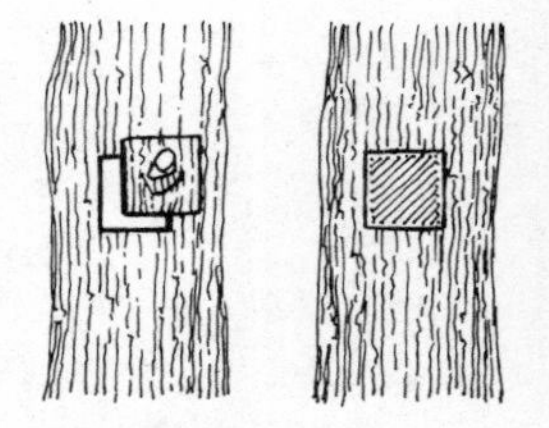
PATCH BUDDING

Partite

Divided; cut almost to the base.

Pasture

Land used for the grazing of sheep or cattle and seldom if ever ploughed.

Patch

Small piece of ground, formerly often attached to the cottage of a farm worker; a quantity of plants growing thereon, e.g. cabbage patch.

Patch Budding

A method of grafting used mainly on walnut trees whereby a patch of bark about 1 in (2.5 cm) square and containing a bud is cut from the scion variety and inserted into the place where a blank square of identical size and shape has been removed from the bark of the stock. The bud-patch is then bound securely with an airtight bandage. *See* Grafting.

Patching-out

Planting-out massed seedlings in small groups or patches when they have grown too close together in the seed bed for separation without damage.

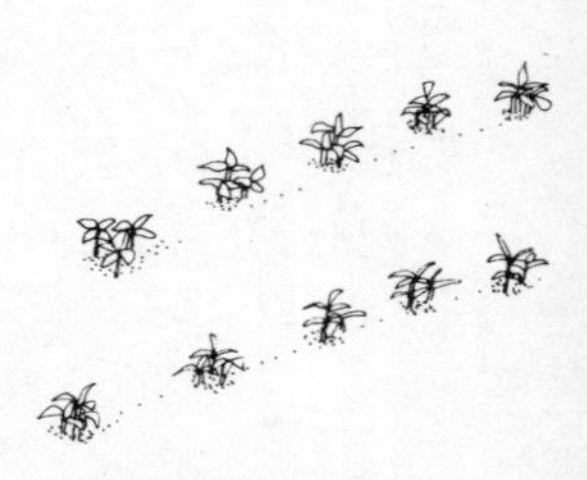
PATCHING-OUT

Patella

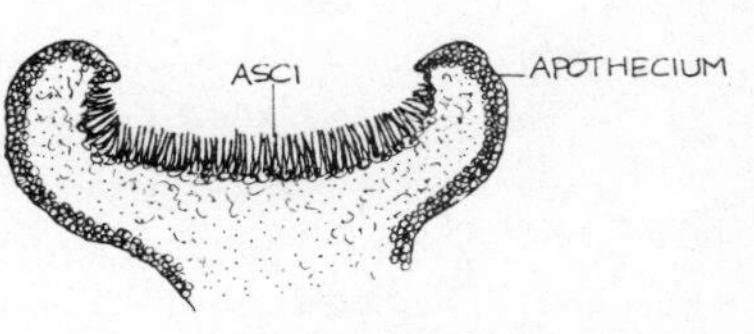

PATELLA (LICHEN)

A saucer-shaped apothecium, as in some fungi.

Patent

Spreading; lax and low-growing.

Path Plants

Plants suitable for planting in the joints between flagstones of paths, e.g. wild thyme (*Thymus serpyllum*). Very few will withstand being continually walked on.

Pathogen

Any living organism that causes disease, especially viruses and some bacteria.

Pathology

The study of diseases.

Patio

Incorrect term frequently used in Britain and USA to describe a paved area or terrace; a sitting-out area. Properly applied, it is the inner courtyard of a house in a hot climate, usually completely surrounded by walls but open to the sky except for slatting or foliage to provide a cool refuge from the sun.

Peach Leaf Curl

See Leaf Curl.

Pea Gravel

Washed gravel graded to about the size of a pea.

Pea Guard

PEA GUARD

Wire mesh formed into a tunnel to cover and protect newly sown peas from attack by birds.

Peat, Peat-moss

Partially decomposed organic matter usually derived from mosses or sedges in boggy or fenland areas. Sedge peats are usually less acid than moss peats; both types contain small amounts of nutrients but these are released too slowly to be of much benefit to plants. Both types of peat are misleadingly called peat-moss in Ireland and USA.

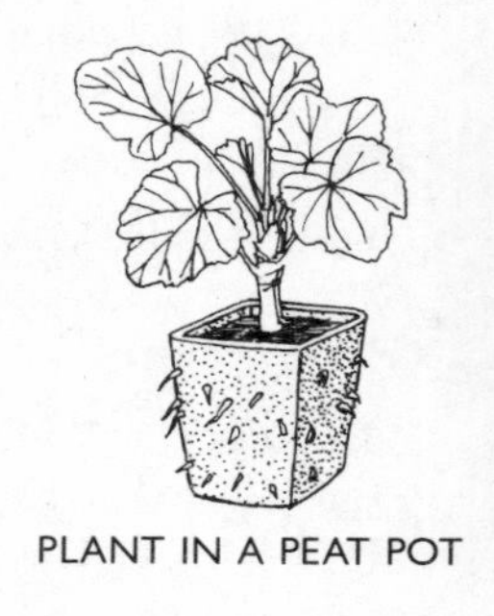
PLANT IN A PEAT POT

Peat Bed

A specially constructed, often raised area, edged with peat blocks and containing acidic peaty soil. Used mainly for growing ericaceous subjects in places where the normal soil is alkaline.

Peat Pot

A pot made from compressed peat. The roots of seedlings or cuttings can penetrate the walls of the pot, enabling the entire unit to be planted out with the minimum of root disturbance. Another type of peat pot is the Jiffy-7 – a pellet of compressed peat enclosed in coarse netting which expands when saturated with water; cuttings or seeds are then inserted, and the roots of the young plants develop through the netting without hindrance.

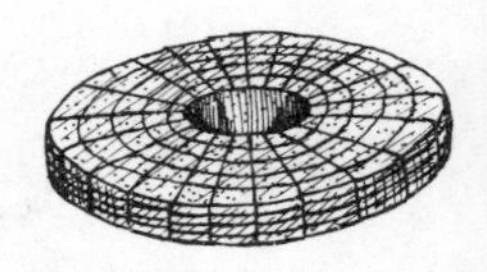
JIFFY-7 BEFORE SOAKING

Peat Wall

The wall of peat blocks enclosing a peat bed.

Peck

(1) *See* Pick, Pickaxe.
(2) Obsolete unit of dry measure, 2 gallons equalling 1 peck.

JIFFY-7 FULLY EXPANDED AND IN USE

Pectic Substances

Important constituents of plant cell-walls and the middle lamella between adjacent cell-walls made up primarily of sugar acids. They are normally insoluble but change into soluble form as the fruit ripens or becomes diseased.

Pectinate, Pectinal

Of leaves having a comb-like structure, or arranged like the teeth of a comb; also the comb-like spines on some cacti. From the Latin *pecten* – a comb.

Pedate

Palmately lobed with the outer lobes deeply cut, or ternately branching with the outer branches forked.

Pedicel

The stalk of a single flower in an inflorescence or cluster.

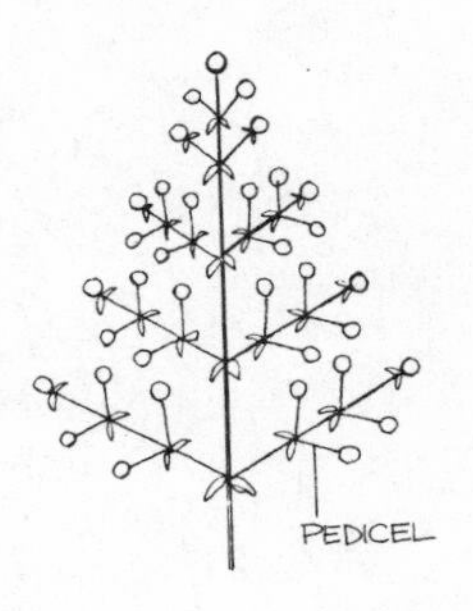

Pedicel Necrosis

The collapse of the flower-stalk which sometimes occurs in peonies and roses and occasionally other flowers.

Pedicellate, Pedicelled

Carried on a pedicel.

Pedology

The science and study of soils and soil structure.

Peduncle

The main stalk of a cluster of flowers, or the solitary stalk of a single flower.

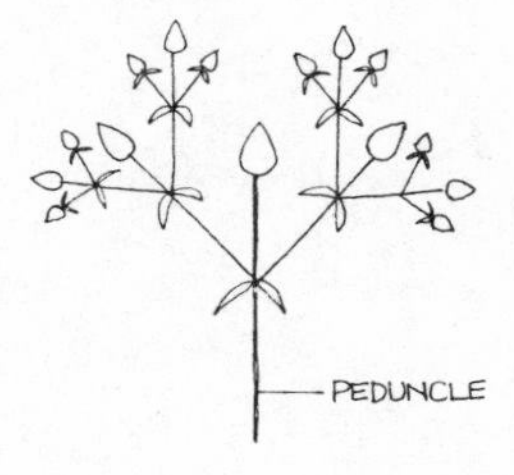

Peel

(1) The rind or outer coat of a fruit.
(2) To strip the bark or rind from a tree or fruit.

Peeler

A plant that impoverishes the soil.

Pelleted Seed

Individual seeds enclosed in a pellet of inert material to which a fungicide may be added. Pelleted seed can be more easily handled and widely spaced so that thinning of seedlings is reduced or avoided; however, germination may be erratic unless the pellets are grown in a moist compost.

Pellicle

Thin membrane or skin; a transparent or translucent film.

Pellucid

Botanically completely or partially transparent.

Peltate

Of a leaf, shield-shaped, with the leaf-stem joining the leaf at or near the centre on the underside, as in the ivy-leaved pelargonium (*P. peltatum*) and the common nasturtium (*Tropaeolum majus*).

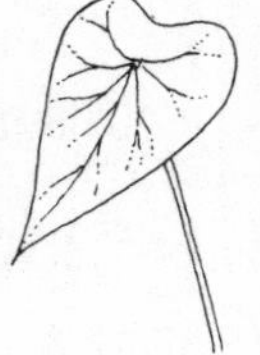

PELTATE

Peltate-palmate

Having long narrow lax foliage with stem attached at the centre of the back.

Pendant, Pendent, Pendulous

Drooping; hanging; dangling; overhanging; suspended from the top.

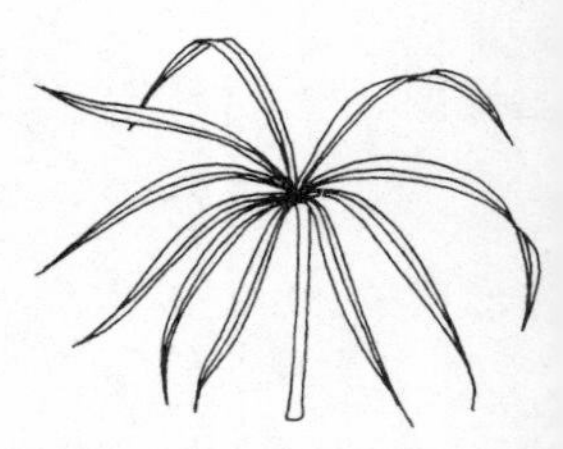

PELTATE-PALMATE

Penicillate

Having a tuft of hairs or a small brush or brush-like growth; brush-shaped.

Penniform

Feathery.

Pensile

Pendulous, hanging down.

Pentamerous

Of flowers having five parts or members; in multiples of five. Sometimes written 5-merous.

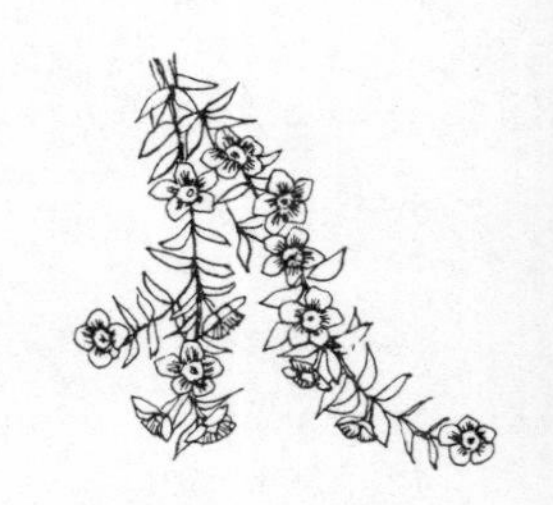

PENDANT *Browallia*

Pepo

Fruit of the melon and cucumber family, a large many-seeded berry formed from an inferior ovary, usually with hard epicarp. Squashes and gourds are of the same family.

Perch

An obsolete unit of length measuring 16½ ft (5 m), the name derives from the length of the measuring rod used in earlier times to calculate land areas. *Same as* Pole and Rod.

Perennate

To survive from season to season; to live over from one year to another.

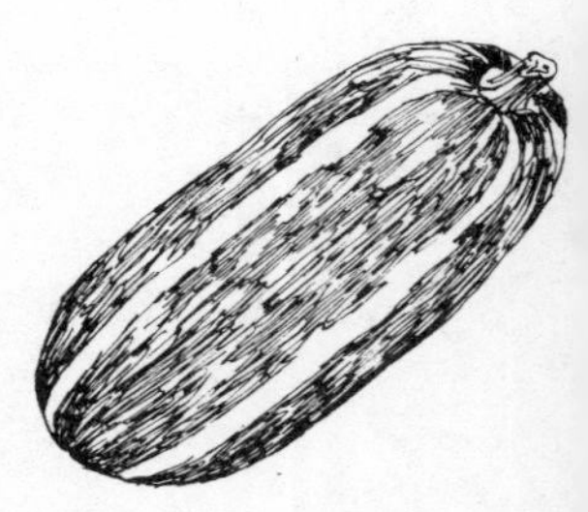

PEPO (MARROW)

Perennial

A plant that lives more than two years or three seasons and normally flowers annually. Many die down during the winter but the roots are unaffected by frost and new growth appears as the weather improves and the temperature rises. The term is usually applied to non-woody plants.

Perfect

A flower with both male and female parts.

Perfoliate

Paired leaves which encircle the stem so that the stem appears to pass right through the leaf, as in the young leaves of *Eucalyptus*.

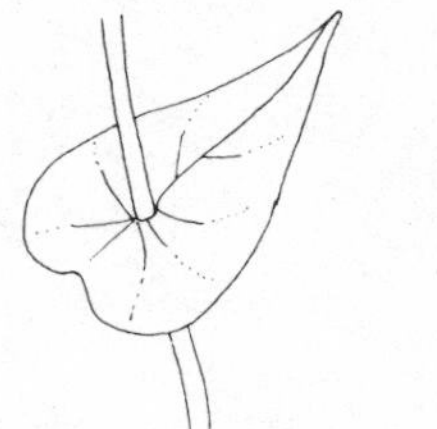

PERFOLIATE ALTERNATE

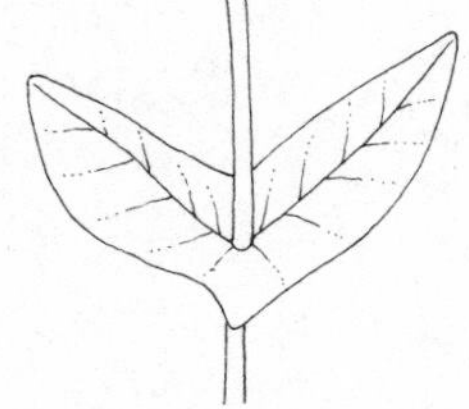

PERFOLIATE OPPOSITE

Pergamentaceous

Like parchment, crisp and papery.

Pergola

Arbour or covered walkway formed of growing plants, trees, etc. trained over a rustic frame of trellis, wire or timber.

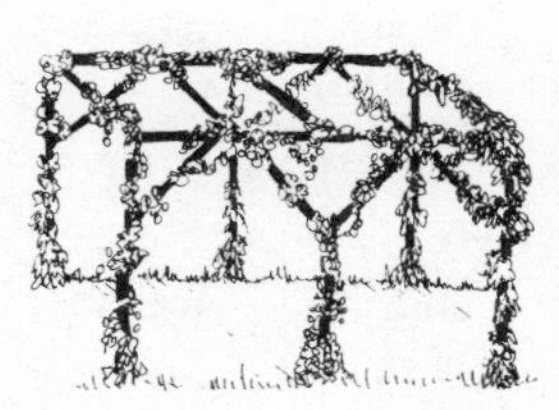

PERGOLA

Perianth

A collective term used to describe the envelope of external floral parts, including the calyx and corolla. The combined sepals and petals of a tulip are called the perianth.

Perianth Segment

One section of the perianth resembling a petal; sometimes called a tepal.

Pericarp

The wall of a ripened ovary; the seed vessel, the outer skin of a fruit.

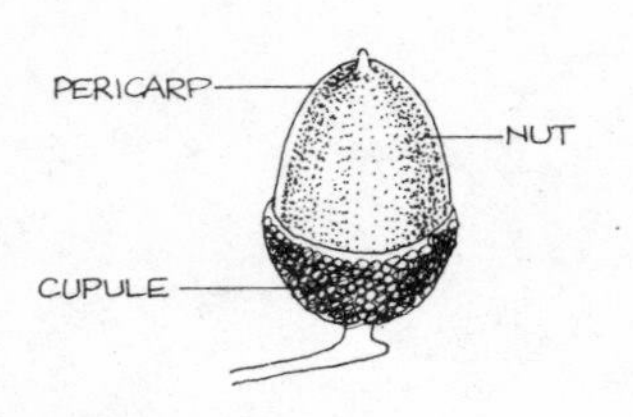

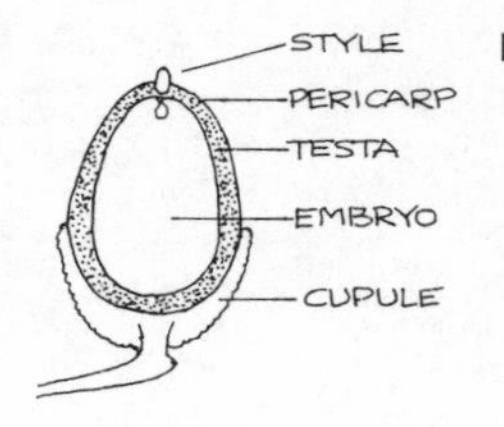

PERICARP (ACORN)

Peridium

The outer envelope which encloses the spores of some fungi.

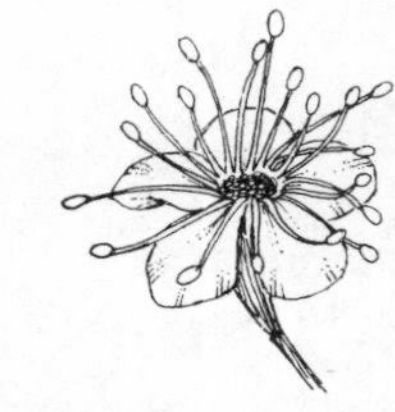

PERIGYNOUS FLOWER OF *Prunus laurocerasus*

Perigynous

Having the receptacle developed as a disc or open cup, with sepals, petals and stamens borne on its margin, as in *Prunus*.

Perisperm

Nutritive tissue in a seed derived from the nucellus.

Peristome

In mosses and other lower plants the peristome is the teeth-like fringe around the mouth of a capsule.

Perithecium

Cup-shaped, spherical or flask-shaped receptacle enclosing the spores of certain fungi.

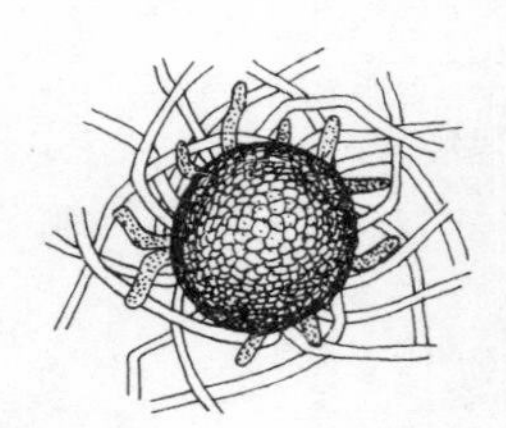

PERITHECIUM OF ERYSIPHE *Graminis*

Perlite, Perlag

Volcanic or other vitreous rock in the form of small globules. Totally inert and sterile, suitable mainly for aeration of soils and composts, and having good water-retaining properties.

Perpetual

A plant with a long flowering period, in bloom more or less continuously, such as the greenhouse carnation. The spinach beet is often described as a 'perpetual' and can yield usable foliage over a very long season.

Persistent

Remaining after the usual time of falling, as of beech leaves in winter.

Personate

Said of a lipped corolla closed by an upward bulge of the lower lip, as in *Antirrhinum*. Sometimes called the palate.

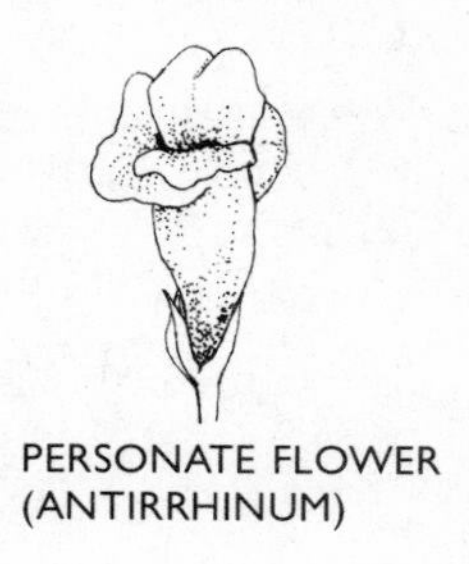

PERSONATE FLOWER (ANTIRRHINUM)

Peruvian Guano

A valuable fertilizer found on the coasts of Peru, rich in phosphates and ammonia, the excrement of sea birds.

Pest

Any living creature which damages plants, except for fungi, bacteria and viruses, which are diseases.

Pesticide

A chemical substance for destroying or controlling pests, especially insects.

Pestology

The study of harmful insects and other pests, and methods of control.

Petal

A non-reproductive part of a flower – a flower leaf. Part of the corolla of a flower surrounding the reproductive organs, often coloured to attract insects.

Petal Fall

The stage in the development of fruit trees or bushes when the flowers shed their petals.

Petaloid

Having the appearance of a petal, usually applied to modified stamens, as in double peonies, etc.

Petiolate

Having a petiole or leaf-stalk: *cf.* Sessile.

Petiole

A leaf-stalk joining the leaf to the plant stem, usually an extension of the central rib of the leaf.

Petiolule

The petiole of one leaflet of a compound leaf.

pH

A symbol, followed by a number, used to indicate the degree of acidity or alkalinity of solutions or soils, related by formula to a standard solution of potassium hydrogen phthalate. pH7 is neutral: lower figures denote the degree of acidity, higher of alkalinity; thus a soil with a pH value of 6 would be acid and pH8 would be alkaline.

Phanerogam

A flowering plant, one that has both pistils and stamens.

Phloem

That part of the vascular bundle by which nutrients are transported within the plant. The softer part of the fibro-vascular tissue compared to the xylem or woody part.

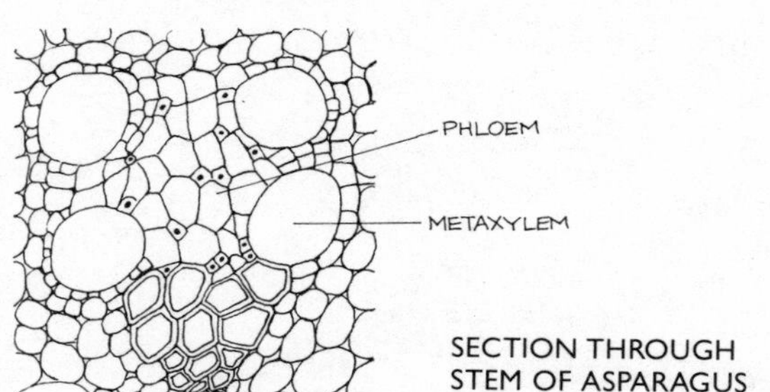

SECTION THROUGH STEM OF ASPARAGUS

Phosphorus

A non-metallic chemical element found in all animal and vegetable organisms. The 'P' symbol of NPK formulae, one of the essential plant foods mainly responsible for healthy root development.

Photoperiodismic

Plants unaffected by day-length. *Also called* 'Day-neutral'.

Photosynthesis

The construction of complex compounds by the chlorophyll apparatus of plants through light-energy, combining carbon dioxide from the atmosphere with hydrogen from water in the soil to form sugars. More simply, the process by which carbon dioxide is converted into carbohydrates by chlorophyll under the influence of light.

Photosynthetic Pigments

The plant pigments which collect the light energy during photosynthesis. The green pigment chlorophyll is the main receptor but other pigments also absorb light energy.

Phototaxis

The movement of a cell or a unicellular organism in response to light.

Phototropic

The growth of plant organs in response to light: *cf.* Geotropic, Heliotropic.

Phylloclad, Phylloclade

A branch with the form and functions of a leaf; a flattened stem or branch functioning and often appearing as a leaf. Plants adapted to dry conditions are usually phylloclads, such as members of the cactus family (Cactaceae).

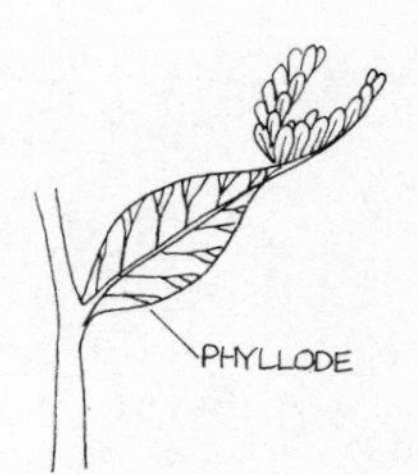

Phyllode

A petiole with the appearance and function of a leaf-blade; a flattened leaf-stalk.

Phyllotaxy

The developing arrangement of leaves on a stem. In verticillate phyllotaxy two or more leaves arise at each node, in distichous phyllotaxy each node bears a single leaf, and in spiral phyllotaxy each node carries one leaf but they are arranged spirally down the stem.

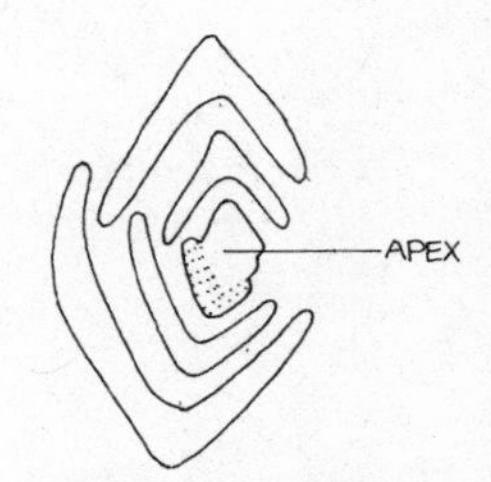

SECTION THROUGH A DISTICHOUS PHYLLOTAXY NEAR THE APEX

Phylogeny

The evolutionary development of a genus or species, or their parts; also applies to larger groups, e.g. families.

Phylum

One of the main divisions of the plant kingdom encompassing organisms of the same general form, (the other main subdivisions in descending order being class, order, family, genus and species).

Physiology

The area of biology involved in the vital functions of plants (and animals), including nutrition, respiration, reproduction, etc.

Phyto-

Prefix denoting a plant.

Phytochrome

A pigment present in small quantities in many plant organs. In its active form it controls many plant processes, including seed germination and the initiation of flowering.

Phytogeography

The study of the distribution of vegetation worldwide.

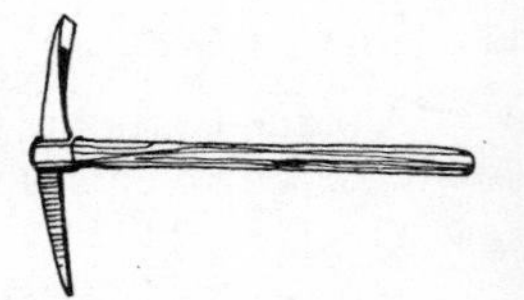

PICK

Pick

(1) To gather fruit, flowers, etc.
(2) Abbreviated form of pickaxe, sometimes called peck.

Pickaxe

A tool with stout ash handle attached to a steel bar, pointed at one end and chisel-shaped at the other, the handle passing through a central hole in the bar. Used for breaking up hard ground, etc. Similar to a mattock but the chisel blade is narrower.

PICOTEE (TULIP)

Picotee

A flower which has a darker edging to the petals, as in some carnations, tulips, etc.

Pig Manure

Not so rich as other animal manures but provides humus and is a valuable fertilizer if left to rot for a year before use. If applied fresh, the ground should not be planted for at least six months.

Pightle

A small plot or enclosure of land; a croft in Scotland.

Pigment

The natural colouring matter of a tissue.

Pilcorn

A type of oat cultivar in which the seed and husk do not adhere to each other.

Pileus

The cap of a mushroom, toadstool or similar fungus.

Piliferous

Hairy; having hair or hairs.

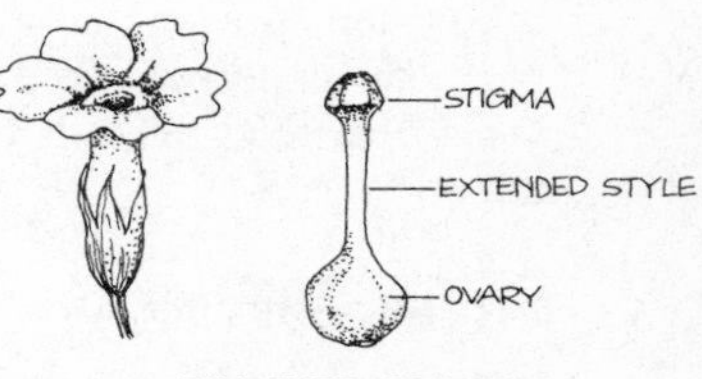

PIN-EYED PRIMULA

Piliferous Layer

The area of root epidermis extending for (usually) less than $\frac{1}{2}$ in (12 mm) behind the root-tip that bears the root-hairs.

Piliform

Hair-shaped.

Pilose

Hairy; having soft, sparse, and moderately long hairs, often covering the leaf surfaces and making them appear downy or having a silvery sheen.

Pin-eyed

Of primula flowers, in which the stigma develops at the mouth of the flower: *cf.* Thrum-eyed.

Pinching, Pinching-out

The action of stopping a plant by removing the growing tip, thus forcing the development of side-shoots.

PINCHING

Pine Needle

The leaf of a pine tree. These collect around the bole of the tree and can be used as a mulch to conserve moisture and suppress weeds. Pine needles decay very slowly because of the high resin and turpentine content.

PINE NEEDLES

Pinery

A sunken greenhouse or similar building for growing pineapples, usually in the gardens of wealthy gentry. Popular in the eighteenth and nineteenth centuries.

Pinetum

A collection of mainly coniferous trees; an arboretum devoted principally to coniferous subjects.

PINK BUD STAGE

Pink Bud

A stage in the development of the buds of fruit trees, essential knowledge for the correct timing of spraying programmes.

Pinna, *pl.* Pinnae

The primary division or leaflet of a pinnate leaf of a fern frond.

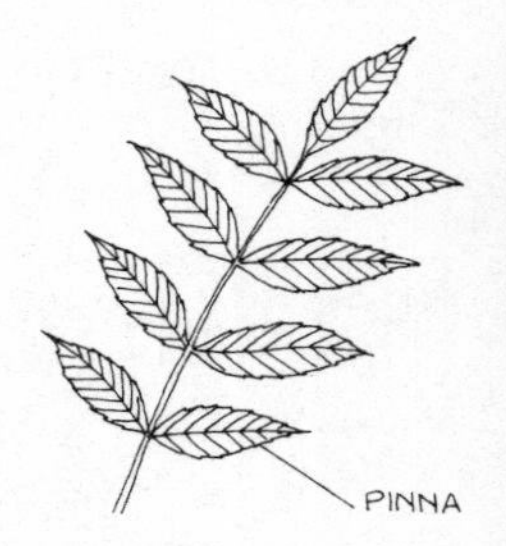

COMMON ASH

Pinnate

Of a leaf, constructed in the manner of a feather with the parts arranged on both sides of an axis in pairs, as in the leaves of the ash (*Fraxinus excelsior*).

Pinnatifid

Of a leaf, divided or parted in a pinnate manner, nearly or about halfway down. Similar to pinnate but divided into lobes, not leaflets.

Pinnatipartite

Of a leaf, pinnately divided or cut rather more than half-way or almost to the midrib; between pinnatifid and pinnatisect.

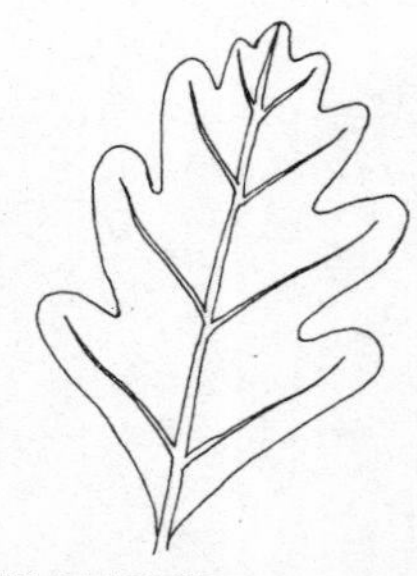

PINNATIFID

Pinnatisect

Of a leaf, cleft almost down to the midrib in a pinnate manner.

Pinnule

One of the individual leaflets of a bipinnate leaf.

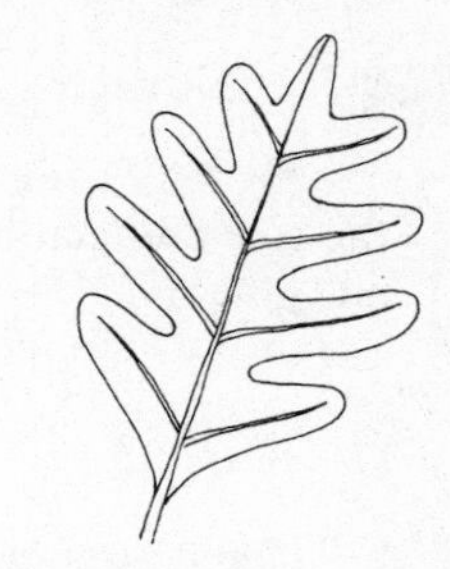

PINNATISECT

Pioneer Species

The first species to colonize a barren area.

Pip

(1) A single blossom of a flowerhead or bloom.
(2) The seed of apple, pear, orange, etc.
(3) A single stem of lily-of-the-valley (*Convallaria majalis*).

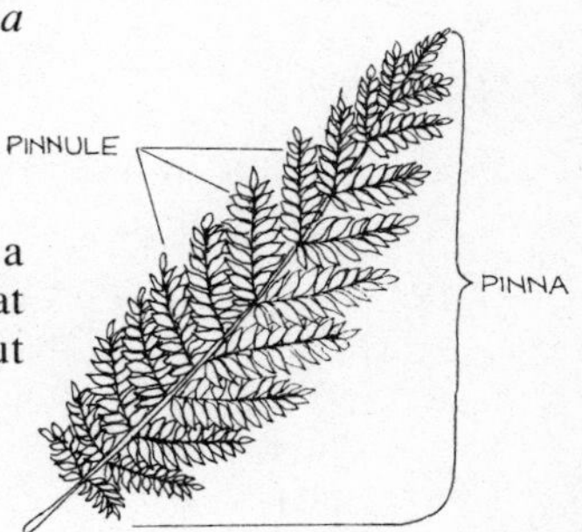

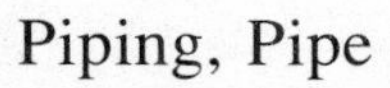

Piping, Pipe

A cutting taken from pinks, carnations, etc. by pulling a young non-flowering shoot away from the mother plant at a joint and inserting it into the rooting medium without trimming.

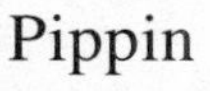

Pippin

Any variety of apple raised from seed.

Piriform

Pear-shaped.

Pistil

The female reproductive organs of a flower, consisting of ovary, style and stigma. It may consist of a single carpel (simple pistil) or a group of carpels (compound pistil).

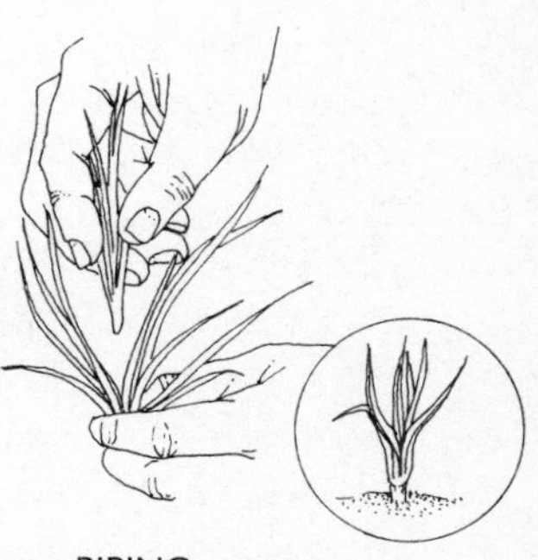

PIPING

Pistillate

Having a pistil but no functional stamens; a female flower.

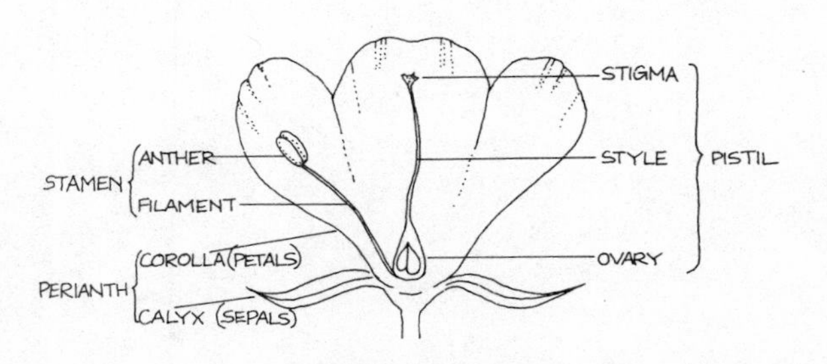

PARTS OF A FLOWER

Pistillode

An abortive pistil, or a vestigial pistil present in some staminate flowers.

Pit

(1) A small depression or cavity on the skin of fruit.
(2) A fruit stone such as that of a cherry; or to remove such a stone.
(3) A natural hole in the ground or one excavated for disposal of rubbish, storage of vegetables, etc.
(4) A sunken, usually unheated garden frame popular in the nineteenth century for raising cucumbers, melons, etc.

PITCHER PLANT (*Nepenthes*)

Pitcher

(1) Modified leaf formed like a pitcher or jug.
(2) A type of plant bearing such leaves, often secreting a fluid to attract and digest insects.

EPIDERMIS
PITH
CORTEX

SUNFLOWER STEM (TRANSVERSE SECTION)

Pitchfork

See Hayfork.

Pith

(1) Spongy cellular tissue in the stems and branches of dicotyledonous plants.
(2) The lining of the rind of oranges, lemons, etc.

Placenta

The part of a carpel that bears the ovules; part of the ovary wall to which the ovules are attached.

Placentation

The arrangement of the placenta within the ovary and the manner of its attachment to the ovary wall.

THE MAIN FORMS OF PLACENTATION

PARIETAL

MARGINAL

AXILE

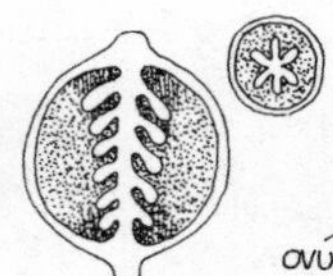

FREE CENTRAL

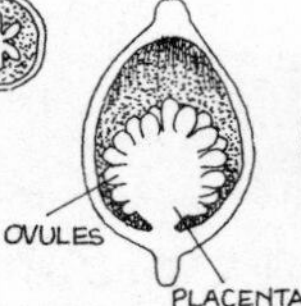

FREE BASAL

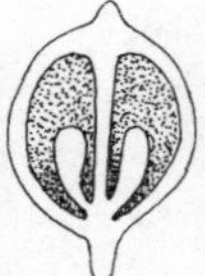

BASAL ERECT

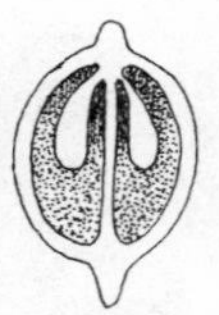

APICAL PENDULOUS

Plagiotropism

The tendency for the growth response of a plant to be orientated at an angle to the force of the stimulus concerned; thus the growth of lateral branches is at an angle to the force of gravity.

Plain

When applied to foliage, a smooth unindented leaf-margin.

Plane

(1) A tree of the genus *Platanus*, includes sycamore, buttonwood, etc.
(2) Flat.

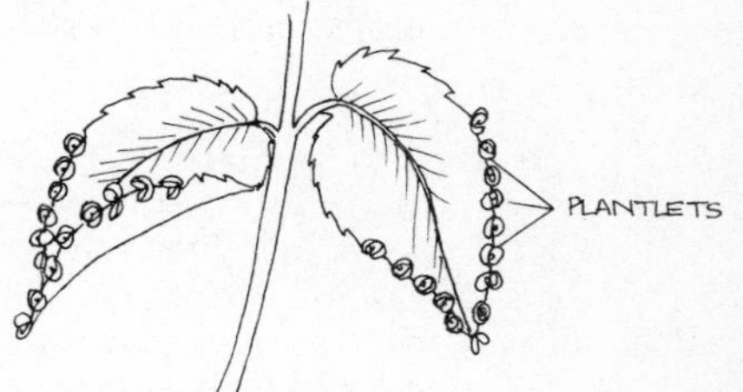

Bryophyllum daigremontianum SHOWING PLANTLETS ON LEAVES

Plant House

Usually a very large greenhouse.

Plant Lice

A general term covering a variety of sap-sucking insects including aphids, blackfly, etc.

Plantlet

Tiny plant produced on the leaves of certain plants, including many bryophyllums such as *Kalanchoe tubiflora*. The plantlets fall off and root in the pot or gravel of the greenhouse bench: *cf.* Propagule.

Plantsman

Originally a nurseryman or florist but now used mainly to describe an expert gardener or one considered to be an authority upon the growing of plants.

Plashing

A form of hedging whereby branches are bent down and interwoven to encourage density of new growth: *cf.* Hedging, Pleaching.

Plasmolysis

Excessive loss of water from a plant cell causing the protoplasm to shrink away from the cell wall. Persistent plasmolysis will cause the plant to wilt.

Plastid

A differentiated granule in protoplasm bounded by a membrane. Plastids containing pigments are called chromoplasts, colourless plastids are leucoplasts.

Pleaching

The horizontal training of branches and twigs of trees, etc. to form a screen or windbreak: *cf.* Hedging, Plashing.

Pleated

Folded lengthwise several times, as in some buds.

Pliable

Bending or bendable; supple; yielding.

Plicata

Bearded iris flowers having a base colour of white or yellow with stippled or feathered margins of contrasting colour.

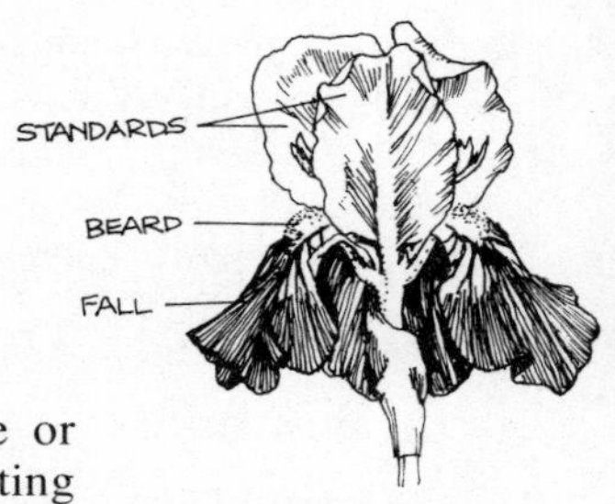

BEARDED IRIS

Plicate

Pleated or folded like a fan, as the leaves of the snowdrop (*Galanthus nivalis*).

Plot

A small area of land usually for the growing of vegetables, etc., sometimes called an allotment.

Plough, PLOW (USA)

One of the earliest of agricultural implements, consisting of a pointed blade pulled through the soil by a man, animal or machine to create furrows and turn over the soil.

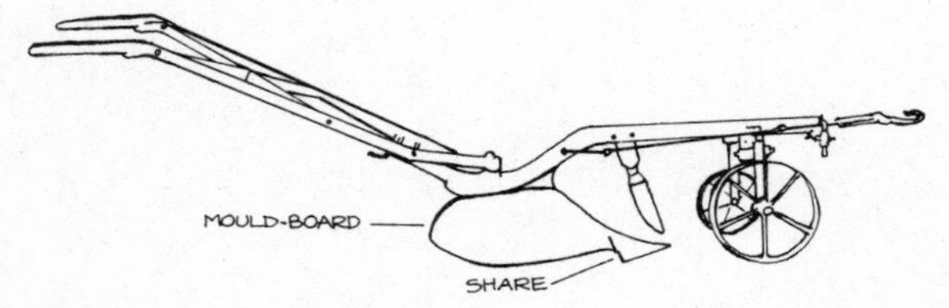

PLOUGH

Plume

A feathery inflorescence.

Plumose, Plumous, Plumate

Feathery; feathered; plume-like.

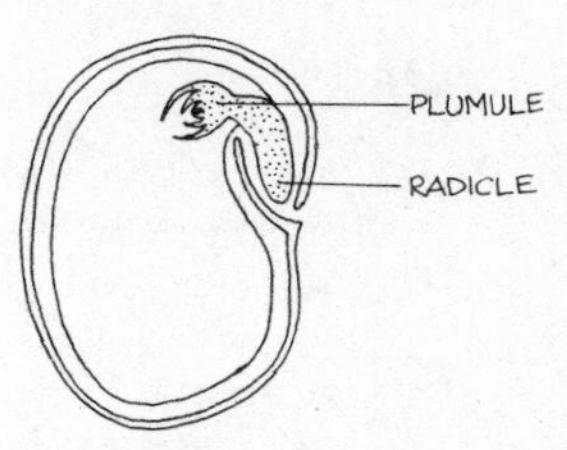

SECTION OF A PEA SEED

Plumule

The rudimentary stem of an embryo plant in a seed, and also in the germinating seedling.

Plunge

To sink a pot plant up to the rim (or nearly so) in soil, cinders, sand or similar materials. This method is often used to prevent the roots from drying out or to keep them cool in hot weather when the subject could be damaged by excessive heat round the roots. In commercial practice, bulbs required for Christmas flowering are plunged earlier in the year to ensure adequate root-formation before forcing can commence.

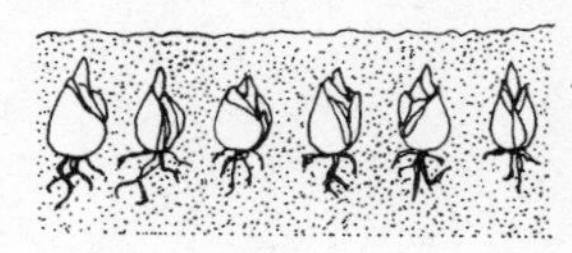
PLUNGED POTS AND BULBS

Pneumatophore

An upward-growing respiratory root in swamp plants such as mangroves.

RESPIRATORY ROOTS OF *Sonneratia alba* IN A MANGROVE SWAMP

Pod

The outer protective shell of peas, beans and other leguminous plants.

Podsol, Podzol

Stratified soil from which various materials have been leached, especially from the upper layers.

Pole

(1) A straight slim stem or branch of a tree often used as a support for protective netting, part of a rustic structure, etc.
(2) An old unit of linear measurement: *see also* Perch and Rod.

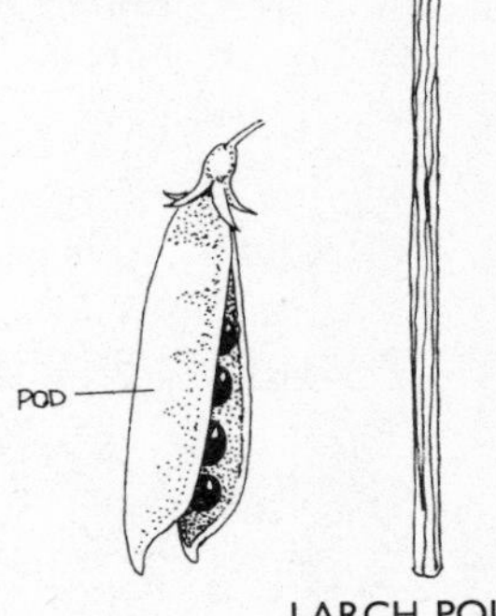

LARCH POLE

Polesaw

A short curved pruning saw attached to the end of a long pole, used for trimming high branches.

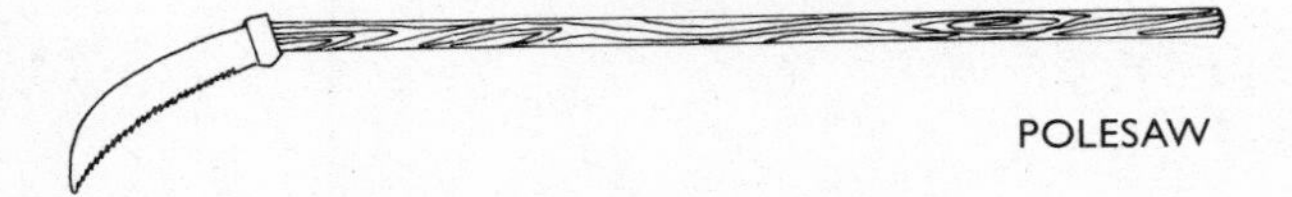
POLESAW

Poll

To cut off the top of a tree or plant; more severe than pollarding.

Pollard

To cut the branches of a tree right back to the trunk at frequent intervals in order to encourage the production of young shoots. The method is also used occasionally on ornamental trees to produce 'mop-heads'. The timber from oak trees that have been pollarded is much valued for furniture-making.

CRACK WILLOW (*Salix fragilis*) POLLARDED TO PROVIDE STRAIGHT POLES FOR HURDLES

Pollen

The male sex cells carried on the anthers of most flowering plants. When these are deposited on the stigma the flower is said to be pollinated and fertilization of the ovule normally ensues. More than one pollen grain may grow a tube but only one will successfully fertilize a single egg.

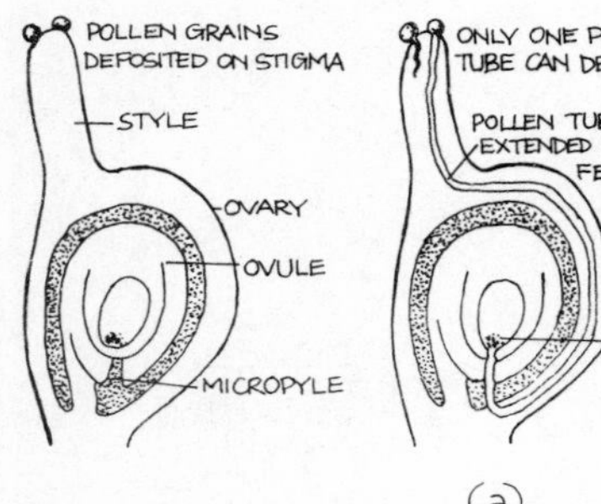

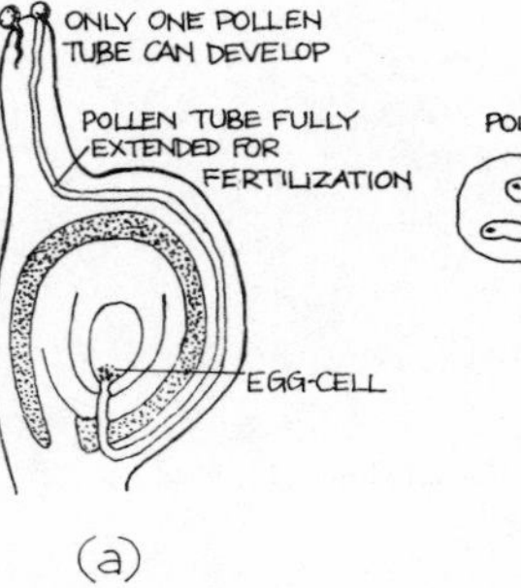

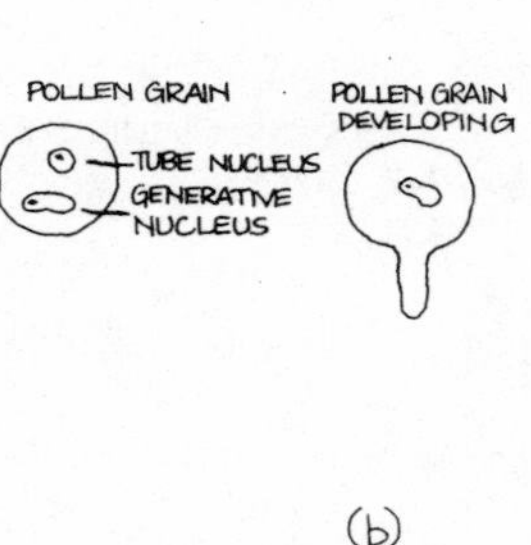

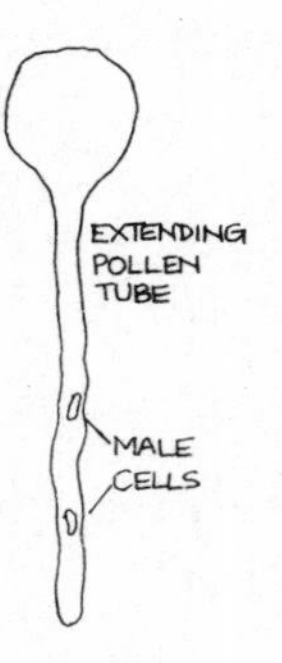

(a) POLLEN GRAINS DEPOSITED ON STIGMA, POLLEN TUBE EXTENDED (b) DETAILS OF POLLEN GRAIN

Pollen Sac

The receptacle in an anther where the pollen is formed.

Pollen Tube

An outgrowth of a pollen grain through which the male gametes are transported to the ovule.

Pollination

The transfer of male pollen to the female stigma, normally performed by insects. Flowers can also be hand-pollinated, or wind-pollinated.

Pollinium, *pl.* Pollinia

Pollen grains adhering together in a mass and not individually enclosed in sacs. A feature of many orchids.

Polyadelphous

Having stamens with filaments united in several bundles: *cf.* Diadelphous, Monadelphous.

Polyandrous

Having a large number of stamens.

Polycarpic

Fruiting many times, or year after year.

Polygamia

The outdated Linnaeus classification of plants; having male, female and hermaphrodite flowers.

Polygamodioecious

Having male and bisexual flowers on one plant, and female and bisexual flowers on a separate plant.

Polygamous

Having both unisexual and bisexual flowers on the same plant.

Polygamy

The occurrence of male, female and hermaphrodite flowers on the same plant or different plants of the same species.

Polygonal

Having many angles, certainly more than four.

Polygynous

Having numerous pistils, styles or stigmas.

Polyhybrid

A cross between parent plants that differ in several heritable characters.

Polymerous

Having many parts or segments in a whorl.

Polymorphic, Polymorphous

Of an organism that occurs in many forms, such as leaves of different shapes appearing on one plant.

Polypetalous

Having a corolla of separate petals, e.g. roses.

Polyphyletic

Arising from mixed ancestry.

Polyphyllous

Having the perianth leaves free.

Polyploid

Having more than twice the usual haploid number of chromosomes.

Polypodiaceae

The family to which most ferns belong, having stalked sporangia with a vertical annulus.

Polypody

Any fern of the genus *Polypodium*.

Polysepalous

Having a calyx of separated sepals.

Polystyrene

An inert, thermoplastic material capable of absorbing air and water. In its granulated form it is sometimes incorporated into soilless potting composts.

Polythene

A thin thermoplastic film used as a protective covering for semi-tender subjects out of doors, and for greenhouse insulation. It is available in transparent or opaque white form, and there is also a black product which can be used for weed-control and mulching. Plant pots and other types of container are made from thicker polythene and extruded in a variety of shapes, sizes and colours.

Pomace

(1) The pulp of crushed apples before and after the juice has been extracted for cider-making.
(2) Fish-refuse used as a fertilizer after the oils have been extracted.

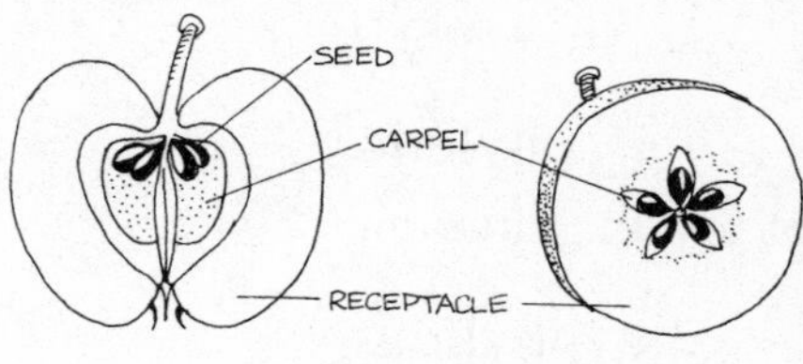

POME (APPLE)

Pomato

A tomato grafted on to a potato plant.

Pome

A fruit constructed like an apple, the enlarged fleshy part enclosing the core formed from the carpels and enclosing the seed. Also applicable to some other members of the rose family (Rosaceae).

Pomiculture

Fruit-growing, especially apples, usually as a commercial venture.

POMPON DAHLIA

Pomology

The science and study of fruit-growing, especially apples.

Pompon, Pompom

Small spherical flower or flowerhead. Derived from a French word meaning 'tuft', it is often applied to a type of dahlia or chrysanthemum but can refer to other flowers.

Pond

Small area of still water produced either naturally or artificially.

Pool

An area of water, usually in a garden, whether for growing plants, swimming or purely to enhance.

Pool Liner

A sheet of tough plastic used for making a pool impervious, or a plastic moulding in a variety of shapes for the same purpose.

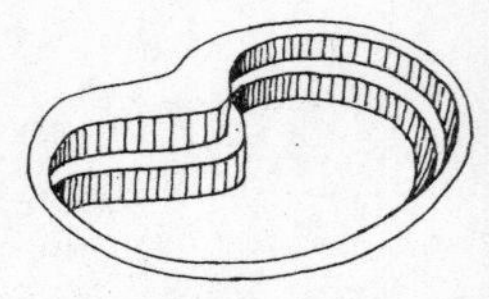

GLASS-FIBRE PRE-FORMED POOL LINER

Pore

A minute passage or interstice with a more or less circular aperture in a leaf or stem permitting the exchange of oxygen and carbon dioxide between the plant and the atmosphere: *cf.* Stoma.

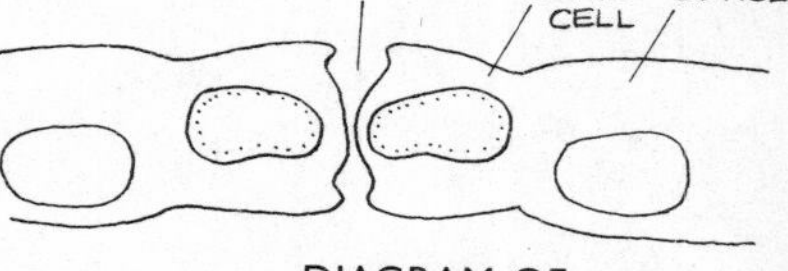

DIAGRAM OF A LEAF PORE

Porrect

Extended forward and upward.

Post-emergent

A type of selective weedkiller applied after the crop concerned has germinated and emerged above the soil.

Posterior

On or at the back; on the side next to the axis.

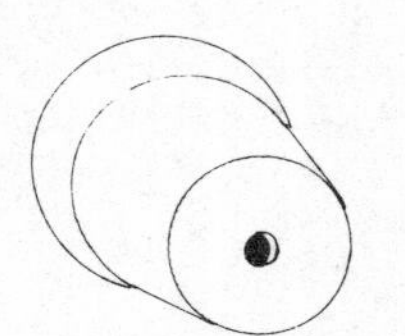

BASE OF A NORMAL CLAY POT

Posticous

Outward; extrorse.

Posy

A small bunch of flowers.

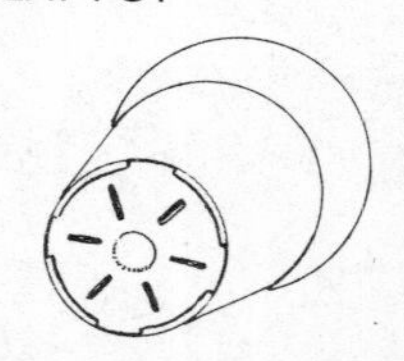

BASE OF A MODERN PLASTIC POT

Pot

A container in which plants may be grown, available in clay, concrete, peat or plastic, in sizes from 1 in (2.5 cm) to several feet, ornamental or plain.

Pot-bound

A pot-grown plant whose roots have completely filled the pot and is therefore ready for potting-on. Some plants flower best when pot-bound, provided they are adequately fed. *Also called* Root-bound.

POT-BOUND

Pot-herb

A herb used for culinary rather than medicinal purposes, such as sage and parsley.

Pot-in-Pot

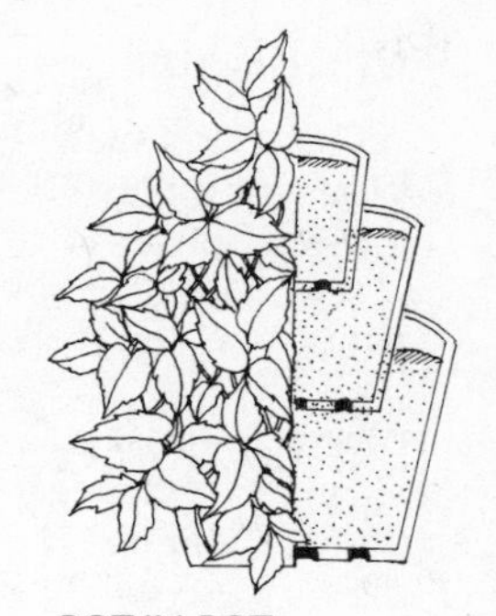

POT-IN-POT

A method of producing an impressive cascade display using mainly creeping or trailing plants. A small pot is set inside a medium pot and the pair set inside a large pot, so that each is slightly elevated above its lower neighbour, forming tiers. Soil is placed in, under and around them and cuttings inserted at every level. When the plants develop, the foliage will completely hide the pots and an attractive cascade effect will be produced, especially if the plants used are restricted to one variety.

Pot-Layering

See Air-layering.

Pot Plant

POT PLANT
(*Begonia rex*)

Any plant grown in a pot can obviously be described as a pot plant, but commercially the term applies to those plants produced in quantity every spring and early summer for growing-on in pots, as distinct from plants grown for planting-out in the garden.

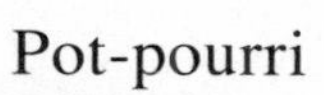

Pot-pourri

An aromatic mixture of dried flowers, leaves and spices, used to scent a room or clothing.

Pot Shard

A piece of broken clay pot formerly used for crocking.

Potassium, Potash

The ‘K’ symbol (Kalium) of the NPK formulae, mainly responsible in plants for flower and fruit development. Originally called kanit or kanite. The word potash derives from the early crude method of leaching wood-ash and evaporating it in clay pots – hence pot-ashes.

Potato-planter

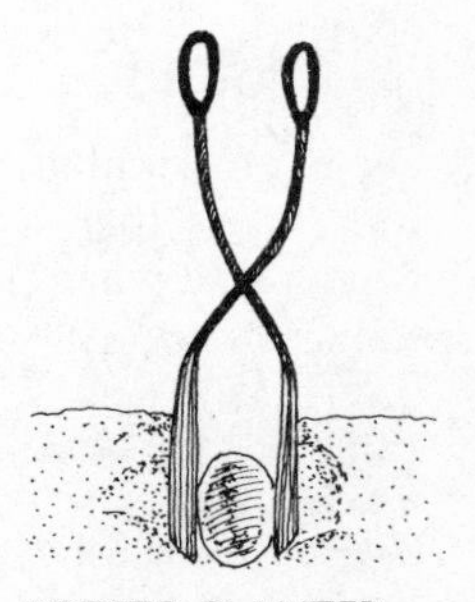

POTATO-PLANTER

An ingenious gadget for planting tubers and bulbs without stooping. It consists of a pair of tongs with flattened blades which can be separated after inserting in the soil, creating a hole into which the tuber can be dropped. As the implement is withdrawn, the soil falls in, covering the tuber.

Potting

Transferring seedlings from the seed bed or rooted cuttings from the propagator into small pots. Most losses occur at this stage through careless handling of the tiny plants, resulting in damage to the young stems which allows the entry of pathogens.

Potting Bench

A table reserved for potting-up plants, usually in a potting shed or greenhouse. Most potting benches have three raised sides to prevent the potting medium spilling, and quite often a shelf above and below to house pots, labels, tools, etc.

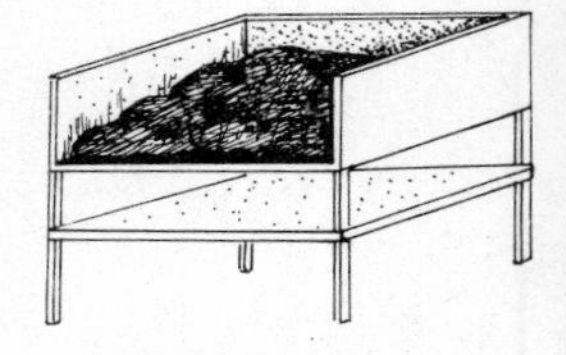

POTTING BENCH

Potting Compost

See Compost.

Potting-on

Transferring an established plant complete with its root system into a larger pot to allow more space for root development.

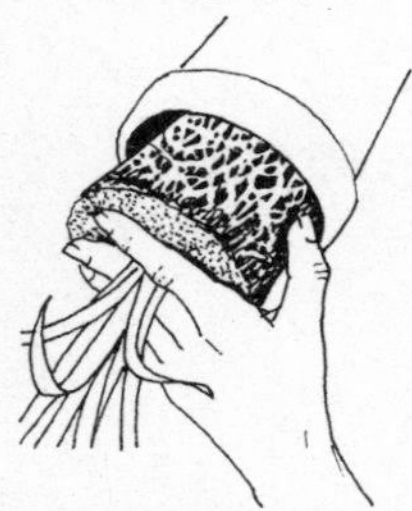

1 REMOVE PLANT FROM ITS POT

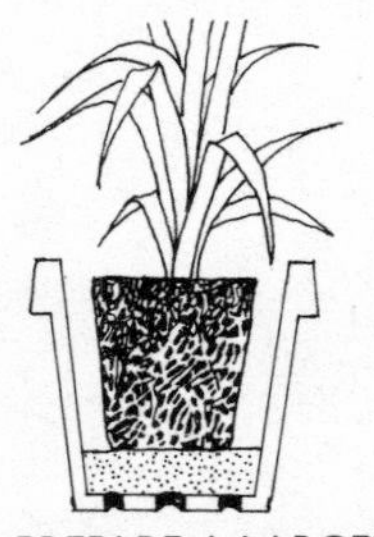

2 PREPARE A LARGER POT WITH SOIL AT BASE – DO NOT DISTURB THE ROOTS

3 CENTRE THE PLANT, FILL POT WITH SOIL, TAP ON BENCH TO SETTLE, WATER CAREFULLY

POTTING-ON

Potting Shed

A building (often attached to one end of a greenhouse) used for all potting operations and for storing garden tools, seeds, etc.

Potting Soil

Before the days of the John Innes Composts and soilless mixtures, this was the sieved soil sold by nurserymen for potting, usually dug from a corner of the nursery and often unsterilized.

Potting Stick

A piece of smooth wood, usually 1 × 1 in (2.5 × 2.5 cm) or 2 × 2 in (5 × 5 cm) and about 12 in (30 cm) long, formerly used to ram down the soil around a newly potted plant to eliminate air pockets. Modern practice frowns upon the use of potting sticks or any similar method of compacting the soil; a few gentle taps of the base of the pot on a bench is all that is required to settle a plant, air spaces in the soil being essential for optimum development.

POTTING STICK

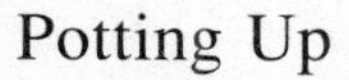

Potting Up

See Potting.

Pottle

Small wicker or woodchip basket for soft fruit such as raspberries; rarely seen today, being largely superseded by cardboard or plastic products.

POTTLE

Potworm

Small whitish worm, often found in pots containing unsterilized soil. The worms can seriously damage roots if present in large numbers. Control by watering with derris or potassium permanganate.

Pouch

Purse-like seed vessel, or bag-like cavity in a plant.

Pre-emergent Weedkillers

Chemical weedkillers that are applied to the ground after the crop has been sown but before it has germinated or any growth has appeared above the surface.

Pre-sprouting

Germinating seeds on blotting paper or similar absorbent material. Same as chitting but this is mainly applied to larger subjects such as potato tubers.

Precocious

Premature; flowering or fruiting earlier than normal.

Predator

In horticulture, an insect that preys upon another. The ladybird (ladybug in USA) eats vast quantities of greenfly, and the chalcid wasp feeds on whitefly; both should be encouraged in greenhouse and garden. Wasps, although generally disliked, are an asset to the gardener since they kill a great many harmful insects, especially in the spring and early summer.

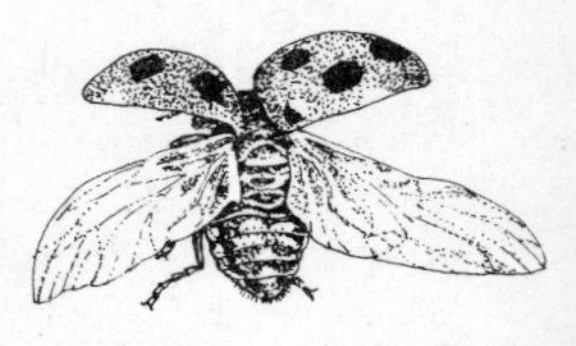

PREDATOR (LADYBIRD)

Premorse, Praemorse

Ending abruptly as if bitten off, as the leaves of the tulip tree *Liriodendron tulipifera*.

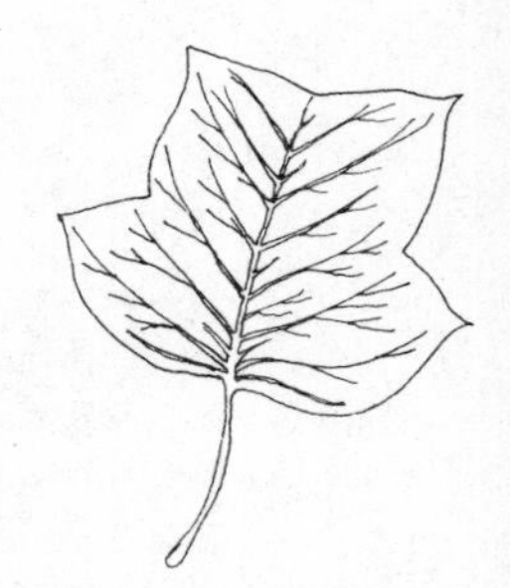

PREMORSE LEAF

Prepared

Bulbs which have been subjected to special treatment before sale so that they will flower earlier than normal. Many hyacinth bulbs are prepared so that they will flower in time for Christmas.

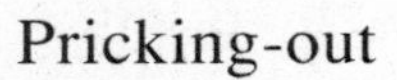

Pricking-out

Transferring young seedlings from their initial seed-bed to more spacious accommodation; so called from the old system of pricking small holes in the soil where the tiny plants were to be inserted.

Prickle

A sharp point growing from the epidermis of a plant, otherwise termed spine or thorn.

PRICKLE OF HAWTHORN

Primary

Belonging to the first stage of development.

Primine

The outer coat of an ovule.

Primitive

(1) Appearing in the earliest or very early stage of development.
(2) Type of single flower with narrow, widely spaced petals.
(3) Features inherited unchanged from distant ancestors.

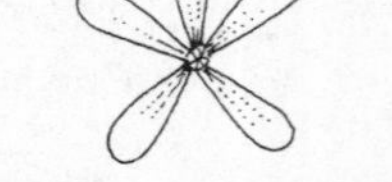

PRIMITIVE FLOWER

Primocane

A biennial shoot, particularly of a bramble during its first year of growth.

Procumbent

Lying along the ground; trailing but not rooting; prostrate.

Proliferation

The production of shoots that may eventually become new plants.

Proliferous

(1) Producing abnormal buds or offshoots, such as plantlets on leaves, fruits on fruits, shoots out of flowers, etc.
(2) Bearing progeny as offshoots.

Prolific

Fertile; producing in abundance. Often used with reference to fruit trees, potato crops, etc.

Prone

Prostrate; lying flat along the ground.

PRONG CULTIVATOR

Prong Cultivator

A long-handled tool, similar to a hoe but having three or more arched prongs instead of a blade.

Prop Root

A rigid aerial root that emanates from the stem, descends to the ground and helps to support the main stem, as in some trees.

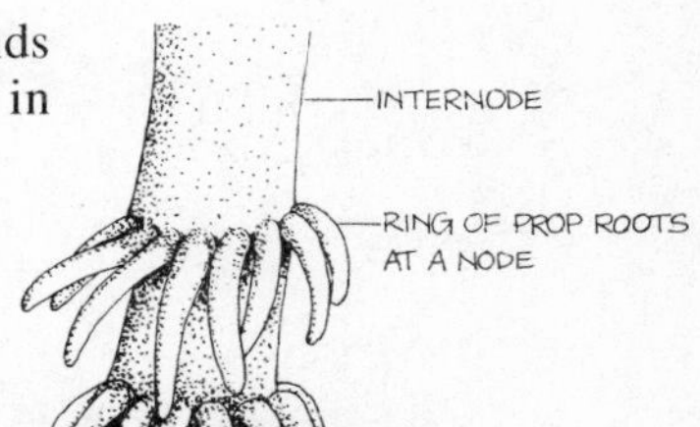

PROP ROOTS OF MAIZE

Propagating Case

See Propagator.

Propagating Frame

Normally an unheated outside structure of glass and wood used for raising young plants that need slight protection.

Propagating House

A greenhouse or part of a greenhouse set aside entirely for the raising of young plants, whether from seed or cuttings; often specially equipped with heated benches and misting sprays.

Propagation

Methods by which plants are increased in number, such as grafting, division, cuttings, seeding, air-layering, budding, etc.

Propagator

(1) Usually a heated glass or plastic covered box-like structure for the raising of seedlings, cuttings, etc.
(2) A person responsible for the production of young plants or seedlings.

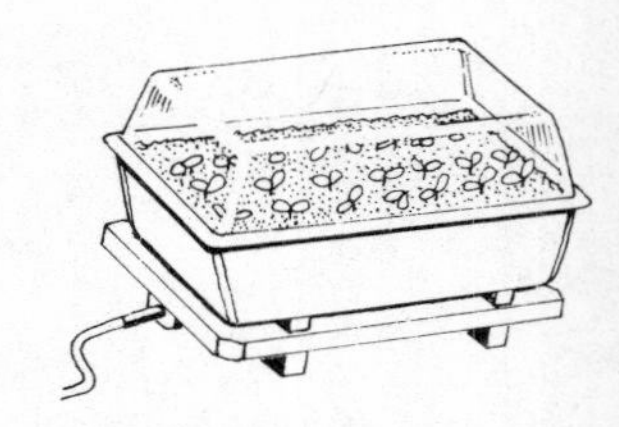

SIMPLE ELECTRIC PROPAGATOR

Propagule

A plantlet or offset produced naturally by an adult plant without assistance or special treatment, such as *Tolmeia menziesii* which produces buds that develop and form new plants at the base of the leaf-blades: *see also* Plantlet.

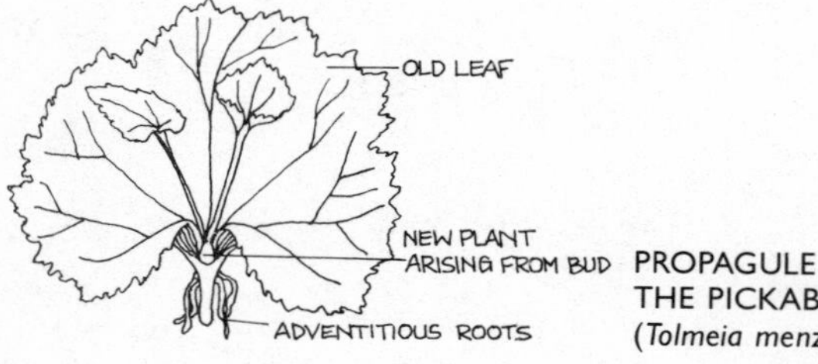

PROPAGULE
THE PICKABACK PLANT
(*Tolmeia menziesii*)

Prostrate

Lying flat along the ground; also sometimes applied to spreading plants such as the pelargonium 'Prostrate Boar'.

Protandrous

The ripening of the anthers of a flower before the stigmas are ready to receive pollen: *cf.* Protogynous.

Protein

Any of a class of organic compounds forming a major part of all living organisms.

Prothallus, *pl.* Prothalli

The liverwort-like gametophyte stage of a fern *(Pteridophyte)*. In some pteridophytes a single prothallus will carry both male and female sex organs, while in others there are separate male and female prothalli.

Protogynous

Of a flower in which the stigma is receptive before the anthers of the same flower have reached maturity: *cf.* Protandrous.

Protoplasm

The viscous living matter within plant cells. The protoplasm is surrounded by a thin cell membrane which is part of the cell, and then by a non-living cell-wall which separates one cell from another.

Protoxylem

The first part of the xylem to be formed.

Provenance

The place of origin.

Pruinose

Having a frosted appearance, a powdery bloom or coating, or a protective waxy secretion.

Pruner

(1) Any implement used for pruning including secateurs, tree-loppers, etc. It can also apply to specially shaped knives used in pruning.
(2) A person employed or engaged in the work of pruning.

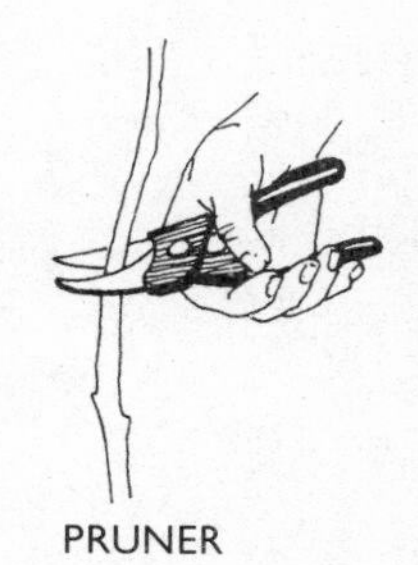
PRUNER

Pruning

The cutting back of plants including trees, bushes and shrubs for four main reasons:
(1) To improve the quality or quantity of the fruit or flowers;
(2) To remove dead or diseased parts;
(3) To open up the plant to allow more light and air to reach the centre, thus reducing the risk of botrytis;
(4) To alter the shape of the plant or tree.

Pruning Knife

A small knife with a blade shaped for special pruning applications; many have fold-away blades.

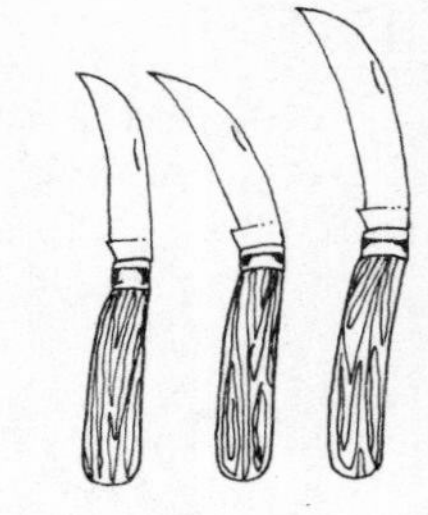
PRUNING KNIVES

Pseudobulb

A swollen growth of bulb-like appearance, but not a true bulb. Common in some orchids.

Pseudocopulation

The attempted copulation by male insects with a part of a flower which resembles the female of the insect species, as in some orchids.

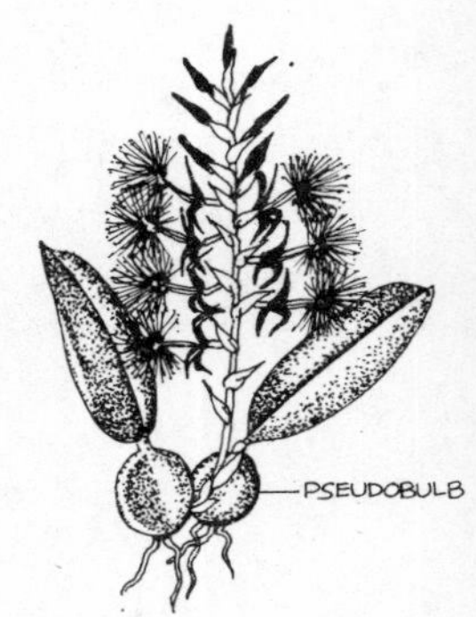

AN ORCHID WITH PSEUDOBULBS (*Bulbophyllum barbigerum*)

Psilocybes

'Magic mushrooms', originally from Mexico, containing the hallucinogenic drug psilocybin. Frequently confused with the peyote (*Liphophora williamsii*), which is a cactus from Mexico and southern USA, also hallucinogenic.

Pteridophyte

A member of one of the main divisions of the vegetable kingdom – ferns, etc.

Puberulent, Puberulous

Feebly or minutely pubescent; clothed with minuscule soft downy hairs.

Pubescent

A general term for hairy, although in fact it should be reserved for those leaves covered with dense short soft hairs, adpressed to the leaf surface.

PTERIDOPHYTE (*Pteris cretica*)

Puddling

(1) A practice in which the roots of plants such as brassicas and leeks are first dipped into a thick mixture of soil and water (thin mud) before being transplanted into their permanent positions. It is probably not much used today but was normal practice in Victorian times and said to prevent invasion by cabbage fly, especially if a little calomel was added to the water.

(2) The compressing or treading of wet clay to form a waterproof lining for a pond or pool.

Puffer

A dusting device usually incorporating some form of bellows for applying powdered fungicides and pesticides to plants.

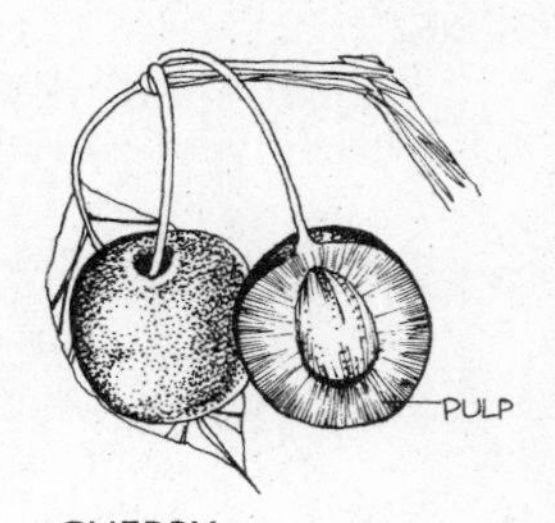

CHERRY

Pullulate

To sprout or develop by sprouting, as of a seed or bud.

Pulp

The fleshy part of fruit, often the edible part.

Pulse

The edible seed of leguminous plants such as peas, beans, etc.

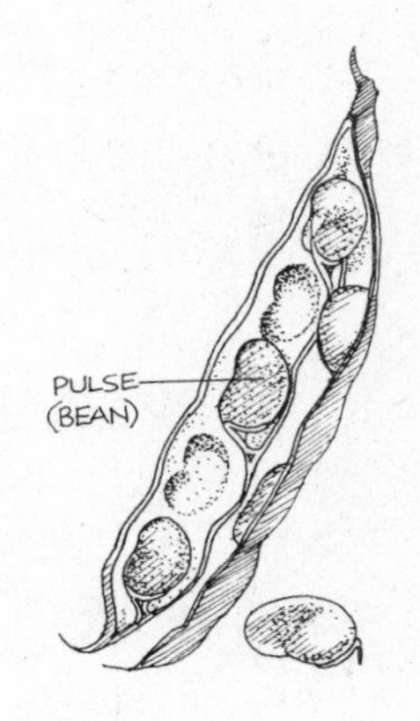

Pulverulent

Literally, dusty, as applied to the powdery coating on the stems and leaves of some plants. It occasionally occurs in specific names such as *Primula pulverulenta*.

Pulvinate

Cushion-shaped, forming a dense low tuft or dome.

Pulvinus

A cushion of tissue, especially one at the base of a leaf or leaflet, which brings about movement by its changes of turgidity. In the sensitive plant (*Mimosa pudica*) the pulvinus is responsible for the folding of the leaves at night-time or when the plant is touched.

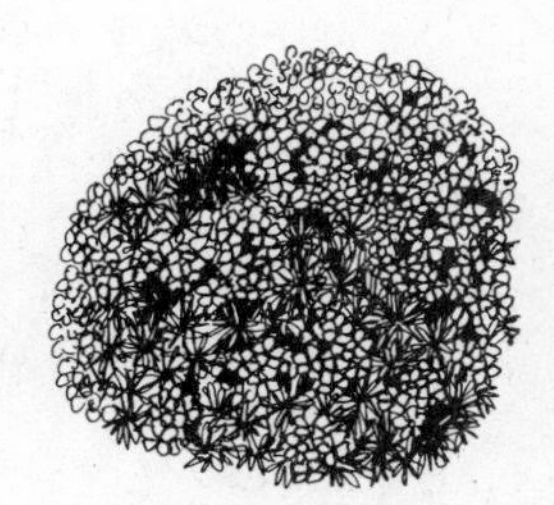

PULVINATE THRIFT (*Acantholimon juniperifolia*)

Punctate

Dotted or pitted, usually shallowly. Apples and pears may be punctate.

Puncticulate

Diminutive of punctate; minutely pitted or dotted.

Punctum

A speck or dot, such as a spot of colour or a small depression on a surface.

Puncture

A small hole or perforation, especially in a leaf, caused by a biting insect or grub.

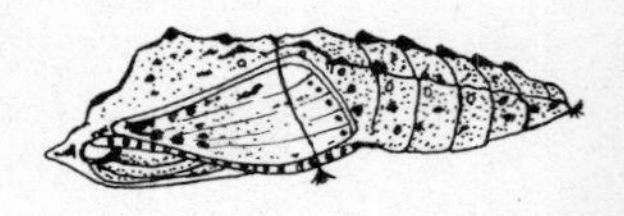

PUPA OF LARGE WHITE BUTTERFLY (*Pieris brassicae*)

Pungent

(1) Acrid to taste or smell.
(2) Sharp; ending in a hard sharp point or tip.

Punk

Rotten wood or the fungus that grows thereon; an expression mainly used in USA.

Pupa

See Chrysalis.

Pustular, Pustulate

Blistered or apparently so; warty; pimply.

Pustule

A blister or blister-like formation.

PYRAMID

Pyramid

(1) A type of fruit tree formerly grown mainly for pears, but dwarf pyramids are now often used in commerce and small gardens for apples.
(2) Any plant with a natural cone-like habit.

Pyrene

A fruit stone, or the nutlet in a drupe as in the cherry, or the 'seed' of a raspberry drupelet.

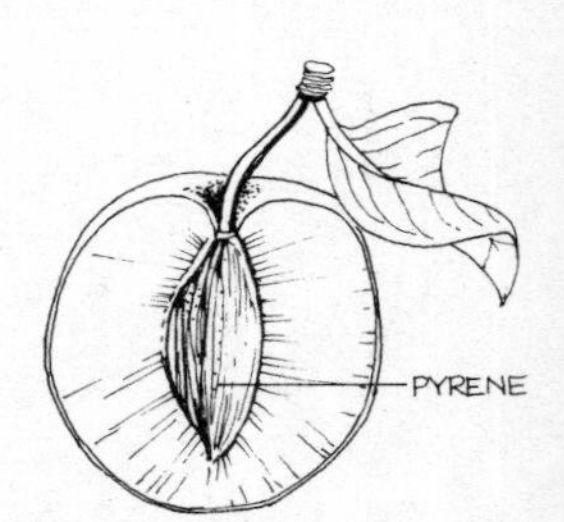

CROSS-SECTION OF A PLUM

Pyrethrum

Insecticide powder made from the pulverized heads of some species of *Pyrethrum*.

Pyriform

Roughly pear-shaped.

Pyxis, Pyxidium

A seed capsule that opens by a transverse circular split, the top coming off as a lid, as in the scarlet pimpernel (*Anagallis arvensis*).

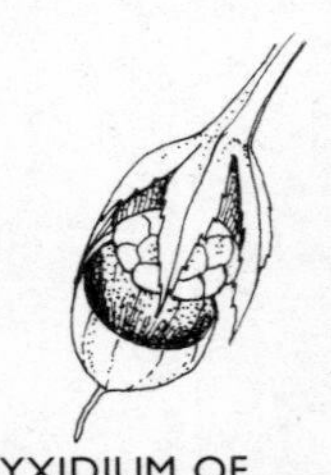

PYXIDIUM OF SCARLET PIMPERNEL (*Anagallis arvensis*)

Q

Quadrat

A square frame of metal or wood, or the area of vegetation enclosed by it, used to calculate the density of species growing on a certain area of land. Both the device and the area it encloses are referred to as a quadrat.

Quadrat Chart

A chart of squared paper on which the results of counting the species in a quadrat are marked, used mainly to record the investigation of large areas. A rectangle of land is marked out with tapes, then divided into smaller squares repeatedly until manageable-sized quadrats are produced and numbered to correspond to the squares on the quadrat-chart.

Quadrigeneric

A hybrid whose ancestry includes four different genera.

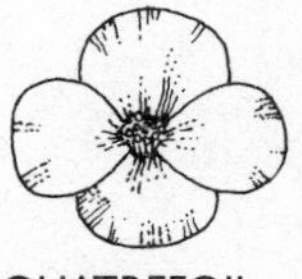

QUATREFOIL FLOWER (*Begonia rex*)

Quadripartite

Of a leaf, deeply cleft into four parts.

Quag, Quagmire

A boggy or marshy area of land; fen; slough; quaking bog.

Quarender

Early type of red apple common in Somerset and Devon in Victorian times.

Quassia

An old-fashioned insect deterrent and bird repellent manufactured from the wood of the *Picrasma excelsa* tree.

It has been largely superseded by modern chemical products but is still sometimes combined with derris as a safe insecticide, harmless to all warm-blooded animals, including humans.

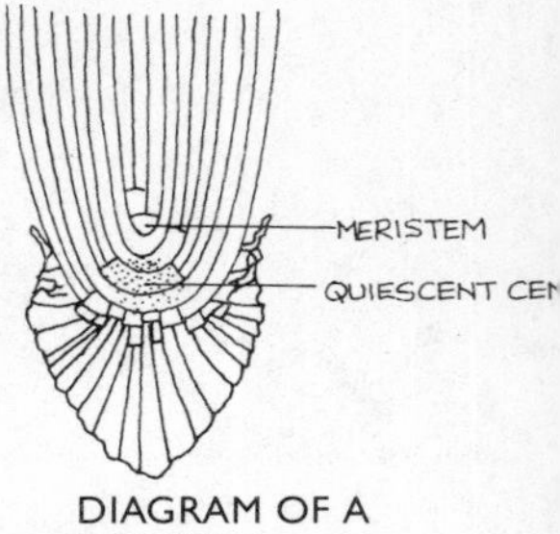

DIAGRAM OF A ROOT-TIP

Quatrefoil

A four-petalled flower, or a leaf consisting of four leaflets.

Queach

A thicket or dense hedge.

Queening

A type of apple, seldom seen today.

Quercetum

A group or collection of oak trees; oak woodland or forest. A plantation of oak trees.

QUILLED OR CACTUS DAHLIA

Quicklime

Oxide of lime, the product obtained by burning chalk or limestone in a kiln.

Quiescent Centre

The central core of a root-apex which lies between the root-cap and the meristem.

Quilled

Of petals which are tubular for all or much of their length, such as certain dahlias often called cactus-dahlias.

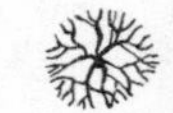

QUINCUNX PLANTING

Quincunx

An ornamental method of planting trees one to each corner of a square with a fifth in the centre; a favoured system of planting fruit trees in the past but now largely discontinued.

Quinsy-berry

Old country name for the blackcurrant, probably derived from its use in folk medicine to relieve the symptoms of throat infections.

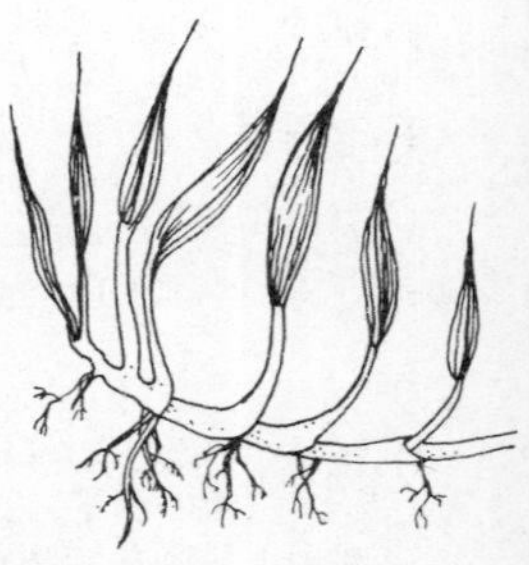
QUITCH-GRASS

Quitch-grass

Alternative name for couch-grass, a troublesome weed with creeping rootstocks, also called quack-grass.

R

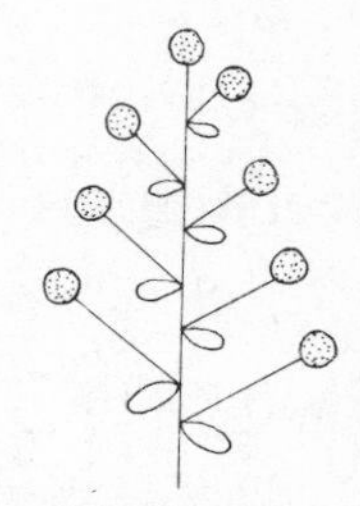

RACEME

Race

A vague term for a group of plants connected by a common ancestry; a subdivision of a species.

Raceme

A flowerhead on which the individual flowers are carried on short stems of approximately equal length and borne on an unbranched main stalk. In most cases the flowers open from the base upward as in the foxglove (*Digitalis*) but in some cases, including hyacinth, they open almost simultaneously.

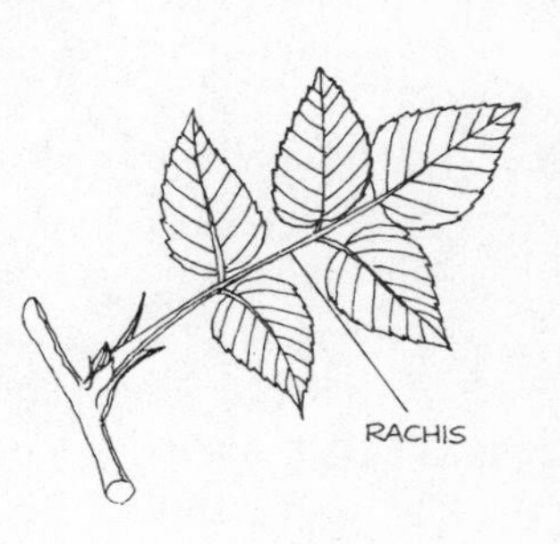

Racemose

Generally any inflorescence capable of indefinite prolongation; arranged like a raceme.

Rachis

(1) An axis carrying flowers or leaflets.
(2) The axis of a pinnately compound leaf or frond. *Same as* Rhachis.

Radial

Of flowers with similar petals which radiate symmetrically from a common central point. *Same as* Actinomorphic and Regular: *cf.* Zygomorphic.

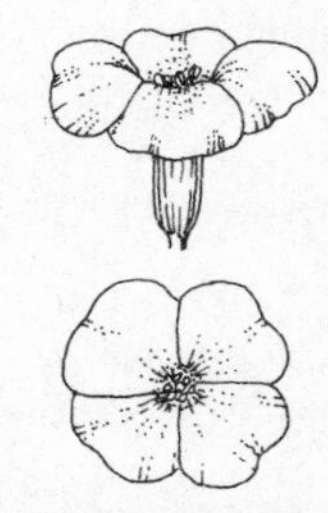

RADIAL FLOWER OF WALLFLOWER (*Cheiranthus*)

Radiate

Spreading from a common centre point.

Radical

(1) Basal leaves; those proceeding from the base of a stem or from a rhizome or crown.
(2) Arising from or near the root.

RADICAL LEAVES GRASS-OF-PARNASSUS (*Parnassia palustris*)

Radicant

Rooting from the stem, e.g. ivy.

Radicate

(1) To plant firmly.
(2) Deeply rooted.

Radicel

A small root or rootlet.

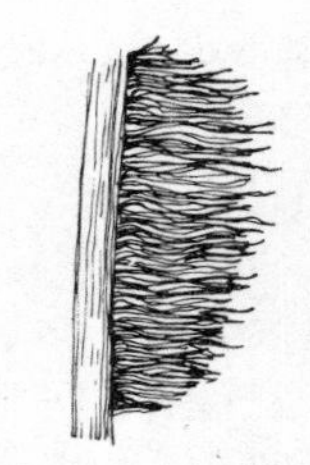

RADICANT ROOTS GROWING FROM THE STEM OF IVY (*Hedera helix*)

Radicle

A rudimentary root or rootlet; that part of a seed which will become the primary root.

Radiculose

Having many rootlets.

Radius

(1) Outer edge of a composite flowerhead, e.g. daisy (*Bellis perennis*).
(2) A radiating branch of an umbel.

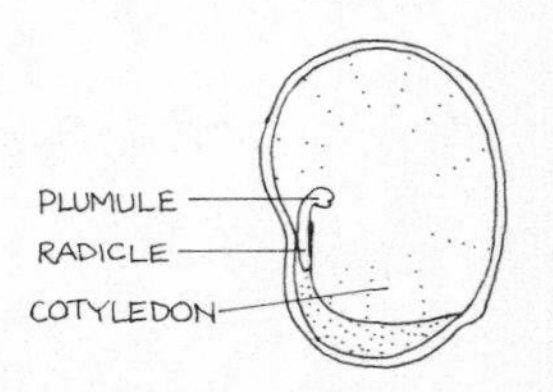

DIAGRAM OF BROAD BEAN SEED

Raffia

Fibre derived from the leaves of the raffia palm which forms a flat strip some 4 ft (1.2 m) or so in length, used in the past for tying and training plants. Now largely superseded by string and plastic strip, it is still preferred by many gardeners for such purposes as grafting and budding. *Also called* Bass.

Raft

A length of bark or similar material, usually oblong, on to which epiphytic plants can be wired. Widely used in the cultivation of epiphytic orchids.

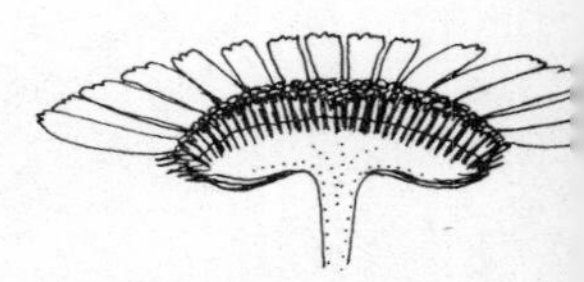

RADIUS OF SHASTA DAISY (*Chrysanthemum maximum*)

Raise

To cultivate or grow to maturity. *Same as* Rear.

RAFT
(*Cryptanthus bromelioides*)

Raised Bed

A plant bed raised above normal ground level and contained inside a wall of peat, bricks, stones or turves, etc. Raised beds are easier for disabled people to manage, and enable a large number of subjects to be grown in a small space.

Rake

A long-handled implement with a number of prongs or metal dowels set into a cross-head at right angles to the handle and used for breaking up the soil surface into a fine tilth ready for sowing. Garden rakes can be made of metal, plastic or stiff rubber; hay rakes are usually much larger and made of wood with metal cross-braces. Small hand-rakes for close work have a short handle and three or four tines bent sharply at right angles.

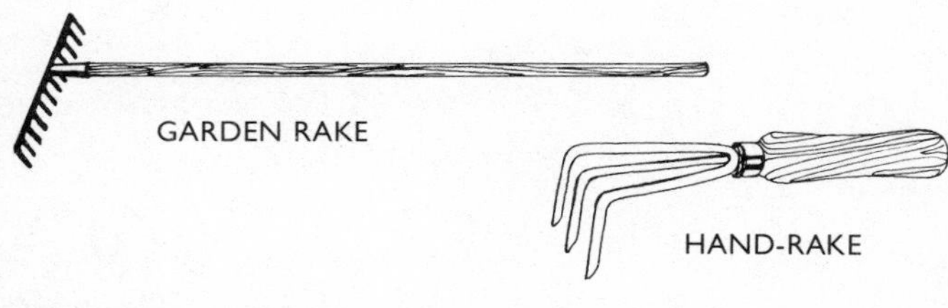

GARDEN RAKE

HAND-RAKE

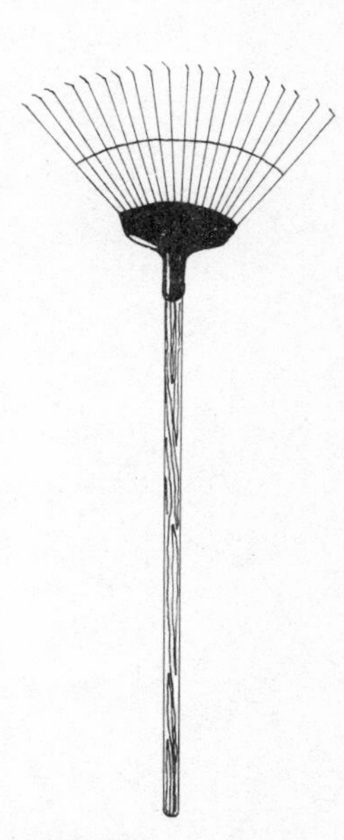

SPRINGBOK RAKE

Rambler, Rambling

Straggling; wandering in a disorganized manner; climbing and sprawling irregularly. A rambling rose is in nature an uncontrolled spreading shrub, whereas a 'climber' tends to grow mainly upward in a more controlled manner.

Ramentum, *pl.* Ramenta

Thin scaly membranous tissue on the surface of leaves and stems.

Ramet

An individual plant of a clone.

Ramification

The arrangement of a tree's branches, the pattern of their spreading.

RAISED BED

Ramifying

Branching; dividing into branches.

Rammer

Another name for a potting-stick, an implement used in the past for compacting the soil around a newly potted plant. The practice is discouraged today by knowledgeable horticulturalists.

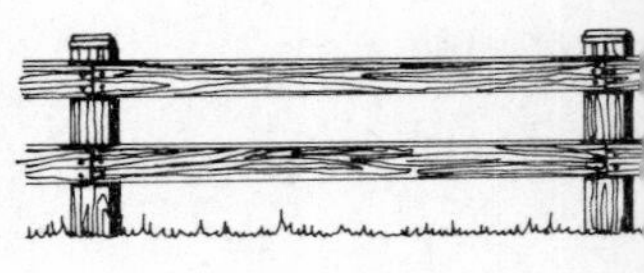

RANCH FENCING

Ramose

Having many branches.

Rampant

Unchecked growth; luxuriant; overwhelming.

Ramulus

A small branch.

Ramus, *pl.* Rami

A branch.

Ranch Fencing

Post-and-rail type of simple fencing.

OPPOSITE RANKS

ALTERNA RANKS

Rank

(1) Vertical row, usually of leaves, the ranks may be opposite or alternate.
(2) Coarse, overgrown, choked with weeds.

Raphe

A seam-like ridge visible on the side of an anatropous ovule continuing the funicle (stalk) to the chalaza.

ANATROPOUS OVULE

Ratoon

(1) A new shoot springing from the base of a plant; a sucker.
(2) To cut a plant down to ground level to encourage the production of new shoots from the base.

Ray

(1) The fringe or outer petals of certain flowers, especially the ray florets of composite flowers: *see also* Ray floret.
(2) Radial strands of living cells in wood for the transport of nutrients.

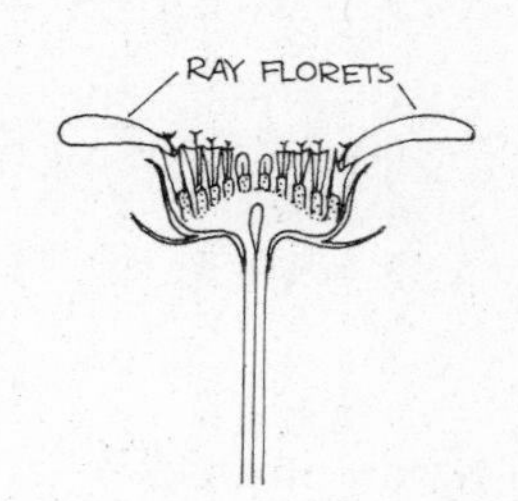

Ray Floret

One of the flowers of a composite flowerhead that form the outer ring, usually with strap-shaped petals as in the daisy family. Sometimes called a ray flower.

Ray Petal

One of the petals of a ray floret.

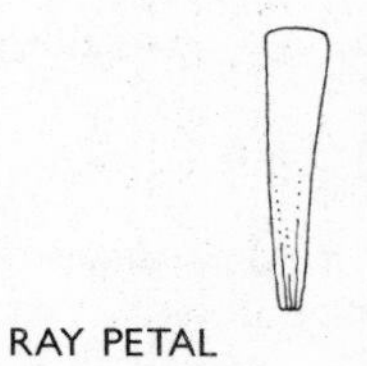
RAY PETAL

Reap

To cut down, as grain; to harvest.

Reap Hook

Another name for a sickle.

Rear

See Raise.

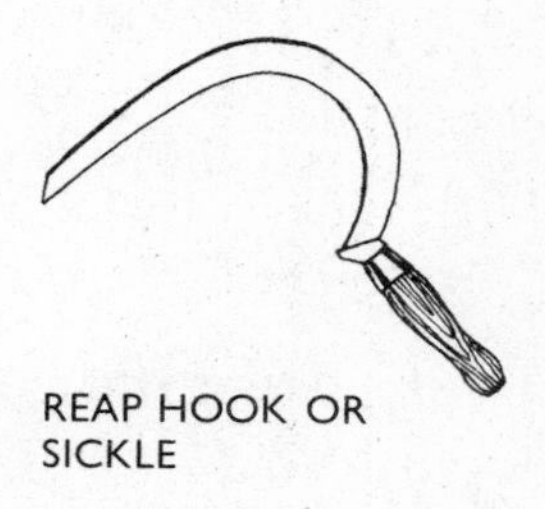
REAP HOOK OR SICKLE

Receptacle

(1) The thickened end of a stalk carrying all the parts of a flower, or supporting the crowded flowers of an inflorescence.
(2) In flowerless plants, a structure carrying the reproductive organs – spores or gemmae.

Recessive

An inherited character not apparent in the organism which inherits it although it may reappear in future generations. A recessive gene controlling such a character needs to be present in both members of a gene pair in order to express the character.

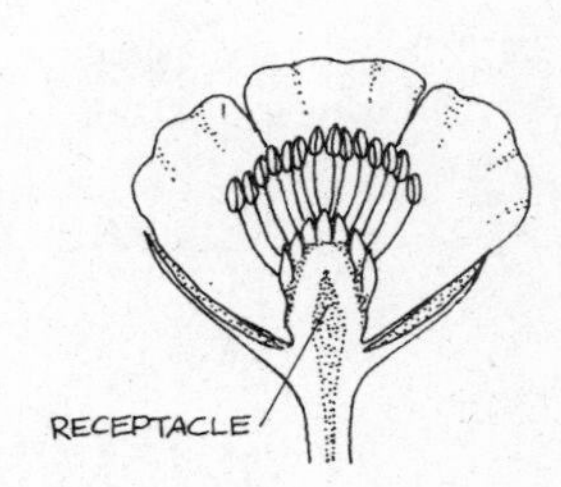

Reclinate

Reclining or bending (usually backward or downward).

Recombination

Genetic process whereby combinations of genes not present in either parent occur in their offspring.

Recumbent

Lying along the ground; flat; reclining.

Recurved

Bent backward; usually applied to types of flowerhead but can equally describe foliage. Recurved petals are gently curved, reflexed more severely so: *cf.* Incurved.

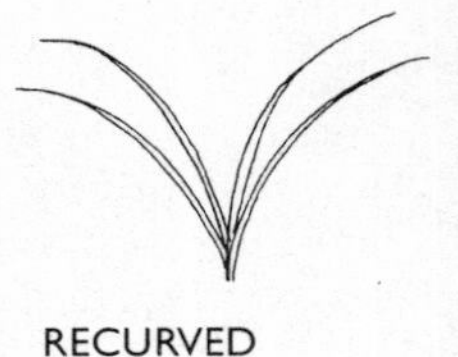

RECURVED

Red Thread

A fungus disease of lawns and grassland which produces irregular discoloured spots followed by reddish thread-like growth. It is usually fostered by a shortage of nitrogen in the soil.

Redolent

Having a strong aroma; strongly scented, not always pleasantly.

Reflexed

See Recurved.

REFLEXED

Regeneration

The growth of new tissues or organs to replace those lost or damaged, the basis of many horticultural propagation methods including various types of cutting.

Regma

A dry fruit that breaks up into single-seeded parts, each of which splits to release a seed, e.g. geranium.

Regular

See Radial.

Regulated Pruning

Otherwise known as saw-pruning. It involves removing all the crowded, crossing or intertwining and drooping branches, together with any weak growth.

Rejuvenation Pruning

A severe form of maintenance pruning whereby all the main branches are cut back to force new shoots to emerge.

Remontant

Flowering again; applied to a plant which flowers in flushes throughout the season, as some roses; also to strawberry plants which produce a second crop of fruits in the autumn.

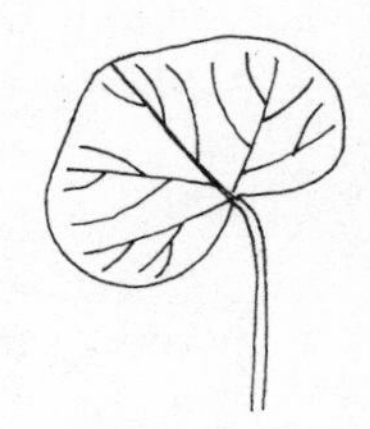

RENIFORM

Reniform

Kidney-shaped, usually applied to leaves.

Repand

Slightly wavy; undulating; usually refers to the margin of a leaf or petal.

Repand-denticulate

Slightly wavy and finely toothed or notched, usually applied to leaves but can include petals.

Repellent

A chemical product formulated to repel pests from plants, mainly used against birds and small mammals such as cats and dogs.

Repent

Growing along or just below the surface of the ground; creeping type of growth.

REPLUM

SILIQUA OF WALLFLOWER

Replum

A partition in a fruit formed by the ingrowth of the placentas.

Repotting

Transferring a plant from a small pot into a larger one.

Reproduction

The generation of new individuals by sexual, non-sexual or vegetative means.

Reptant

Crawling or creeping over the surface of the ground.

Reserve Border

A nursery bed or area where replacement plants are raised or stored until required.

Residual

Leaving a residue; applied to types of weedkiller that remain active in the soil over a long period.

Resinous

Covered in, producing or containing resin or a similarly sticky substance, e.g. horse-chestnut buds.

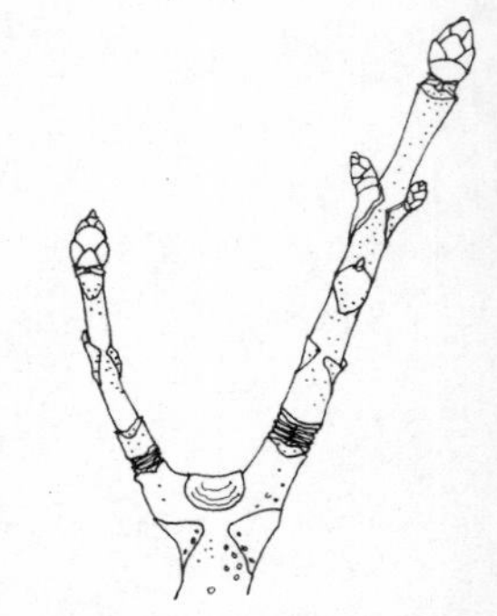

HORSE-CHESTNUT BUDS ARE RESINOUS

Resistant

Having powers or properties of resistance. Some plants are selected or specially bred to be resistant to specific diseases, e.g. rust on chrysanthemums. This renders them better able to ward off an attack but does not mean they are immune, a term applied only to a plant which is totally protected, whether naturally or by breeding.

Respiration

The process whereby plants absorb carbon dioxide from the atmosphere and expire oxygen during the day, absorbing oxygen and releasing carbon dioxide during darkness. Respiration is essential for the metabolism of the sugars created by photosynthesis to be converted into energy for the growth and wellbeing of the plant.

Resting Period

The phase, usually during the winter months, when some plants are either completely dormant or apparently so.

Resupinate

Upside down or inverted by twisting, as in the flowers of many orchids.

Retaining Wall

Any wall constructed to hold back soil, whether it be of brick, stone or any other material. Retaining walls can be

RETAINING WALL

a feature of a garden created on sloping land and terraces, or may simply separate one property from another.

Retarding

Specially treating plants to bring them into flower later than they would naturally. Hormone retardents are often used commercially to delay the flowering of fruit trees to avoid frost damage.

Reticulate Venation

Finely netted or net-like. In the case of the leaves of *Pelargonium peltatum* 'White Mesh' the virus emphasizes the reticulate venation which is already present. In grasses, where there is no main vein with branches, it is known as parallel venation.

Retrorse

Turned backward or downward.

Retuse

Slightly notched at the apex, as in some leaves and petals.

RETUSE

Reversion

(1) A virus disease of blackcurrants (*Ribes nigrum*) which results in smaller nettle-like coarse leaves, stronger growth and reduced crops.
(2) A plant reverting to its original form, as frequently occurs with intensely bred hybrids.

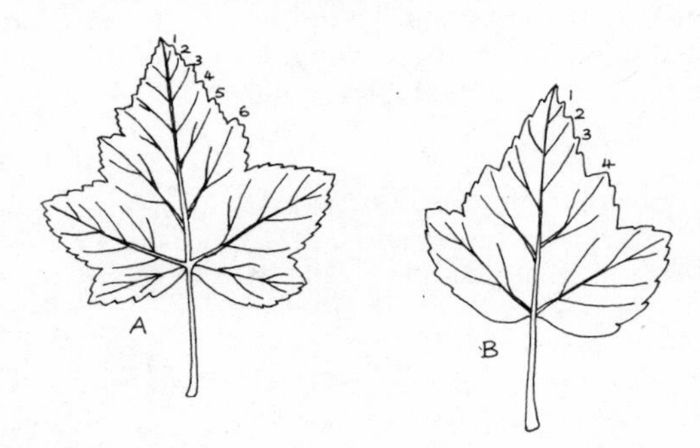

BLACKCURRANT LEAVES
A. HEALTHY
B. REVERTED

Revert

To return to its original state, as when a plain green leaf appears on a variegated plant.

Revolute

Rolled backward, outward or downward, usually applied to leaves or petals.

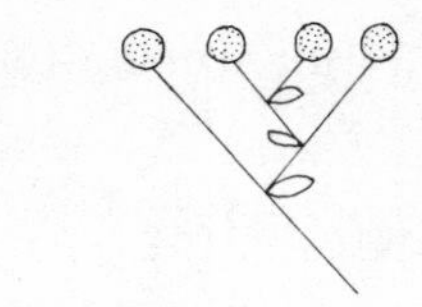
RHIPIDIUM

Rhachis

See Rachis.

Rhipidium

A fan-shaped cymose inflorescence, normally flattened in one plane, as in the iris species.

Rhizobium

A type of bacterium found in the roots of leguminous plants connected with the fixation of nitrogen. One of the many beneficial bacteria.

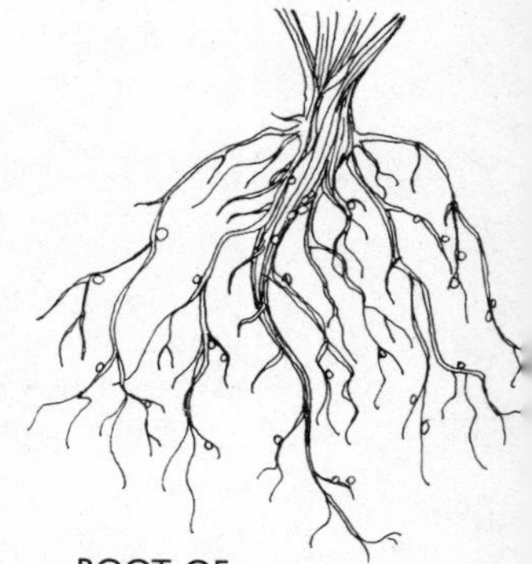
ROOT OF LEGUMINOUS PLANT SHOWING ROOT NODULES CONTAINING RHIZOBIUM

Rhizoid

(1) Anatomically different from a root but having a similar function and appearance.
(2) A delicate and usually colourless outgrowth in algae, etc. responsible for anchoring the plant to the substrate and absorbing water and mineral salts.

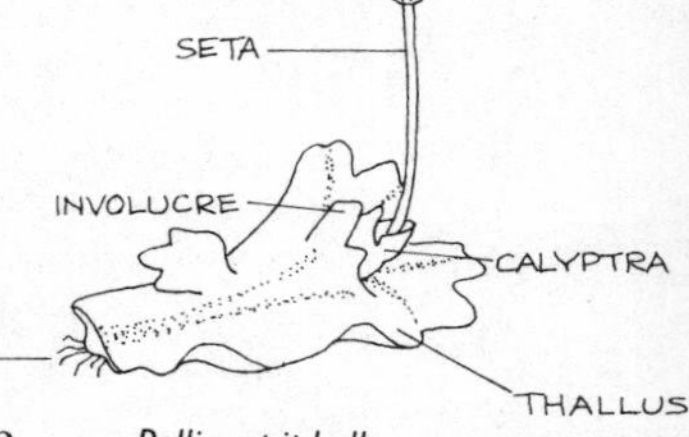

Pellia epiphylla

Rhizomatous

Having or producing rhizomes.

Rhizome

Normally a horizontal creeping stem lying on or under the ground from which shoots will arise and roots descend; a type of rootstock. The bearded iris is an example of an underground rhizome.

Rhizoplane

The surface of a root and the soil in immediate contact with it.

Rhizosphere

The soil surrounding the root system of a plant, usually subject to considerable microbiological activity.

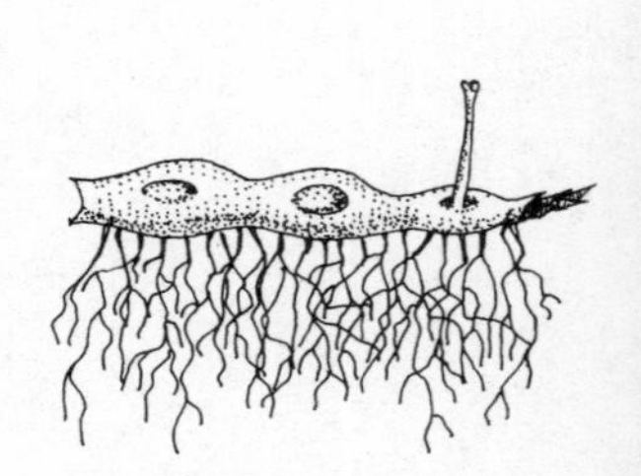
RHIZOME OF SOLOMON'S SEAL (*Polygonatum multiflorum*)

Rhomboid, Rhomboidal

Of a leaf shaped roughly like a parallelogram with opposing acute and obtuse angles.

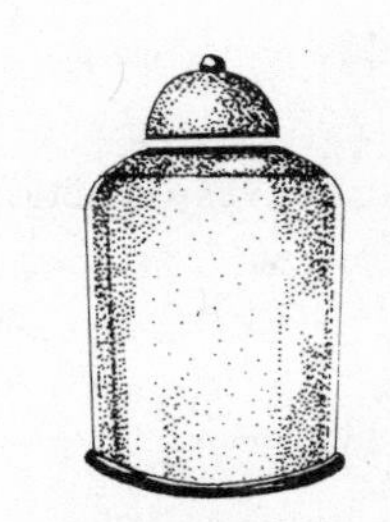
RHUBARB POT

Rhubarb Pot

A fairly tall earthenware vessel, open at the bottom and the top and fitted with a lid. Victorian gardeners placed them over rhubarb crowns in early spring to produce long tender stems. Today, rhubarb is usually forced commercially in light-proof sheds, but pots can occasionally be seen in private gardens, and are much sought by collectors of antiques.

RIDDLE

Rib

The primary vein in a leaf; can also be a ridge on a stem or fruit.

Riddle

Any type of sieve, but usually applied to the large coarse-meshed, often rectangular kinds used in commercial operations.

Ridger

A type of hoe with a V-shaped blade used as a miniature plough to create a furrow through the soil.

RIDGER HOE

Ridging

(1) A method of exposing soil to the elements over winter, thereby improving the texture.
(2) Raising soil in ridges for the planting of early crops.
(3) Drawing-up soil around the haulm of potatoes to prevent frost damage, or around the stems of celery to blanch them: *cf.* Trenching. *Same as* Earthing-up.

Rind

(1) The peel of a fruit or vegetable.
(2) The bark of a tree or plant.

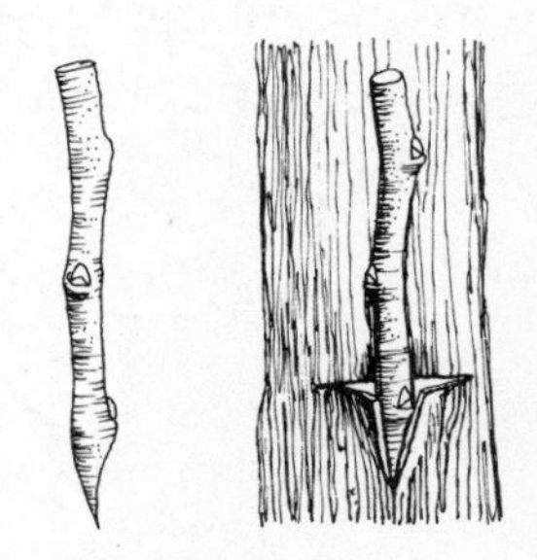
RIND GRAFTING

Rind Grafting

A method of inserting a scion under the bark or rind of a tree. Crown-grafting is a form of rind grafting. *See* Grafting.

Ring Culture

A method of growing certain types of plant, e.g. tomatoes, by which they are encouraged to produce a dual root system, one for nutrients, the other for water. Very popular 30–40 years ago but not so well thought of today.

RING CULTURE OF TOMATOES

Ringing

A method of restricting the growth of trees, particularly fruit trees, to increase cropping. A ring of bark not more than $\frac{1}{4}$ in (6 mm) wide is removed from the trunk about 2 ft (60 cm) above ground level in spring; this will callus over as the season progresses. A wider ring may well cause the death of the tree. In partial ringing, instead of a full ring being cut around the tree, two half-rings are removed on opposite sides of the trunk and about 6 in (15 cm) apart.

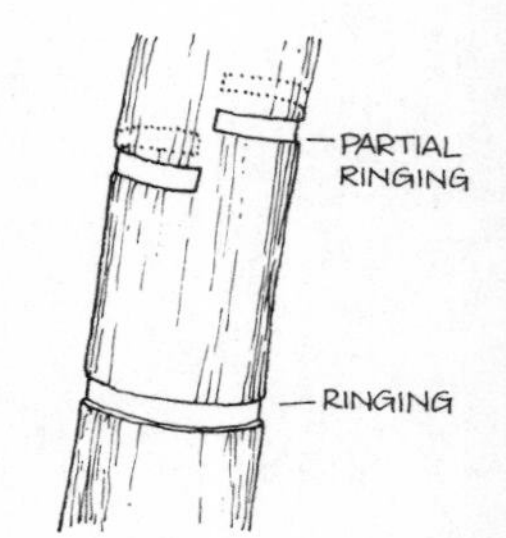

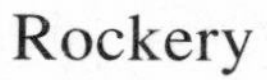

Rockery

Artificial (rarely natural) mound of rough stone, rocks and soil for growing rock plants, alpines, heaths, etc.

ROCKERY

Rod

(1) Slender, straight, round shoot growing on a tree (often from the base), and harvested for a variety of purposes including fishing rods, supporting canes, etc.
(2) An old unit of linear measurement: *see also* Perch and Pole.

RODS

Rogue

(1) A mutant; a plant which does not conform to type in either colour or habit.
(2) An inferior plant among a group of seedlings, or one that is different from the rest.

Rogueing

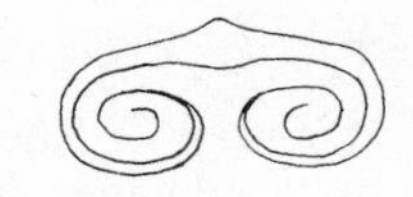

ROLLED

The removal and destruction of diseased, distorted, infected or sickly plants; especially applied to the careful checking of seedlings.

Rolled

Of leaf margins rolled inward towards the midrib; sometimes caused by leaf-roller caterpillars or certain other pests.

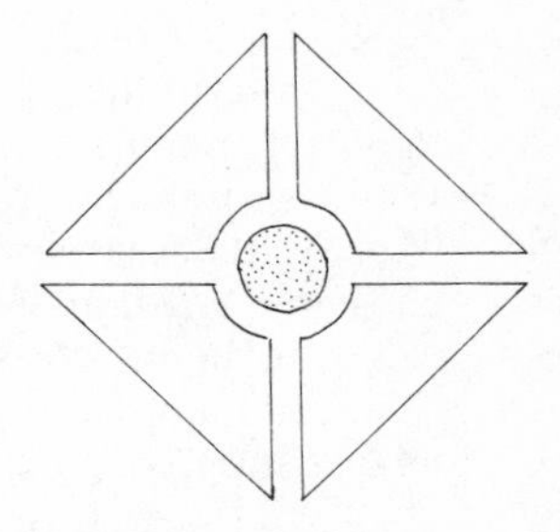

ROND POINT

Rond Point

The centre point of a formal garden layout from which avenues radiate. It is usually a circular flowerbed or paved area.

Roof Garden

A garden created on a flat roof, often above a block of flats or a departmental store, to which all materials and supplies have to be elevated.

Root

Usually the underground part of a plant. The main, easily visible part of the root-system is chiefly for anchoring the plant to the soil; nutrients and water are absorbed by the root-hairs which are generally invisible without the aid of a microscope.

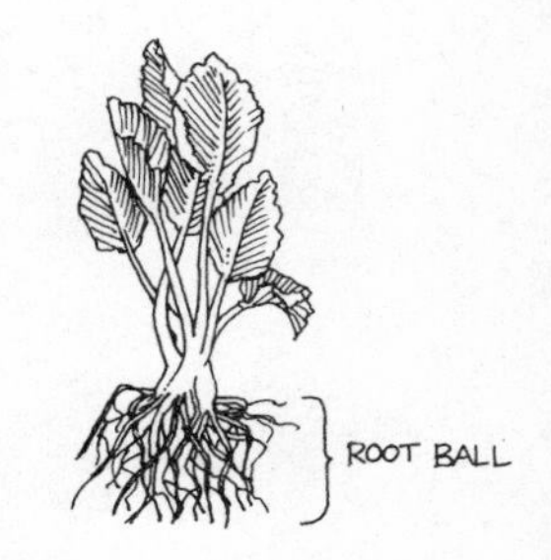

Root Aphid

Sap-sucking insect which attacks the roots of some vegetables and flowers. Control is by watering the plants with a suitable insecticide.

Root Ball

The matted roots of an established pot plant, tree, shrub or any other plant, including the soil enclosed by the roots.

Root-bound

See Pot-bound.

Root-cap

A cluster of cells at the tip of a root protecting the meristem.

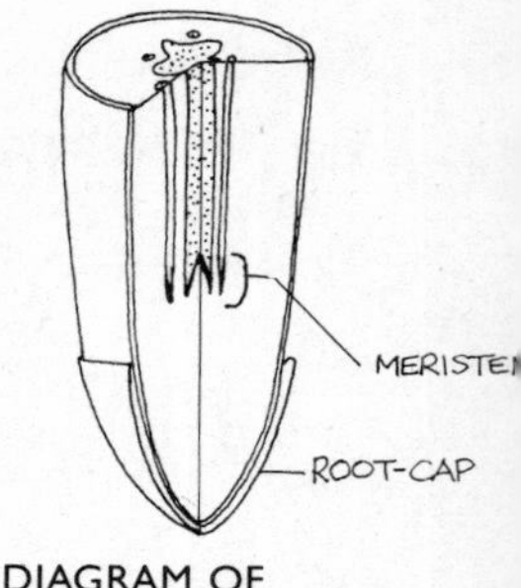

DIAGRAM OF ROOT TO SHOW ROOT-CAP

Root-climber

A plant that climbs by means of roots rather than tendrils such as ivy (*Hedera*) or Virginia creeper (*Ampelopsis hederacea*).

Root Crop

Plants grown for their roots; many vegetables, including potatoes, turnips, carrots, etc. Onions are not a root crop because the roots are not eaten.

Root-cuttings

Roots cut into segments from which new plants will develop. Mint (*Mentha*) is often propagated by this method.

Root-dipping

Immersing the roots of young plants in a solution of insecticide to protect against certain pests. Brassicas, for instance, are dipped or 'puddled' as a protection against club root disease.

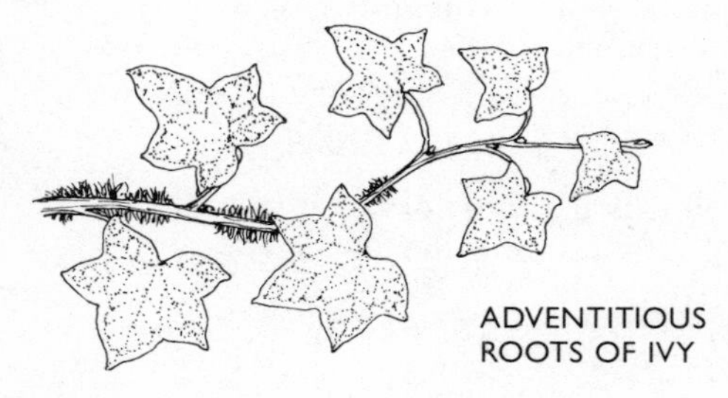
ADVENTITIOUS ROOTS OF IVY

Root Division

A method of increasing stock of a chosen plant by breaking off pieces of root from the parent plant. An old root can also be split into several sections, each eventually becoming a new plant.

ROOT-DIPPING OF BRASSICAS

Root-hairs

Extremely fine hairs which grow near the root tips by which the plant absorbs water and nutrients, usually invisible to the naked eye.

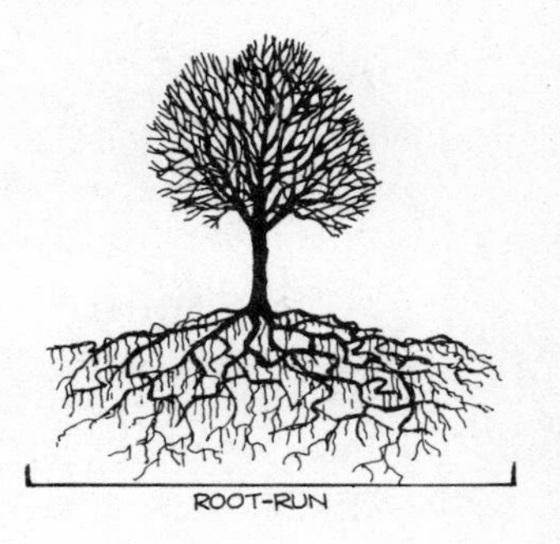

Root-knot Eelworm

A pest which attacks the roots of a range of plants including cucumbers and lettuce grown under glass, causing small swellings on their roots, so that the plants wilt.

Root Maggot

A general term occasionally used to describe the larvae of any fly which attacks roots.

Root Restriction

A method of encouraging more prolific flowering or fruiting, forcing the plant to restrict its growth and concentrate its energies elsewhere. Some subjects do better if their roots are restricted, including many bulbs and several kinds of pot plant.

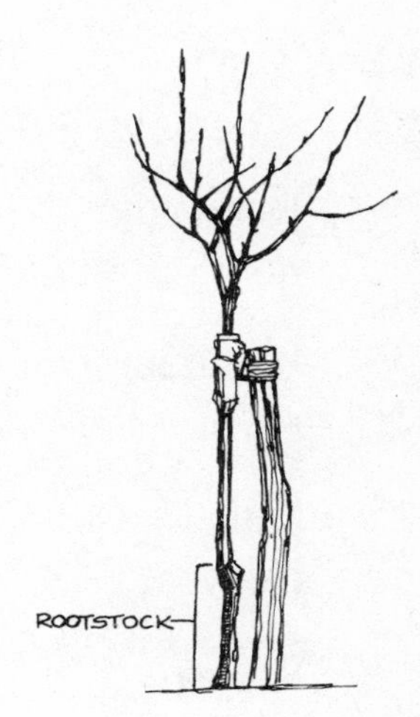

Root-rot

Condition often caused by excessive watering and waterlogged soil but may also be the result of attack by several kinds of disease organisms such as fungi.

Root-run

The area of soil occupied by the roots of a plant.

Rootstock

(1) An underground stem or rhizome.
(2) The lower part of the trunk of a fruit tree used entirely for its root system, on to which a fruit-bearing variety is grafted.
(3) Sometimes erroneously used to describe the crown and root system of a very short, compact plant.

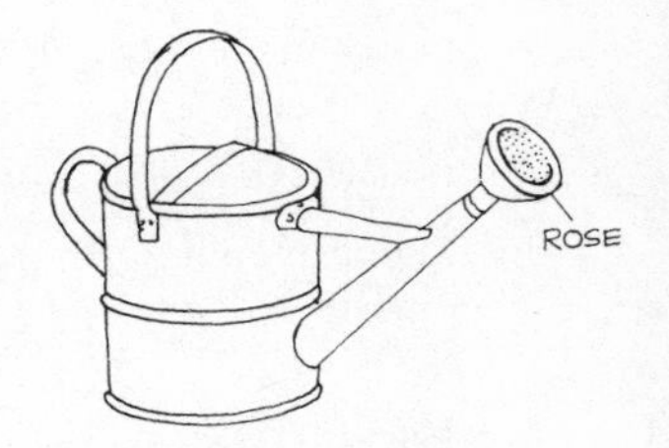

Rosary

A rose garden, or a rose bed in a mixed garden; a favourite expression in Victorian times but seldom heard today.

OVAL ROSE

Rose

The perforated spray-head attached to the spout of a watering can, or to the lance of a hose, to produce a fine shower of water.

ROUND ROSE

Rosette

Any formation of leaves radiating from a crown and usually close to soil-level, e.g. dandelion (*Taraxacum officinale*); also sometimes used to describe the corolla of a flower.

DAISY (*Bellis perennis*)

Rostellum

Short beak-shaped outgrowth on the stigma of certain flowers, e.g. many orchids and violets.

Rostrate

Having a beak, or beak-like in form.

RIBWORT (*Plantago lanceolata*)

TWO COMMON EXAMPLES OF ROSETTE LEAF FORMATION

Rostrum

A beak or beak-like part.

Rosulate

In rosette form.

Rot

Decay; decomposition; putrefaction.

Rotate

A wheel-shaped corolla consisting of petals joined, especially at the base, forming a flat disc.

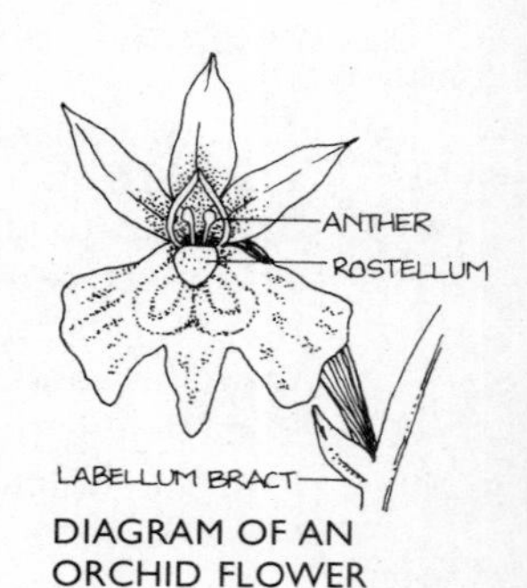

DIAGRAM OF AN ORCHID FLOWER

Rotation

Successional cropping of land to ensure maximum fertility.

Rotenone

The active ingredient in the insecticide Derris.

Rotting Agent

A compost heap additive used to accelerate the speed of decomposition.

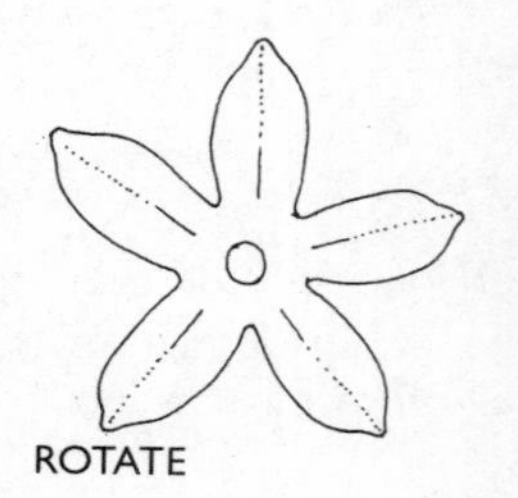

ROTATE

Rotund

Almost circular; rounded; nearly spherical.

Rough Leaves

The first true leaves produced on a seedling following the cotyledons. Cotyledons always have smooth plain margins, but the true leaves may be indented or toothed.

A ROTUND UTRICLE

Rubbery Wood

A virus disease of apple trees which causes the shoots to droop like a weeping willow. The variety 'Lord Lambourne' is rather prone to rubbery wood disease.

Rubiginous

Rust-coloured or rusted.

Ruderal

Of a plant growing in waste places or among rubbish near habitation, e.g. rosebay willowherb (*Epilobium angustifolium*).

Rudimentary

Undeveloped; vestigial; of organs which are non-functional or do not develop beyond the initial stage.

Rugose

Wrinkled or covered with wrinkles; corrugated. Usually applied to leaves with a wrinkled, puckered surface, e.g. a pepper (*Peperomia caperata*).

Rugulose

Finely wrinkled.

Ruminate

Appearing as though chewed or mottled, usually applied to seeds.

Run

(1) A line of plants such as a commercial row of peas, beans, etc.
(2) To bolt or flower prematurely, such as a lettuce which has run to seed.
(3) Plants which produce an abundance of creeping growth are said to run.

Runcinate

Of a leaf, coarsely serrated.

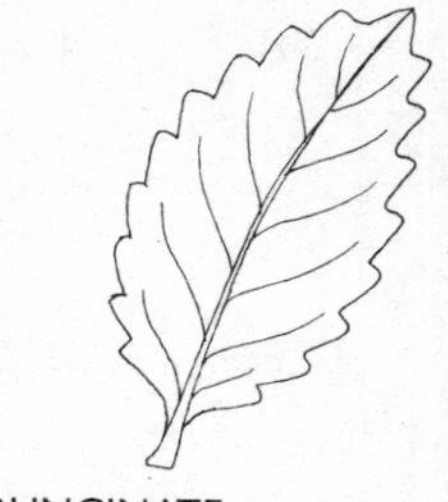
RUNCINATE

Runner

A slender stem extending from a parent plant on the surface of the ground and producing roots at the nodes (and eventually small plants) as with strawberries. A runner differs from a stolon which roots at the tip.

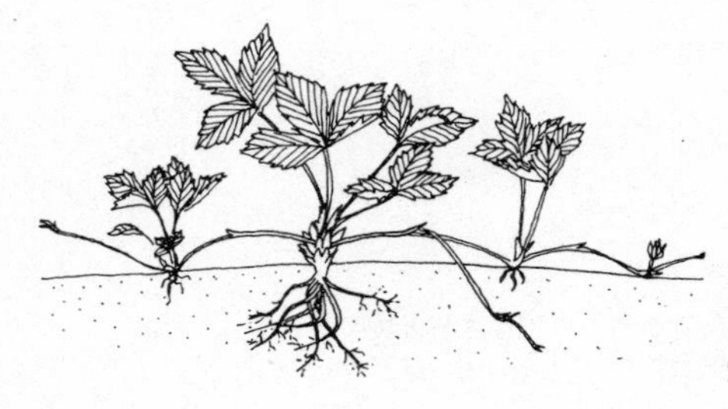
RUNNERS OF STRAWBERRY

Russet

An apple or pear with a naturally rough, brownish skin.

Rust

A plant disease showing as rust-coloured spots caused by fungus spores. A portmanteau word covering a number of different fungi, many of which include specific hosts, such as wheat rust, chrysanthemum rust, pelargonium rust, etc.

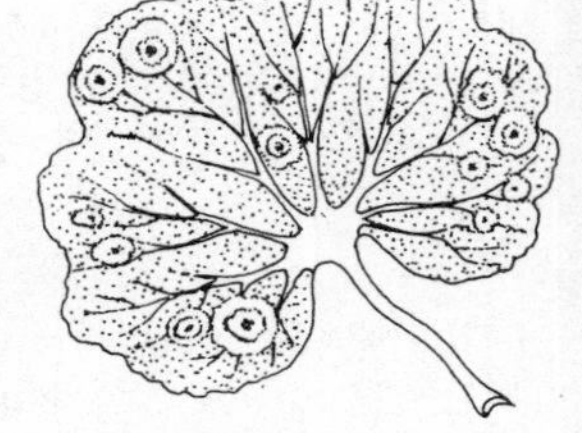
RUST SPORES ON THE UNDERSIDE OF A PELARGONIUM LEAF

Rustic Poles

Timber in its natural state, usually still carrying its bark, used for constructing pergolas, arches, fencing, garden furniture, etc.

RUSTIC SEAT

S

Saccate

Pouched or bag-shaped; enclosed in a sac.

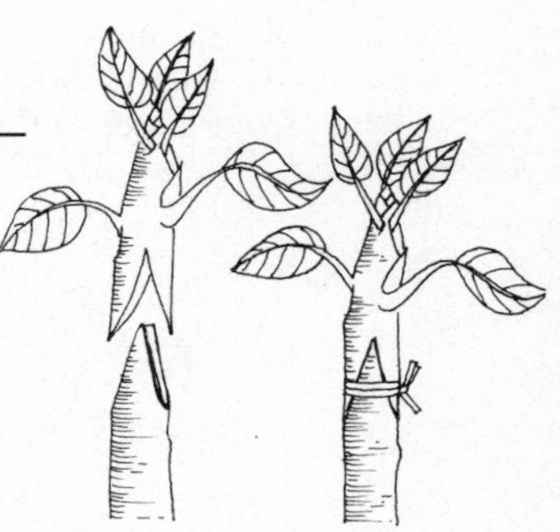
SADDLE-GRAFT

Sacculus, Saccule, *pl.* Sacculi, Saccules

A small sac or pouch.

Saddle-graft

A method of grafting used when the stock and scion are of the same circumference. The end of the scion is notched in the form of a 'V' and the stock shaped to fit it; they are then bound together and sealed. An inverted saddle-graft is when the stock has the 'V' cut in it and the scion is pointed to fit. *See* Grafting.

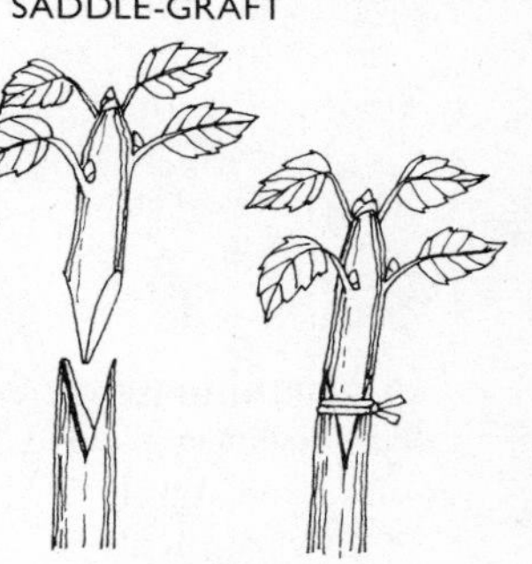
INVERTED SADDLE-GRAFT

Sagittate

Of leaves that are arrow-shaped, roughly triangular with the basal lobes pointing downward.

Salicetum

A group or collection of willows, once popularly called a sally-garden or sally-wood.

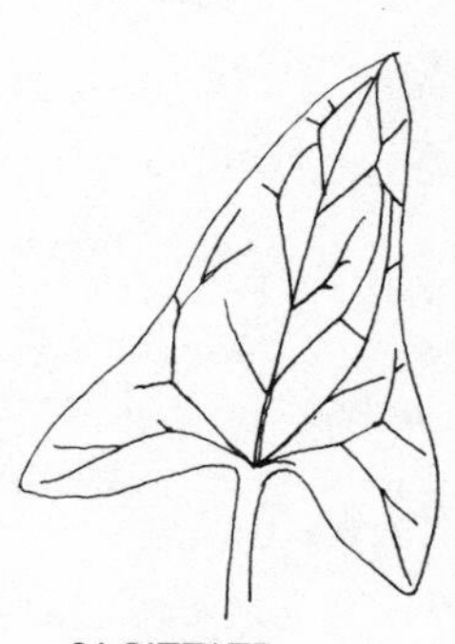
SAGITTATE

Saline

Salty; impregnated with salt as seashore sand, etc.

Salt

Substance formed from an acid when all or part of its hydrogen is replaced by a metal or metallic radical.

Salverform

Normally applied to a corolla having a long tube formed by petals joined at least at the base and limbs spread out flat.

SALVERFORM

Samara

Winged dry indehiscent fruit, single as in elm (*Ulmus* species), double as in maple and sycamore (*Acer* species): *cf.* Key.

SINGLE SAMARA (DUTCH ELM)

Sampler

A young tree or sapling left standing after adjacent trees have been cut down, especially in forestry.

Sand

An ingredient, to a greater or lesser extent, of most soils, which, although having no food value, is nonetheless useful for aeration. If used in potting mixtures it must be sharp sand; soft builder's sand tends to clog and consolidate the soil.

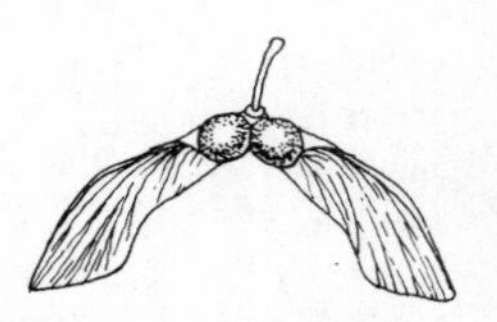
DOUBLE SAMARA (SYCAMORE)

Sandstone

A natural lime-free stone which weathers well in the garden and has good water-retaining properties.

Sap

(1) The fluid or juice in plants.
(2) The soft layers of wood growing between the bark of trees and the heartwood; alburnum.

Sap-sucker

Any insect that sucks the sap of a plant, such as an aphid.

SAP-SUCKING APHIDS

Sapling

A young tree; a tree at any stage before the heartwood has hardened.

Saponins

Group of soap-like compounds found in many plants, often toxic; frequently used medicinally, e.g. digitalin from the foxglove.

Sappy

Lush, soft growth of shoots and foliage, usually the result of excessive nitrogen in the feed. Sappy plants are more than usually vulnerable to pest and disease attack, and flowering or fruiting will be below normal.

Saprophyte

A plant that feeds upon decaying organic matter, lacking chlorophyll, e.g. fungi.

Sapwood

The outer annual rings in tree trunks through which water and nutrients are conducted from the roots upward: *cf.* Springwood, Summerwood.

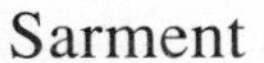

Sarment

A long, weak, flimsy twig.

Sarmentose

Producing runners or stolons.

Sarmentum, *pl.* Sarmenta

A runner or stolon.

Saxatile

A plant which grows among rocks or in rocky, arid situations.

Scab

Parasitic disease of plants causing scab-like rough areas or surfaces. Sometimes seen on potatoes, apples, etc.

A SAXATILE PLANT
(*Saxifraga oppositifolia*)

Scaberulous

Slightly rough or scabrous.

Scabrid

Rough or scabby, somewhat scabrous.

Scabridulous

Slightly or minutely scabrous.

Scabrous

Rough to the touch, harsh, often with projections such as small warts or pimples.

Scalding

Scorching or blotching of the leaves of greenhouse plants in hot weather, caused by the sun's rays falling upon wet foliage.

Scale

(1) A small, usually dry leaf or bract.
(2) An insect of the homopterous family Coccidae which attaches itself to a plant and exudes a waxy protective shield.
(3) A fern, the back of whose leaves are densely covered with rusty-orange scales.
(4) A small leaf attached to and protecting a bud.
(5) The fleshy segments of a bulb, such as a hyacinth or onion.
(6) Part of the cone of a coniferous tree.

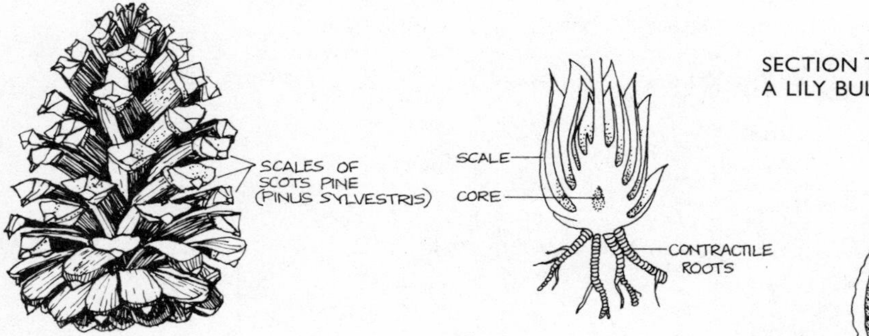

SECTION THROUGH A LILY BULB

Scale Insect

Small insect with a hard protective coat which attacks a wide range of plants both inside the greenhouse and in the garden.

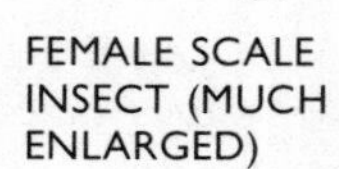

FEMALE SCALE INSECT (MUCH ENLARGED)

Scalloped

Of leaves with crenate margins, having rounded teeth or notches.

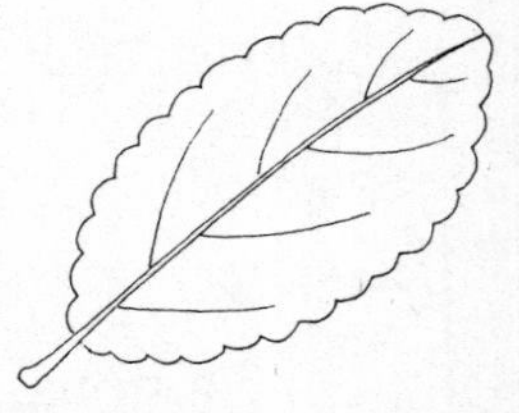

SCALLOPED OR CRENATE

Scandent

Climbing unassisted by tendrils, often by intertwining long flexible stems through other plants.

Scape

A leafless flowerstem arising directly from the soil. It may bear one or several flowers. Foliage may be present but is not attached to the scape.

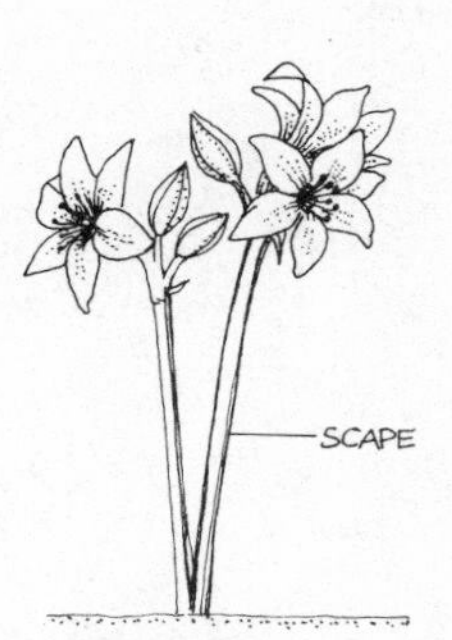

Amaryllis belladonna

Scaphoid

Boat-shaped.

Scar

The hilum on a seed; the mark on a plant, tree, etc. left when a leaf falls. Flower loss or accidental damage can also leave scars.

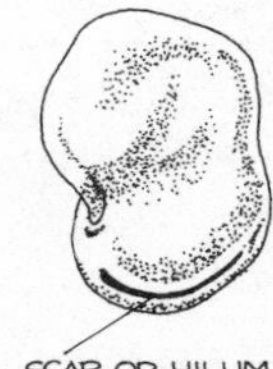
SCAR OR HILUM ON BROAD BEAN

Scarifier

A machine or implement with prongs for loosening the surface of the soil; a particular type of rake used for removing moss from lawns. *See* Rake.

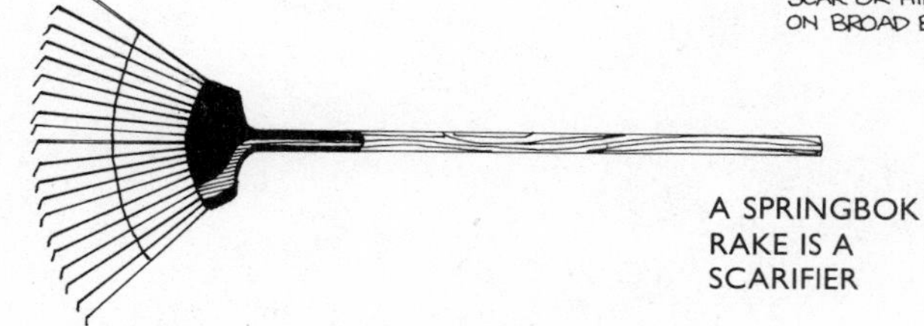
A SPRINGBOK RAKE IS A SCARIFIER

Scarify

To roughen, slightly score or scrape the hard outer coating of a seed before sowing to aid the absorption of moisture and thus accelerate germination.

Scarious

Thin, dry, stiff and membranous. Having no green colouring, often translucent and papery; with a dried-up appearance.

Scatter

To throw (seed) or distribute haphazardly; diffuse in various directions; strew here and there in a disorganized manner.

Schizocarp

A dry, dehiscent fruit which splits into two mericarps, e.g. carrot seed.

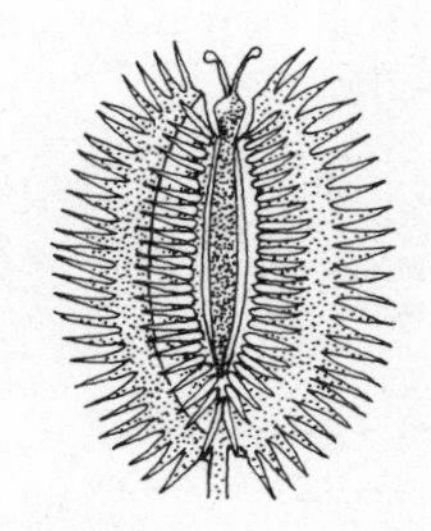
SCHIZOCARP (CARROT)

Schizogeny

The separation of plant cells to form a cavity surrounded by the intact cells, in which secretions accumulate, such as the resins of some conifers.

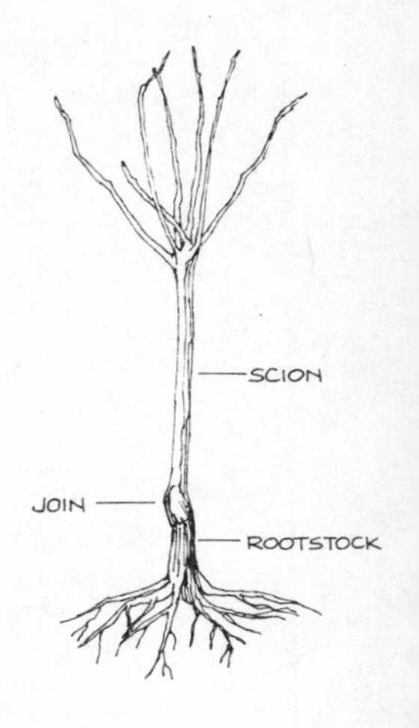

Schizophyta

Plants that multiply by fission, including blue-green algae and bacteria.

Sciarid Flies

Also called fungus gnats (in USA) and peat flies, which can be a pest in greenhouses if not severely controlled. The larvae eat up into the stems of plants, causing rapid collapse. They are especially troublesome when peat-based composts are used.

Scion

A detached shoot, bud or cutting which is grafted on to a rootstock of another variety, as in the case of many fruit trees.

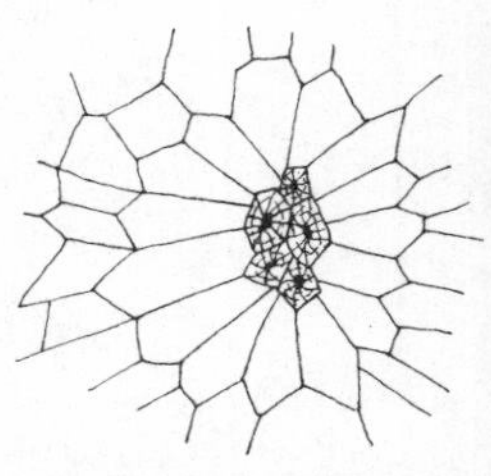

SCLERENCHYMATOUS CELLS IN PEAR FRUIT

Sclerenchyma

(1) Dead plant tissue with thick, woody cell-walls, providing the plant with rigidity and strength, and conducting water. Often encountered in the gritty flesh of pears.
(2) The material of a nut-shell or seed-coat.

Sclerocaulous

Possessing a hard, dry stem.

Sclerogen

The hard lignified matter found on the inner surface of plant cells.

Sclerophyll

A hard, stiff leaf.

Sclerosis

Hardening of the cell-wall by lignification; woodiness.

Sclerotinia Rot

The correct term for stem rot.

Sclerotium

A compacted, usually more or less spherical mass of hyphae.

Scorching

(1) The effect upon foliage of excessive sunlight, especially under glass if the foliage of the plants is wet.
(2) Marks on the foliage caused by chemical fertilizer coming into contact with the leaves.

Scorpioid

A form of cymose inflorescence curved to one side like a scorpion's tail.

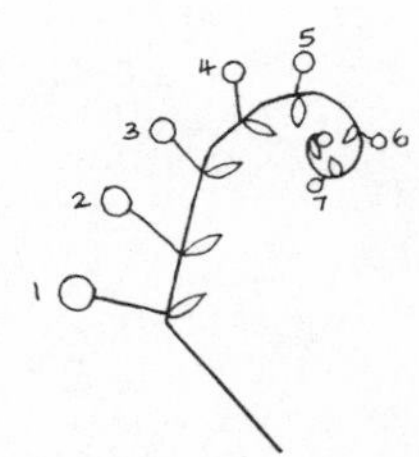

SCORPIOID CYME (THE NUMBERS INDICATE ORDER OF FLOWERING)

Scrambler

A plant with sprawling, creeping habit, usually assisted by having thorns or tendrils, e.g. bramble.

Scree

In nature, the dry debris which collects at the foot and the lower slopes of mountains. In horticulture, any free-draining mixture of coarse gravel, small stones and sharp sand which, together with a little soil, is used mainly as a growing medium for alpines and other plants that require exceptionally good drainage.

Screen

(1) A large sieve or riddle used to separate the coarser from finer particles of materials such as sand, gravel, etc. Usually applied to a sieve with a fairly open mesh.
(2) A fence, hedge or wall, shelterbelt or windbreak. Ornamental screens can be of trellis, wire, rustic work, etc.

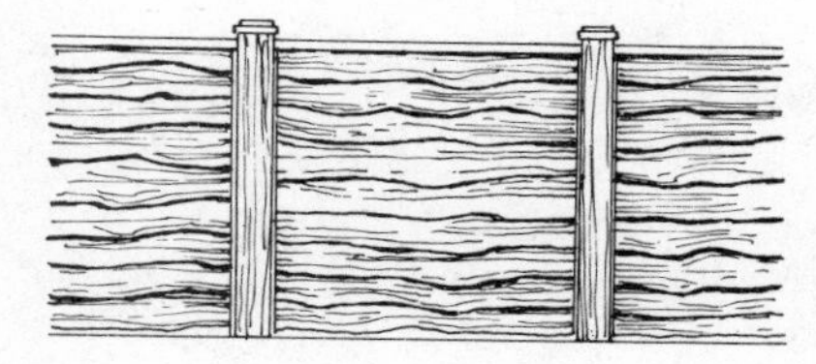
WEATHERBOARD SCREEN

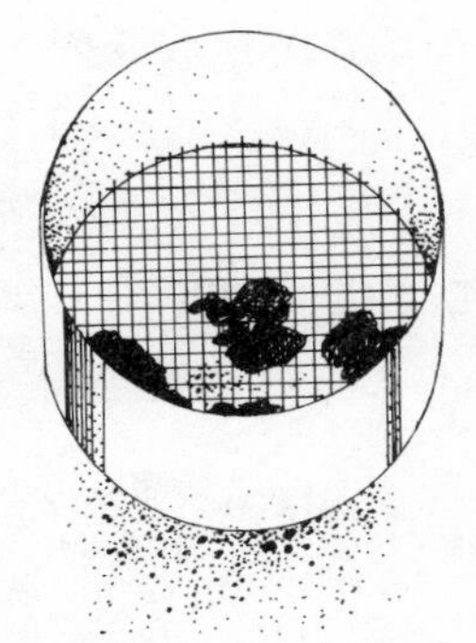
SCREEN

Screenings

Material which has been passed through the screen or sieve, or the residues from screening.

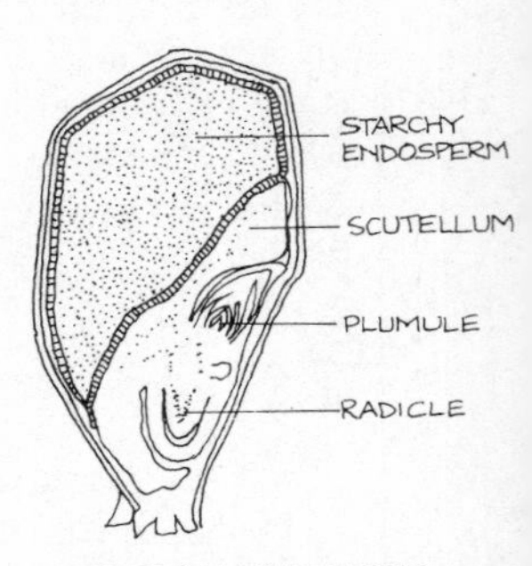

DIAGRAM THROUGH THE EMBRYO OF MAIZE GRAIN

Scrim

A loosely woven cotton material formerly used for shading greenhouses, rather like coarse window-netting; now largely superseded by plastic products.

Scrub

Uncultivated land (often waste land) covered with an assortment of plants but dominated by low shrubs such as gorse. An overgrown area which has reverted totally or partially to nature.

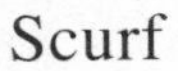

Scurf

The scaly covering of leaves and stems of bromeliads which gives them a silvery appearance; technically known as lepidate leaves or stems.

Scurfy

Covered with tiny bran-like scales.

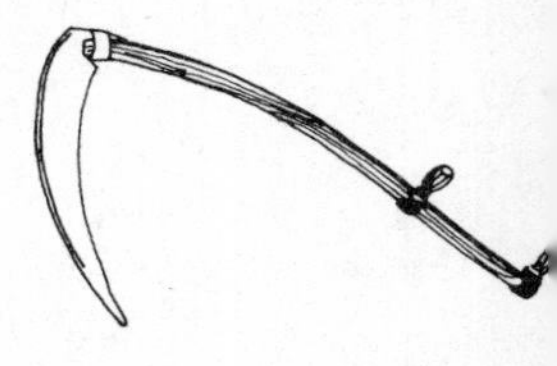

SCYTHE

Scutate

Shield-shaped, usually applied to leaves.

Scutcher

The striking part of a threshing machine.

Scutellum

(1) A cotyledon by which a grass embryo absorbs the endosperm.
(2) A small scale or similar membrane on a plant.

Scythe

Long-handled implement for cutting and reaping, with a narrow slightly curved blade, used with long sweeping strokes.

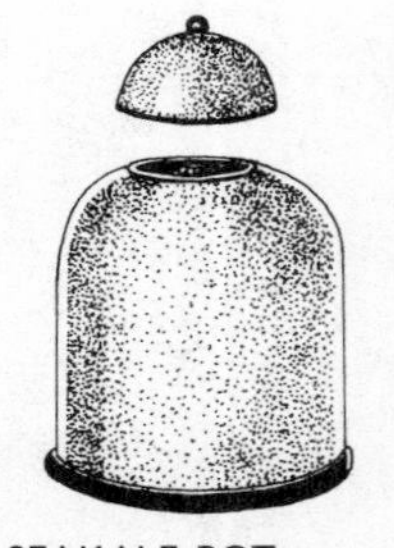

SEAKALE POT

Seakale Pot

An earthenware pot much used in Victorian England for blanching seakale. Smaller than a rhubarb pot but of similar design with removable lid.

Sealing Agent

Grafting clay, grafting wax, etc. applied to freshly cut or damaged surfaces to prevent the entry of fungus spores, wood-boring insects, etc. A sealing agent must be waterproof and elastic so that the plant's natural callusing process can proceed normally underneath.

Seaweed

A high-potash fertilizer, especially the strap-shaped kelps (*Laminaria*). It is best composted for six months before use, or it can be dried and stored indefinitely.

Secateurs

Hand-held pruning shears (by which name they are so called in USA) with scissor-action.

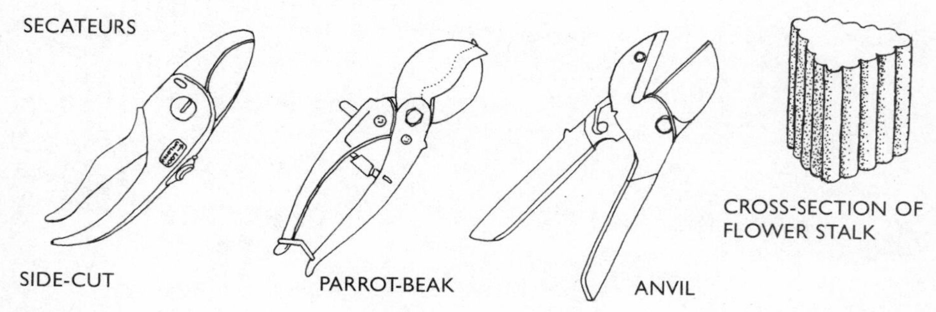

Second Soil

The soil remaining after seedlings have been planted out, or from pots when the plant has died. This soil may contain disease and should not be reused; it should be put on to the compost heap.

Secund

Having all parts turned to the same side.

Secundine

Usually the inner coat of an ovule, rarely the outer and second formed.

Sedge

Any species of *Carex* or other plant of the Cyperaceae family distinguished by its triangular stems and unslit leaf-sheaths. Sedges normally grow in damp areas.

SEDGE (*Carex panicea*)

Seed

(1) Fertilized and ripened ovule of a flowering plant containing the embryo from which a new plant may develop.

(2) To go (or run) to seed, applied to plants which have matured and produced seed; lettuce, etc. which have bolted.

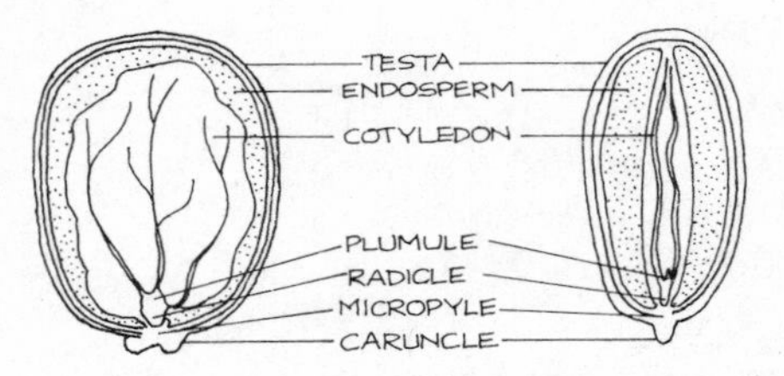

CASTOR OIL SEED (EMBRYO)

Seed Bed

An area of land specially prepared for seed sowing, usually by raking to a fine tilth.

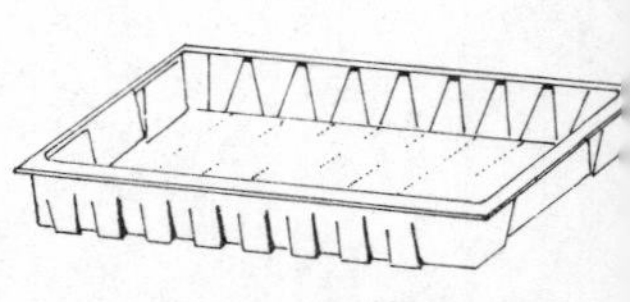

PLASTIC SEED TRAY

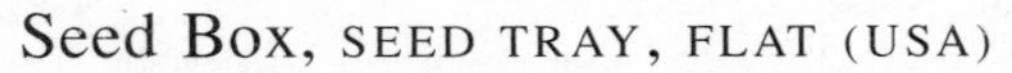

Seed Box, SEED TRAY, FLAT (USA)

A rectangular container in which seeds can be raised. The standard British seed tray is about $14 \times 8\frac{1}{2} \times 2$–3 in (35.5 $\times$ 21.5 $\times$ 5–7.5 cm) deep and they are obtainable in wood or plastic.

Seed Dispersal

The various ways in which plants spread their seeds, such as the 'parachutes' of thistle and dandelion, the 'propellers' of sycamore, and the bristles of goosegrass, etc.

Seed Dressing

Treating seeds before sowing to deter attack by soil-borne and other pests. In the past peas and beans were dipped in paraffin, then coated with red lead to discourage birds and mice, a practice which is now illegal.

Seed Drill

(1) A mechanical implement used for the efficient sowing of seeds, especially when large areas, such as cornfields, are involved. The seeds are contained in a hopper and released in a regulated manner as the

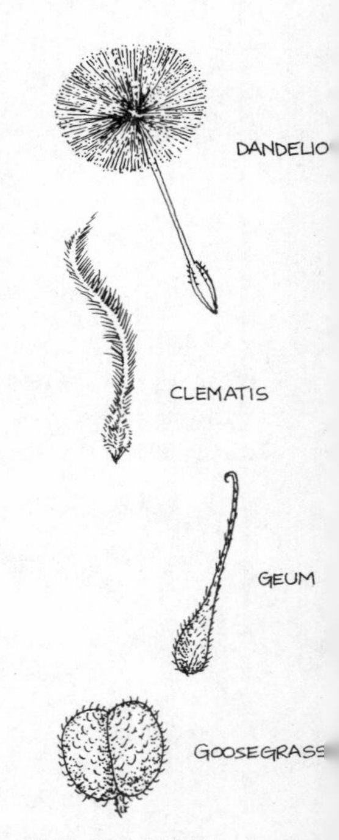

SEED DISPERSAL

wheels rotate. Smaller hand-held gadgets are available for gardeners but these tend to be more hit-and-miss.

(2) A furrow drawn in the soil to receive seeds.

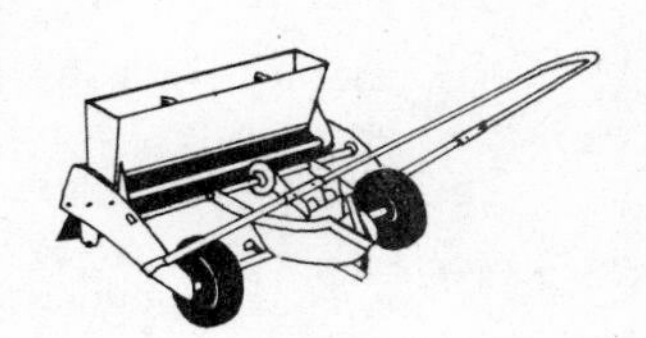
SEED DRILL

Seed Head

Any usually dry fruit that contains ripe seeds.

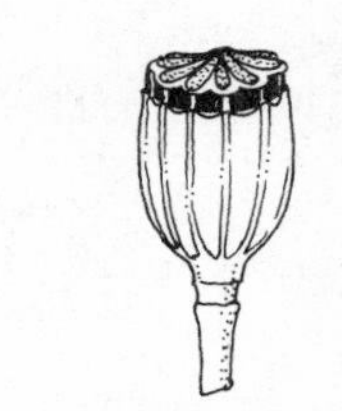
SEED HEAD (POPPY)

Seed Leaf

Also known as cotyledon. The first leaf or leaves produced by a seed after germination. In some plants, e.g. peas, the seed leaves remain below ground.

Seed Stalk

See Funicle.

Seed Viability

A seed's capacity for living or germinating. Very few seeds have 100 per cent viability, and it lessens as the seed ages. The Seeds Act of 1920 requires all vegetable seeds to conform to prescribed minimum germination percentages, but flower seeds are not covered by the Act.

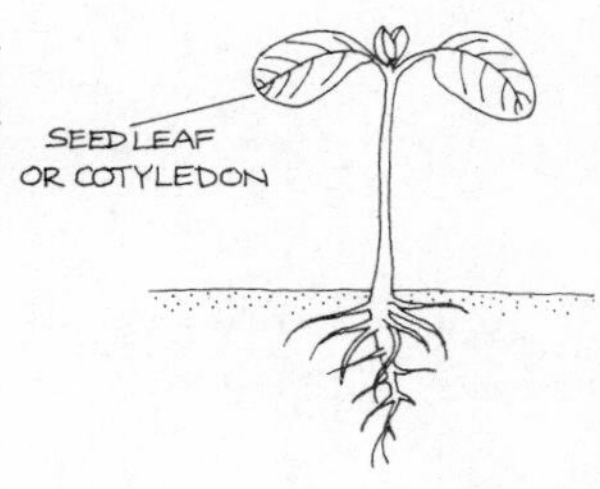

Seedling

(1) The young plant that develops from a germinating seed.

(2) A plant raised from seed as distinct from a cutting or scion.

Segment

A lobe of a leaf or part of a petal, corolla or perianth, not quite separated.

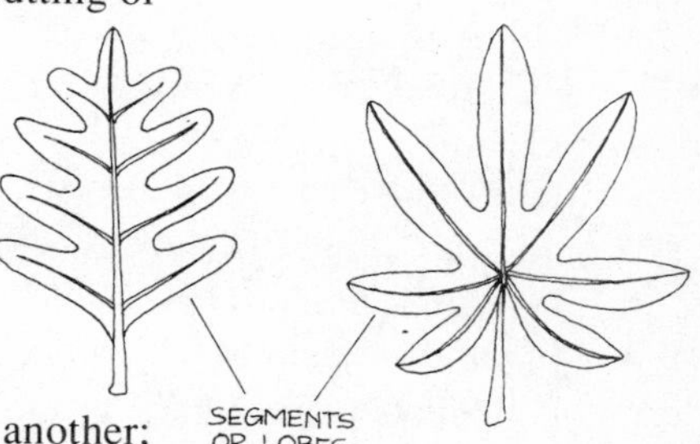

Segregated

Separated from the parent plant or from one another; species separated from aggregate species.

Segregation

The genetic process whereby the two alleles of a gene pair pass into different gametes and therefore into different individual offspring.

Selection

The method used by hybridists to improve the stock of certain plants, whereby they gather seed from only the best plants and are enabled over many years to produce superior strains or varieties. George Russell's work on lupins is an outstanding example.

Selective Weedkillers

Hormone weedkillers which affect the growth-process of selected weeds while being harmless to surrounding plants. One of the most familiar of the selective weed-killers is 2,4–D which kills broadleaved weeds in lawns without affecting the grass. A much wider range is available to the farming community.

Self

A florist's term used to describe a flower with blooms of a single colour throughout, not variegated, veined or shaded. *Same as* Self-coloured.

Self-clinging

Plants which produce aerial roots or adhesive tendrils, and can therefore climb without artificial assistance, e.g. Virginia creeper (*Vitis hederacea*).

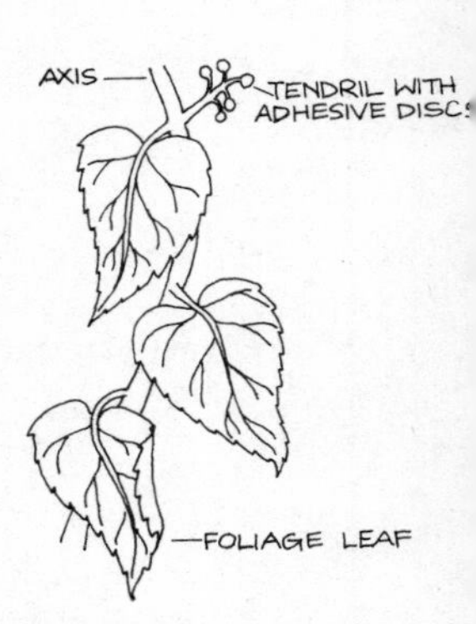

SELF-CLINGING
Ampelopsis

Self-coloured

See Self.

Self-fertile

A plant which does not need another to act as pollinator but is able to pollinate its own flowers: *cf.* Self-sterile.

Self-incompatible

Plants incapable of self-fertilization, often through the incomplete development of the pollen tube.

Self-pollination

The natural pollination of a plant with its own pollen.

Self-seed, Self-seeded

Seedlings produced naturally, usually around or close to the parent plant.

Self-sterile

A plant which is unable to produce fruit without the use of pollen from another. A plant whose flowers are sterile to their own pollen: *cf.* Self-fertile.

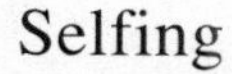

Selfing

Artificially pollinating a plant with its own pollen.

Semen

Seed of flowering plants.

Semi-annual

A plant lasting for half a year only, or one that flowers for this period.

Semi-double

Of a flower having a few more than the normal number of petals; between a single and a double. Actually a contradiction in terms.

THREE DIFFERENT CAMELLIA FLOWERS

Semi-evergreen

A plant which sheds most but not all its leaves in winter, or is deciduous only in extreme conditions.

Semi-parasite

A parasitic plant which does not depend entirely upon its host for survival, having some green (chlorophyll) in its leaves, but extracts water and minerals from its host. e.g. mistletoe (*Viscum album*).

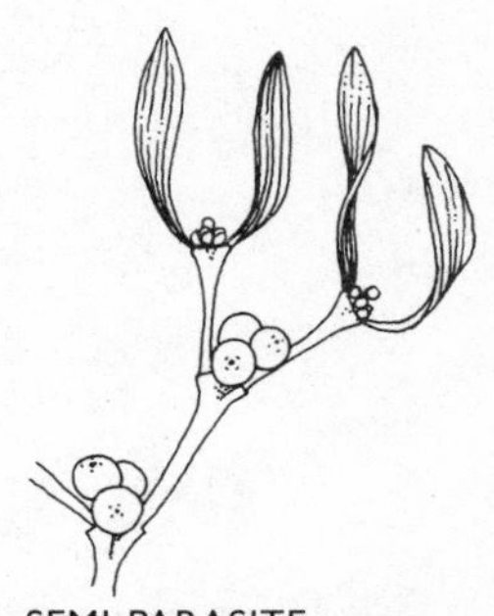

SEMI-PARASITE (MISTLETOE)

Semi-ripe

Type of cutting used for the propagation of many shrubs and similar woody plants such as heaths and heathers.

Seminal

Of seed. Increase by seed or seeding is called seminal propagation, and a structure containing seed is a seminal organ.

Seminal Root

The initial root development from the seed.

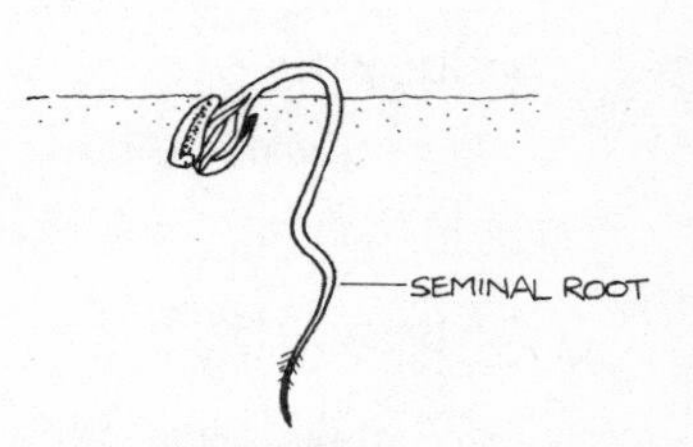

SUNFLOWER SEED

Seminate

To sow or propagate from seed.

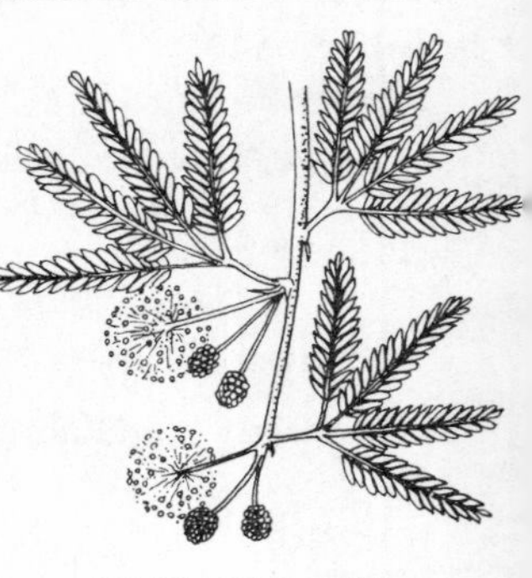

SENSITIVE PLANT
(*Mimosa pudica*)

Senescence

The ageing process in plants (and animals); the changes that occur between maturity and death. Not all parts of an organism age at the same rate, and in deciduous trees the loss of senescent leaves in autumn is a normal process repeated throughout the lifespan of the tree.

Senescent

Old, past its prime, moribund. Said of a stock plant which has been kept beyond its useful lifetime, or a tree which is dying.

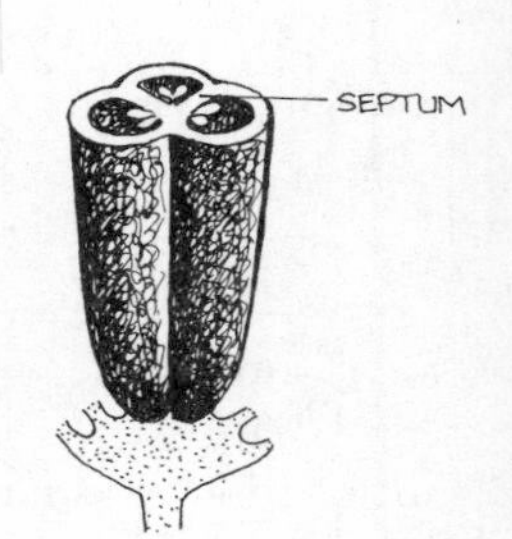

DIAGRAM OF A SEPTATE OVARY (TULIP)

Sensitive

Plants which respond to touch such as the sensitive plant (*Mimosa pudica*); stamens which move when touched by a pollinating insect; the leaves on some insectivorous plants which trap insects coming into contact with them.

Sepal

A part of a flower calyx, usually at the base of the flower protecting the petals, often like a small leaf.

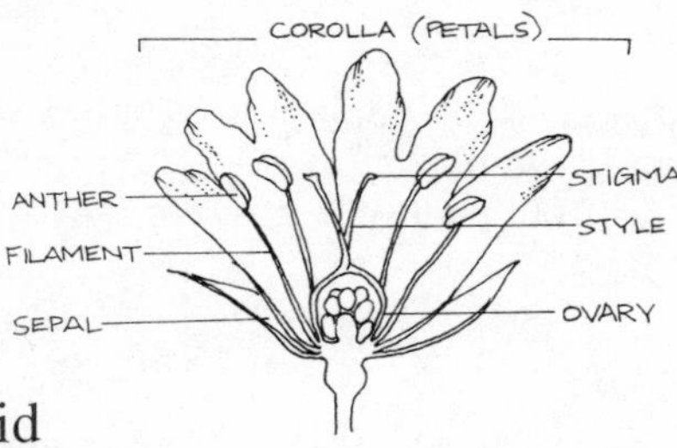

SEPTICIDAL CAPSULE (FOXGLOVE)

Sepaloid

Resembling a sepal, part of a flower calyx.

Septate

Partitioned, divided by partitions. Refers mainly to ovaries.

Septicidal

Fruits that when ripe split open longitudinally through the septa, separating the carpels.

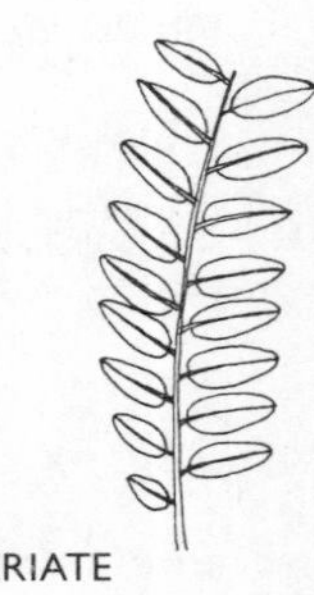

SERIATE

Septum, *pl.* Septa

A partition or dividing wall of ovaries.

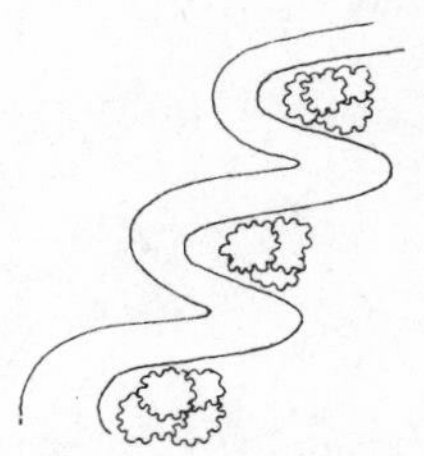

SERPENTINE

Sequestrol

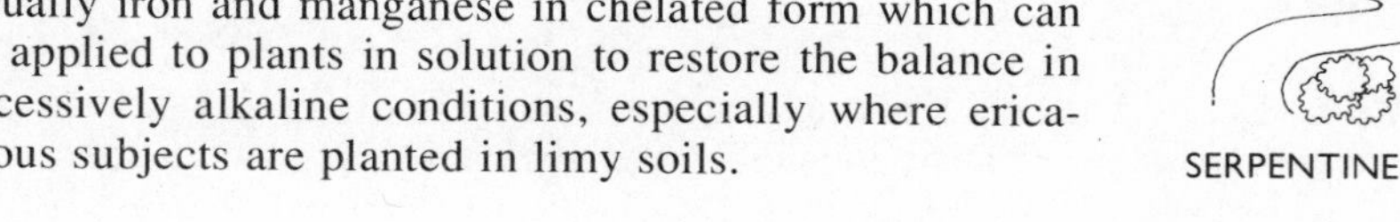

Usually iron and manganese in chelated form which can be applied to plants in solution to restore the balance in excessively alkaline conditions, especially where ericaceous subjects are planted in limy soils.

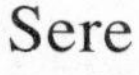

Sere

Withered or dried; desiccated.

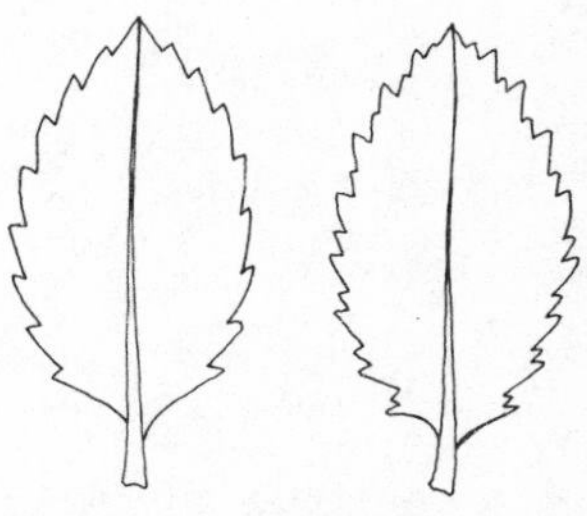

SERRATE DOUBLY-SERRATE

Seriate

Arranged in rows or series, such as leaves on a stem.

Sericeous

Silky, covered with soft silky hairs or having a silky sheen.

Series

A group of similar but not identical plants, linked by some common features. A term usually applied to annuals.

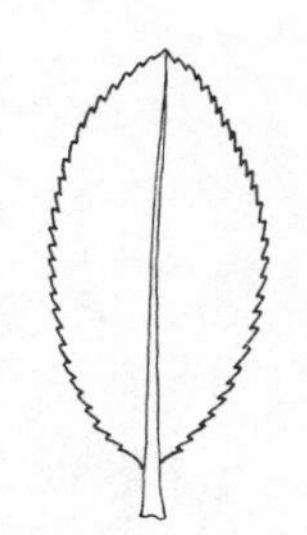

SERRULATE

Serpentine

Snake-like, sinuous, as of walls, borders, paths, hedges, etc. constructed in this manner. Serpentine walls need no buttresses, and provide sheltered spots for tender subjects.

Serrate, Serrated

Applied to leaf margins notched like a saw with forward-facing teeth.

Serrulate, Serrulated

Of leaves that are minutely serrated or notched.

STEM WITH SESSILE FLOWERS

Sessile

Having no stalk; plants with flowers which arise directly from the base, leaves without petioles, stigmas without styles. The oak tree (*Quercus robur*), for instance, bears leaves that are attached directly to the twigs: *cf.* Petiolate.

Set

(1) Onions, shallots, potatoes, etc. used as seed.
(2) Blossom which has been successfully fertilized.
(3) A slip or shoot for planting.

ONION SET

Seta

(1) A bristle or bristle-like structure.
(2) The stalk of a moss capsule which carries the spore.

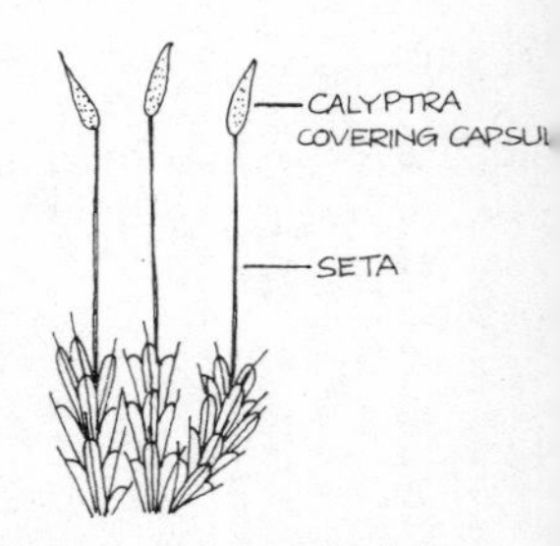

MOSS (*Tortula muralis*)

Setiferous

Having bristles; bristly.

Setiform

Bristle-shaped.

Setose

Covered with bristles or with stiff, straight hairs.

Setulose

Minutely setose, covered with very tiny bristles.

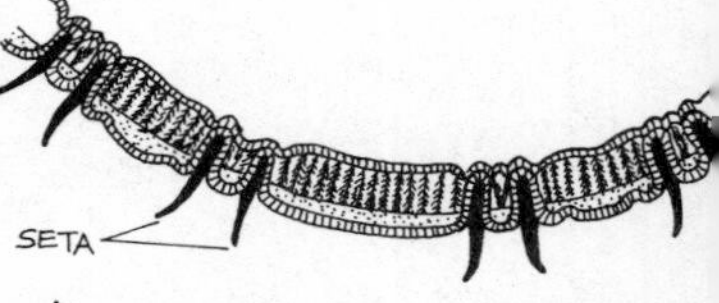

DIAGRAM OF LOWER HALF OF AN EARTHWORM (*Lumbricus*)

Sewage Sludge

The processed residue from a sewage plant which in its raw state is unpleasant to handle and normally only used agriculturally. Dried sewage sludge is odourless, and in the form of a coarse powder can be used as a top-dressing to provide organic matter and a small percentage of chemical fertilizer.

Sexual Reproduction

Most garden plants are naturally reproduced by a form of sexual reproduction in which the pollen from the male anthers is deposited on the female stigma which is connected to the ovary where seeds are eventually produced.

Shaft

Handle of a tool, such as a trowel or spade: *see also* Haft, Helve and Hilt.

Shake

A crack or split in growing timber.

Shanking

A condition of grapes in which the individual stalk of each grape withers, causing the grape to dry up through lack of nourishment.

Sharawadgi

A type of Chinese garden in which geometric designs are avoided in favour of more natural curves and general irregularity.

Shard

A piece of broken pot or similar material used in the past for crocking flowerpots. The term is more widely used in USA to mean crock. *Same as* Sherd.

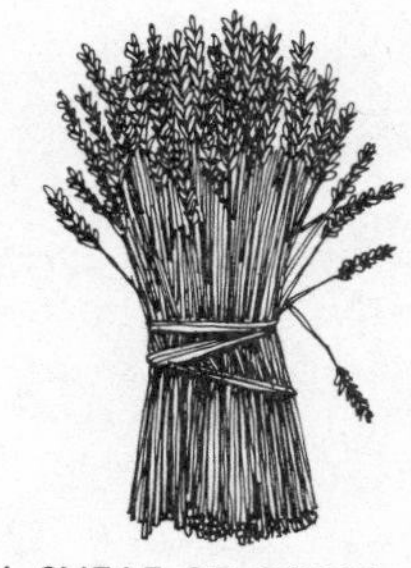

A SHEAF OF CORN

Shaw

An old word for a wood or thicket.

Sheaf

Large bundle of plants bound together; usually applied to harvested cereals, e.g. a sheaf of barley.

Shears

Scissor-like tools in various sizes for several uses. In USA pruning shears are what in Britain are called secateurs.

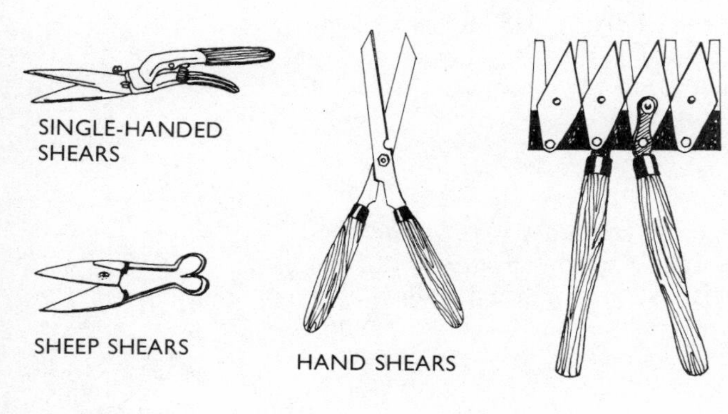

SINGLE-HANDED SHEARS

SHEEP SHEARS

HAND SHEARS

'ASTOR' OR 'FOUR-IN-ONE' SHEARS

Sheath

A roughly tubular and close-fitting enclosure; a clasping leaf-base or similar protective covering usually around a stem. Often a feature of grasses.

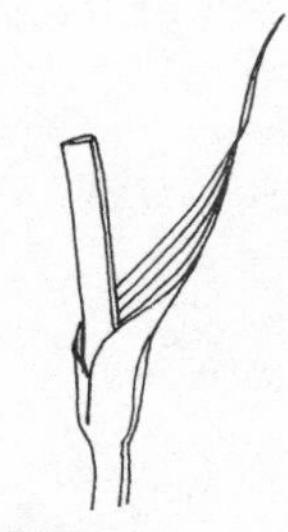

SHEATH

Sheathing

The enclosure of the stem by a sheath-like leaf, as in tradescantia.

SHEATHING

Shell

The hard outer case enclosing a kernel or nut of some kinds of seed or fruit.

Shells

Crushed shells of sea creatures such as oysters and cockles, sometimes added to bulb fibre to reduce the acidity.

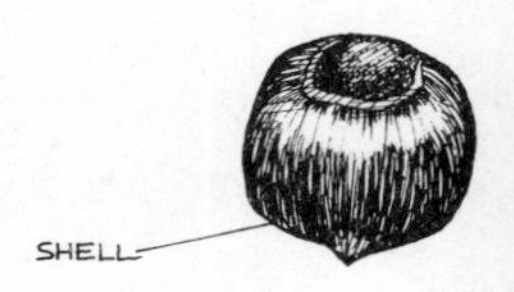

HAZELNUT

Shelterbelt

A row or group of trees or shrubs planted to act as a windbreak and to provide shelter for crops or delicate subjects. To be effective, a shelterbelt must be dense from

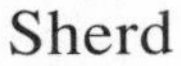

the ground upward, and will then reduce wind velocity for an area of roughly 10 ft (3 m) from the base of the trees for each foot of height. For very small areas, specially designed plastic netting serves the same purpose.

Sherd

See Shard.

Shingle

(1) A rectangular piece of wood, usually cedar, used as roofing material, especially in USA, Canada and Australia.
(2) Small water-worn pebbles, especially those found on beaches.

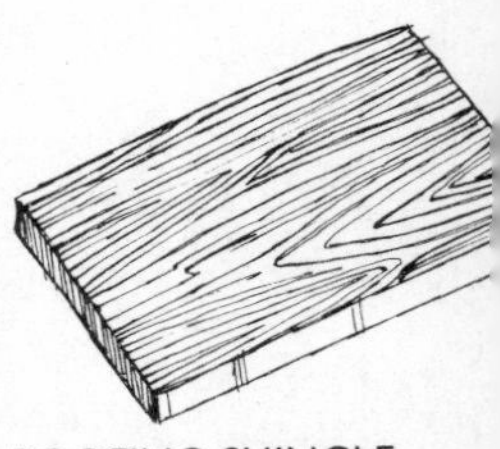

ROOFING SHINGLE

Shock

A number of sheaves of corn (commonly twelve) propped against each other in pyramid fashion to dry before being collected for threshing. A system still seen occasionally in Europe, especially in places inaccessible to a combine harvester, but seldom in Britain today. Also called a stook in some parts of the country.

A SHOCK OF CORN

Shoddy

Wool waste and fabric collected from woollen mills. It can be used as a bulky organic manure which decays very slowly when dug into the soil. Synthetic fibres do not decay and should not be used for this purpose.

Shoot

(1) The growing portion of a plant, such as leaves and stems above ground.
(2) A young branch or sucker.
(3) A side-shoot – one that develops from a main shoot or branch.

Short-day Plant

A plant which flowers only when days are short: *cf.* Long-day plant.

Shot Eye

A rose bud which, after budding on to the stock, produces a shoot in the first summer rather than remaining dormant. Such shoots seldom mature successfully and are usually more prone to frost damage.

Shot-hole Borer

A type of beetle which bores into the branches and trunks of trees, mining the bark with dozens of small tunnels. To date there is no cure; infected trees or branches should be burned.

Shot-hole Disease

A fungus disease, or the summer stage of bacterial canker, both of which affect *Prunus* species. Control is by the use of copper-based sprays at precise times of the year.

Shovel

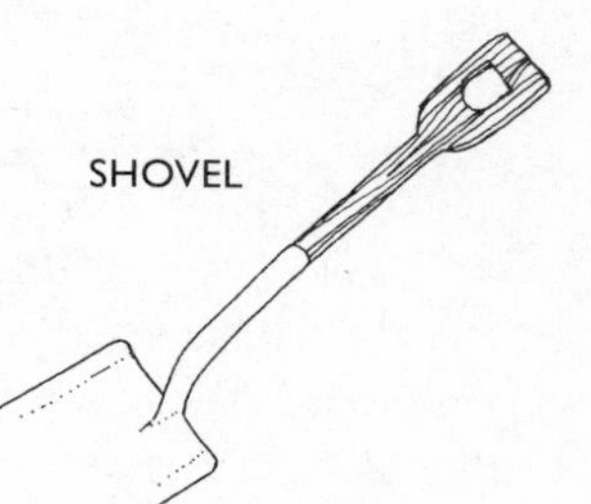
SHOVEL

A spade-like implement for shifting earth, sand, etc. Similar to a spade but usually having a larger concave blade.

Shreds

Nicotine shreds are strips of paper or similar material, saturated in a solution of nicotine and a combustion agent, which are placed in small heaps around the greenhouse and ignited to produce clouds of dense smoke that is lethal to many insects such as aphids. The smoke is dangerous if inhaled and the shreds should not be handled with unprotected hands, and for this reason they are no longer available to the general public.

Shrew, Pigmy Shrew

Small beneficial insectivore, somewhat mouse-like but with a longer, more pointed snout, and, usually, a darker coat. It feeds mainly on slugs, snails, caterpillars and insects and is said to need eight times its own weight of food each day to survive.

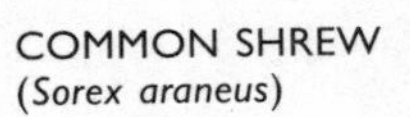

COMMON SHREW
(*Sorex araneus*)

Shrub

A loose term often used to describe a relatively short-growing woody plant producing shoots from the base and having several stems, not a single trunk as a tree. A bush.

Shrubbery

Plantation, border or area of shrubs in a garden.

Shuck

The husk or pod, as of peas. A word not much used today except in USA.

PEA SHUCK

Shy Flowering

Producing few flowers; backward; reluctant to flower.

Sickle

Hand-held reaping hook with a short handle and almost semi-circular blade.

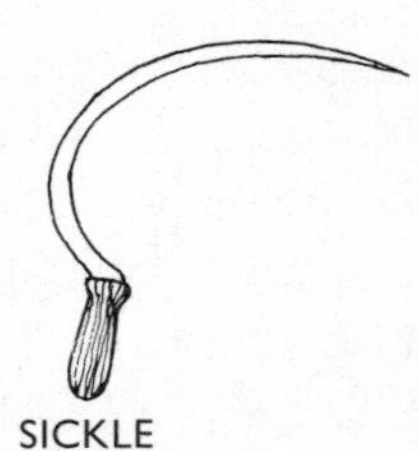

SICKLE

Side-dressing

A top dressing or fertilizer applied alongside growing crops without actually coming into contact with the foliage or stems of the plants.

SIDE GRAFT

Side Graft

A method of grafting whereby the scion is inserted into the side of the stock, and after it has taken the stock is beheaded just above the union. *See* Grafting.

Side-shoot

A lateral growth.

Side-shooting

Removing side-shoots by cutting, pinching or rubbing out, especially on tomatoes when they are to be grown on a single stem.

SIDE-SHOOTING TOMATOES

Sieve

A fine type of riddle consisting usually of a metal or wooden frame with a wire mesh bottom, normally hand-held. Used for separating coarse from finer materials.

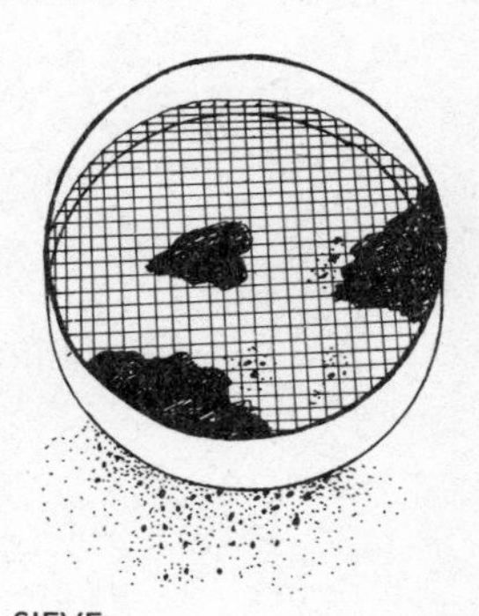

SIEVE

Sieve-plate

The perforated area connecting sieve-tubes in phloem.

Sieve-tube

A conducting element in phloem which is the tissue in the vascular bundles responsible for transporting sugars and other synthesized food materials around the plant.

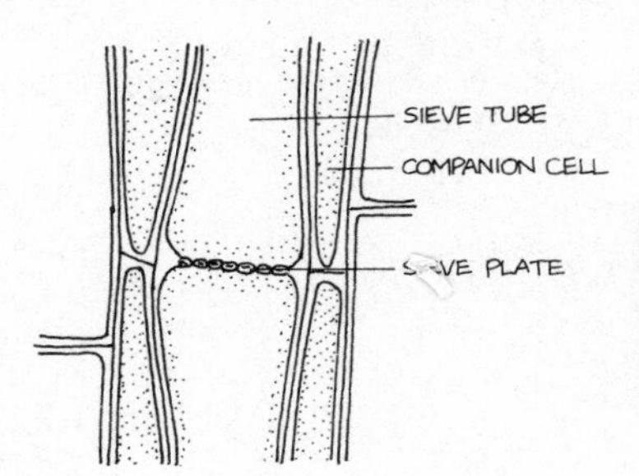

LONGITUDINAL DIAGRAM OF SIEVE-TUBES AND SIEVE-PLATE IN PHLOEM OF PUMPKIN (*Cucurbita pepo*)

Sift

To put material through a sieve; to riddle or sieve.

Sigmoid

Usually taken to mean S-shaped, as the word is derived from the Greek letter *sigma*. This is written as Σ and therefore could also be correctly called C-shaped.

Siliceous

Containing silica, usually in minute particles as in sand, etc.

Silicula, Silicle, Silicule

A short pod of two carpels with a replum. Shorter and often wider than a silique, as in the Cruciferae, e.g. the fruits of *Alyssum*.

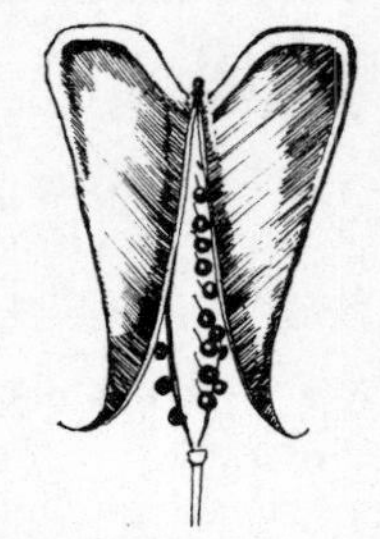
SILICULA OF SHEPHERD'S PURSE (*Capsella bursa-pastoris*)

Siliqua, Silique

A long pod or seed vessel consisting of two carpels with a replum, usually much longer than it is wide, as in cabbage.

Silky

Covered in soft, fine hairs; smooth to the touch.

Silver-leaf

A disease of plum trees and many other kinds of *Prunus*.

SILIQUA (CABBAGE)

Silvery

Having a metallic, whitish lustre.

Simple

Undivided foliage (not compounded into leaflets); unbranched inflorescence arising directly from the main stem.

Single

A flower with the normal number of flowers for the species. Not double.

SIMPLE INFLORESCENCE OF IRIS (*Iridaceae*)

Singling

Thinning out of seedlings.

Sinuous

Wavy, winding, bending in a supple manner; e.g. the stems or canes of raspberry.

Sinus, Sinuate

An indentation of the margin of a leaf or petal in one plane only; wavy-edged, as in some garden pinks. Distinct from undulate which is up and down waviness.

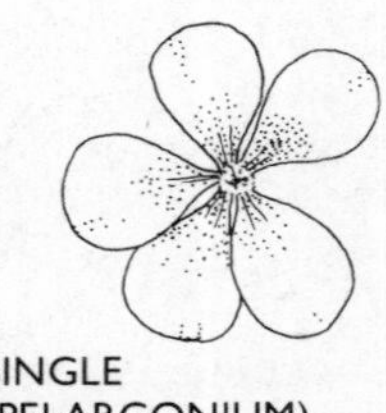
SINGLE (PELARGONIUM)

Siphonet

The honey-tube of aphids consisting of the four-jointed labium which is inserted into the plant tissue after it has been penetrated by the mandibles and maxillae.

Skep

A beehive, usually one made of straw or wicker. Known as a skip in some rural areas.

SINUATE

Skin

(1) The protective membrane covering the stems, foliage, etc. of plants, the epidermis.
(2) To remove the rind or peel of a fruit or vegetable.

Skip

See Skep.

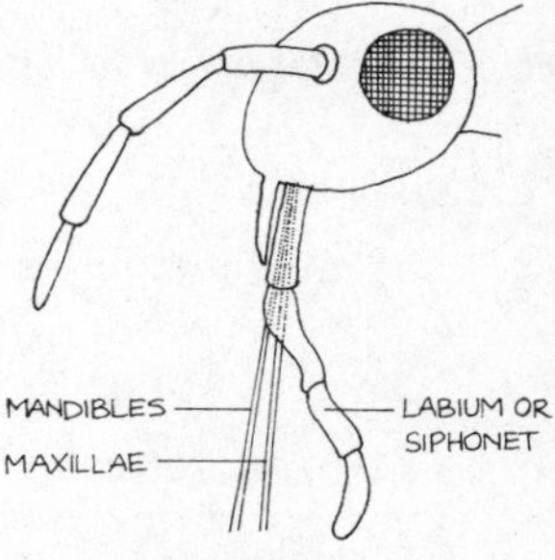

MOUTH PARTS OF AN APHID

Slag

The waste product from iron furnaces which can be used as a general fertilizer; usually marketed as 'basic slag'.

Slashed

See Incised.

Slasher

A type of billhook used for laying and trimming hedges, especially hawthorn and blackthorn.

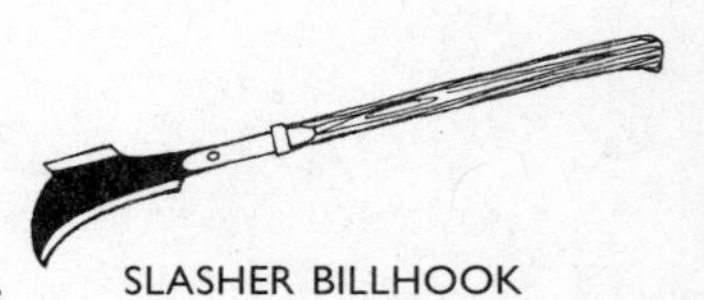

SLASHER BILLHOOK

Sleepy

Of fruit that is past its prime and beginning to rot, especially those that appear to be sound but are rotting below the surface, e.g. pears.

Slender

Slim; willowy; of small girth; not robust.

Slip

(1) A cutting taken from a plant for grafting or planting; often used to describe a shoot pulled off a plant rather than cut.

(2) A piece pulled off a clump of roots consisting of a stem with some roots already attached. Otherwise known as an Irishman's cutting.

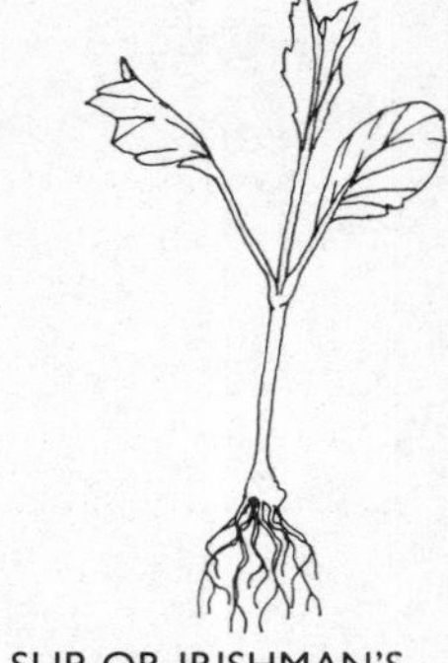

SLIP OR IRISHMAN'S CUTTING

Slitting

(1) Making slits or cuts in turf with a spade or machine to improve aeration; an alternative to spiking.
(2) Making vertical cuts down the bole of a tree in winter to release the restriction caused by hardened bark. Quite often trees will split or crack their bark naturally and not only vertically, as in the cork oak (*Quercus suber*).

Sliver

Strip of wood torn from a tree.

Slow-release

Fertilizers formulated to dissolve slowly in the soil to release their nutrients over a prolonged period.

Sludge

See Sewage sludge.

Slug

Slimy land-mollusc, very destructive, especially to young plants but also to some root crops. They differ from snails in having no visible shell although some have a reduced internal shell.

Slugworm

The larva of a species of sawfly which resemble slugs but have legs. They attack the foliage of plants and can be controlled by a variety of insecticides.

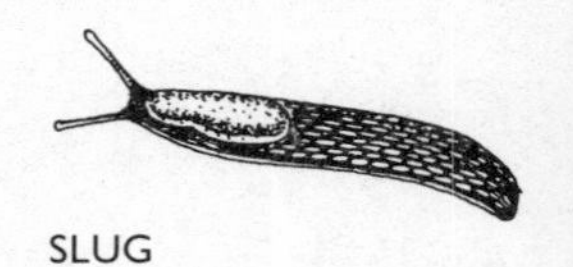

SLUG

Smoke

A chemical fumigant, usually in the form of a small cylinder, cone or pellet which when ignited produces a dense cloud of smoke through which a pesticide is dispersed.

Smother Crop

A dense, fast-growing crop used to deprive weeds of the light and air they need to survive. After maturing, the crop is dug into the soil as a form of green manure. Plants used commercially for this purpose include mustard, etc.

Smudge

To fumigate with smoke; thick choking smoke; a smoky fire, often in a garden or on agricultural land.

Smut

Fungus disease, mainly on cereal crops but also found on some garden plants; the black powdery deposit caused by this.

Snag

A jagged projection on a branch or tree-stump left after cutting or pruning.

Snail

Terrestrial gastropod with normally conical shell, which feeds upon plants and can cause much damage. Control as for slugs. Some water-snails can harm plants in pools, but others can be beneficial by controlling algae.

SNAIL

Snead

Old word for the handle of a scythe.

Snick

Small cut or notch.

Snout Beetle

Weevil, similar in appearance to other beetles but having a distinct beak or snout. *Curculio nucum* attacks certain nuts and has a most pronounced snout, and the vine weevil is a common pest in greenhouses where the black adults eat notches in the foliage of vines and will also attack the tubers of begonias and cyclamens; the creamy-white grubs eat the roots of many pot plants.

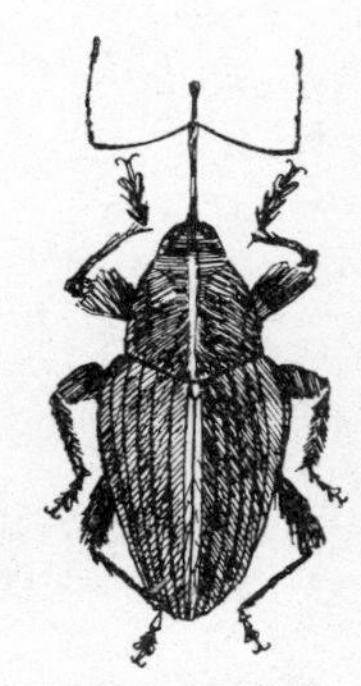

SNOUT BEETLE
(*Curculio nucum*)

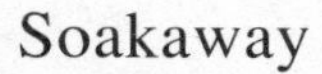

Soakaway

A deep pit filled with rubble or stones to collect water from a drainage system when no other outlet is available.

Soap

A solution of soft soap and water was once used as the main means of controlling aphids and red spider mite. Plants were syringed under considerable pressure and the insects were washed off the plants; an excellent and effective method if performed on a frequent and regular basis. Modern insecticides are less demanding and cover a wider range of pests.

Sobole

An underground (often creeping) stem producing shoots and roots.

Soboliferous

Having soboles; throwing up lateral shoots from ground level; clump-forming.

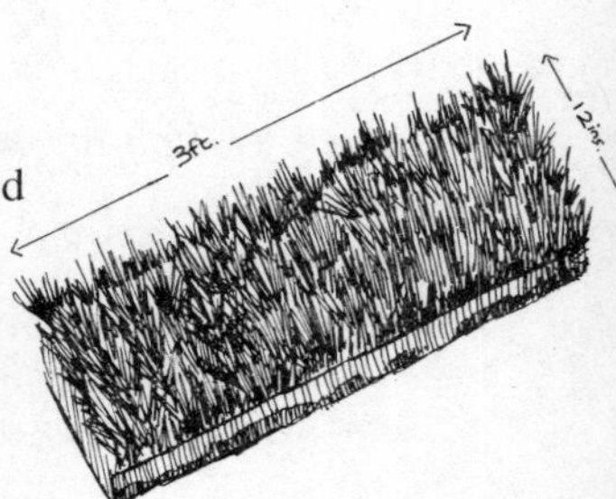

A SOD OF TURF

Sod

See Turf.

Sod-lifter

See Turfing Iron.

Sodding

Turfing, or the laying of turves to make a lawn, especially in USA.

Sodium Chlorate

A persistent general weedkiller used mainly to eliminate weeds from paths and drives; must be kept well clear of plants and trees which can easily be killed. In its pure dry form, it easily ignites by friction or a blow and must therefore be handled with great care.

Soft Fruit

Fruits grown on bushes or canes as distinct from top fruits borne on trees.

Soft Rot

A portmanteau term covering several fungal and bacterial diseases which result in rotting plant tissue.

Soft-stemmed

Of a plant with green and often immature stems: *cf.* Woody.

Softwood Cuttings

The method of propagation most used for perennials, sub-shrubs, alpines and certain conifers. The young green shoots are taken, trimmed and inserted into a propagating bed where they normally root rapidly. Hardwood cuttings are taken from older, riper wood and will take longer to root.

Soil

The upper strata of the ground in which plants grow.

Soil Block, PLUG (USA)

A usually rectangular or cylindrical block of compressed soil, used for sowing seeds individually. Blocks can also be made from peat or peat-based composts.

SOIL BLOCK

Soil Conditioner

Any substance incorporated into the soil to improve its physical structure; can be organic or chemical, can also include substances which improve drainage or aeration whilst providing no additional nutrients, such as sand and grit. Many soil conditioners marketed today are seaweed derivatives or gypsum-based.

Soil Mark

The mark or stain on the stem of a plant indicating the soil level before it was moved or transplanted.

Soil Testing

A method of analyzing the chemical content and the pH of a soil. Crude soil-testing kits are available in most garden shops and these provide a general guide as to the pH but cannot indicate a deficiency of trace or other elements.

Soil Warming

(1) The application of bottom heat to cuttings and seed trays to promote early rooting and germination. They

usually involve the use of thermostatically controlled electric cables, but hot water systems are equally efficient.

(2) Covering the garden soil with black plastic in spring to warm the soil before planting.

Soilless Composts

U. C. Mixtures were originally formulated by the University of California in the early 1940s to provide a standard medium for the raising of pot plants, cuttings and seeds, and were peat-based. In recent years the use of shredded coconut fibre has been substituted for peat with some success and its use is becoming more widespread: *see also* Coconut fibre.

Soilless Cultivation

Any system of cultivation which dispenses with the use of soil, such as hydroponics, straw bale culture, etc. Ring culture of tomatoes is a partially soilless method.

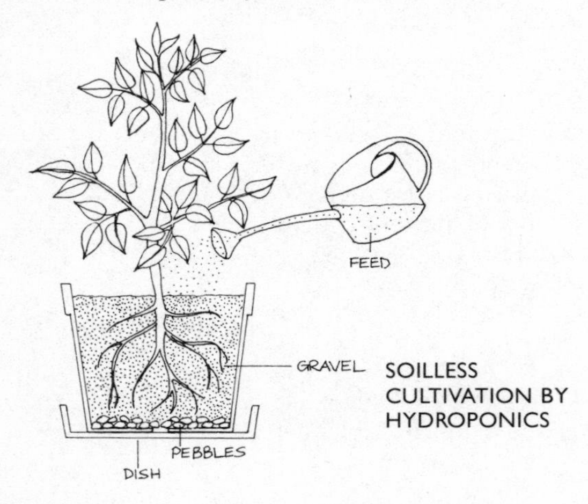

SOILLESS CULTIVATION BY HYDROPONICS

Solitary

Flowers which occur singly; separately; not in clusters.

Somatic

Describes all the cells of a plant (or animal) other than the reproductive cells.

Soot

A black deposit caused by the imperfect combustion of carbonaceous matter. It contains up to 6 per cent of nitrogen but must be weathered for at least six months before use. Said to deter slugs and snails if scattered around choice plants, it is a useful plant food if incorporated into the soil at the rate of 4–6 oz (113–170 g) per square yard or metre, and can be used as a liquid fertilizer if a bag of soot is suspended in a barrel of water, then diluted until it resembles the colour of weak tea.

SOROSIS (MULBERRY)

Sooty Mould

A black mould which grows upon the 'honeydew' excreted by aphids, etc. It is a fungus which blocks the breathing pores of the leaves and distresses the plant. After the insects have been destroyed, the fungus can be removed by washing with soapy water, then rinsing with clean water.

Sorosis

Fleshy compound fruit, as mulberry, pineapple, etc.

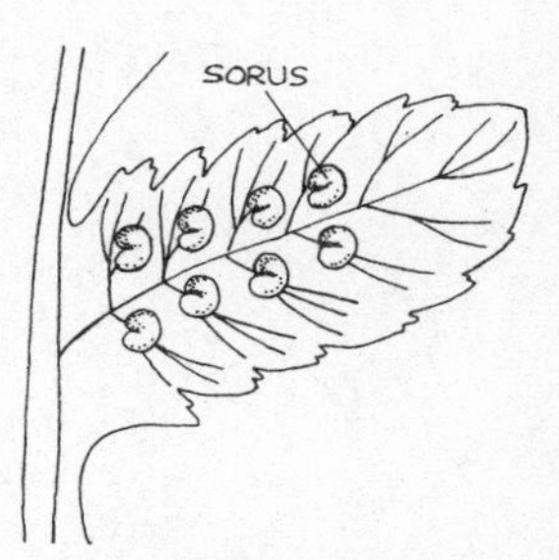

UNDERSURFACE OF A LEAF OF WOODFERN (*Dryopteries*) SHOWING SORI

Sorus, *pl.* Sori

One of the clusters of spore-bearing organs or spore cases found on the underside of mature fern fronds, usually in distinct and regular patterns. Sori are often mistakenly confused with rust.

Sowing

Originally the broadcasting of seed over the soil, often by women following the plough or harrow; now any method of planting seed.

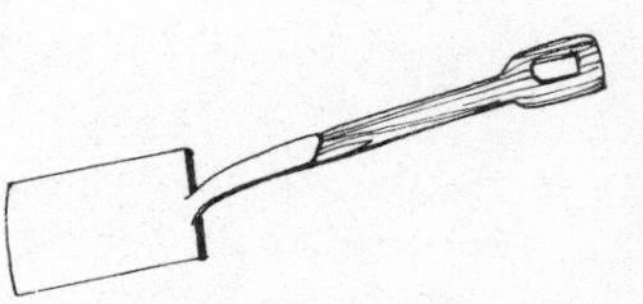

SPADE

Spade

Tool for digging or turning over the soil, usually having a rather flat, rectangular metal blade and wooden or metal handle.

Spading Fork

A broad-tined fork often used for lifting root crops such as potatoes without damaging the tubers.

SPADING FORK

Spadix, *pl.* Spadices

A fleshy flower-spike usually partially enclosed by a spathe, as in the Araceae family, etc.

Spaghetti Watering

A method of watering large numbers of plants in a greenhouse by connecting each one to a narrow tube, this being connected to a thicker tube which takes water from a central supply point; by turning the water on at the control cock, the entire crop can be watered at once. *See* Drip-feeding, Trickle irrigation.

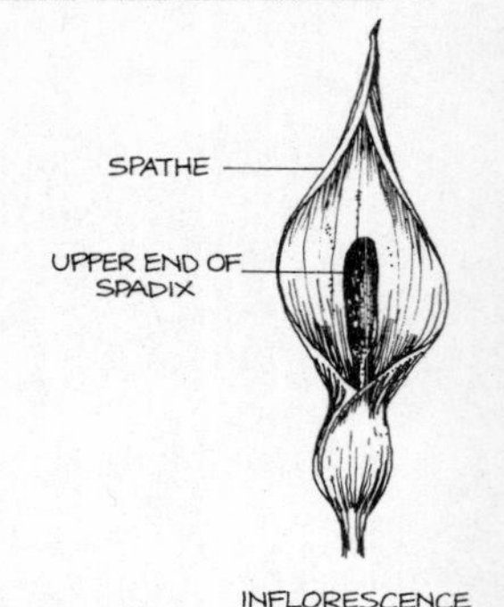

CUCKOO-PINT
(*Arum maculatum*)

Spathe

A leaf or bract, often brightly coloured, enclosing or partially surrounding a flower or flower-cluster such as a spadix.

Spatulate, Spathulate

Of leaves that are broad and rounded at the tip, tapering at the base – spatula-shaped.

SPATULATE

Spawn

(1) The mycelium of fungi, e.g. mushroom.
(2) The tiny cormlets produced by gladioli around the main corm.

Speciation

The development of one or more new species from an existing species.

Species

A group of plants having common characteristics, distinct from others of the same genus. The basic unit in classification. A group of plants which will breed among themselves but not normally with members of another group, and will breed true. (Species is both plural and singular.)

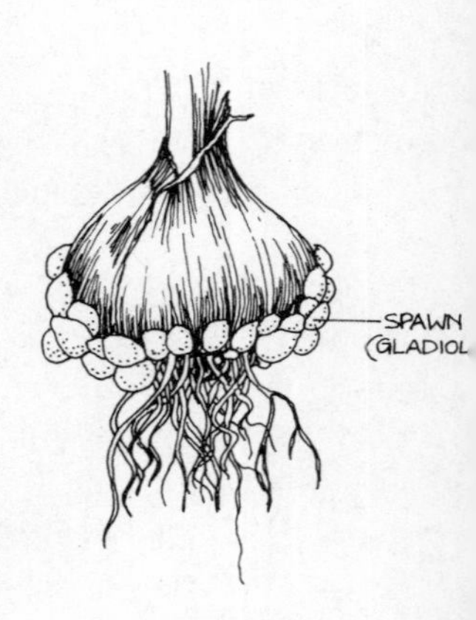

Specific

Relating to or constituting a species.

Specimen Plant

A plant separated from others, usually given special treatment to prove its superiority, or one set into a flowerbed in a focal position.

Speck

(1) A small spot or area of rot in fruit.
(2) A dot or small mark of a contrasting colour, usually on a petal.

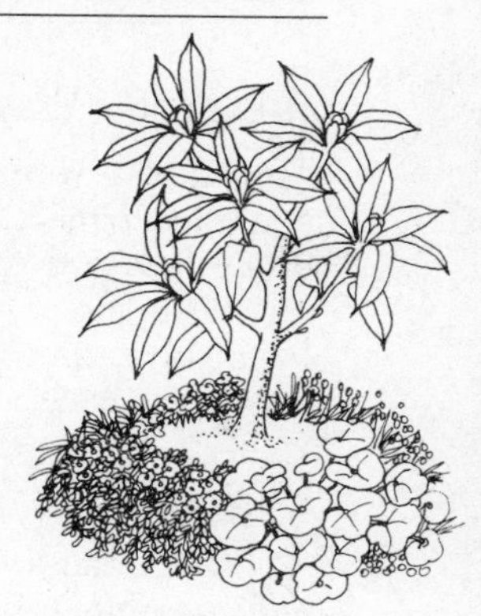
SPECIMEN PLANT

Spent Hops

See Hop manure.

Sperm

The male reproductive cell; seed.

Spermatium, *pl.* Spermatia

A spore-like structure found in some fungi.

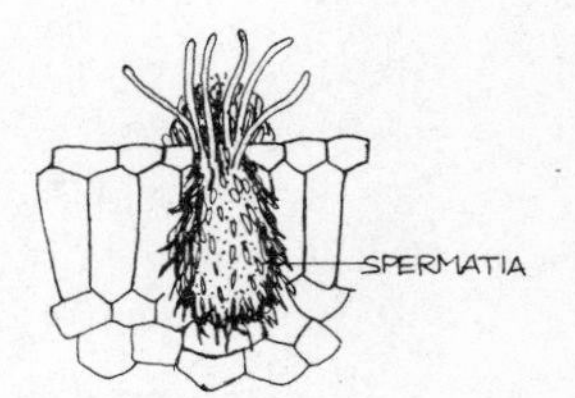

SPERMATIA ON *Berberis*

Spermatophyte

Any seed-producing plant.

Sphagnum

Types of moss which normally grow in damp situations such as bogs and saturated banks. When decomposed in boggy conditions and compressed over several centuries, it becomes sphagnum peat (peat moss in USA) as distinct from sedge peat.

Spicate

Pointed; formed into or carried like a spike, e.g. lupin.

SPICATE (*Lupinus*)

Spicula

A prickle or spicule; a splinter.

Spicule

Slender, pointed, needle-like formation; a prickle.

Spider

(1) Beneficial creature that traps and devours many kinds of flying pest. The term is often wrongly applied to red spider mites which are very destructive.
(2) A type of chrysanthemum with narrow curling petals.

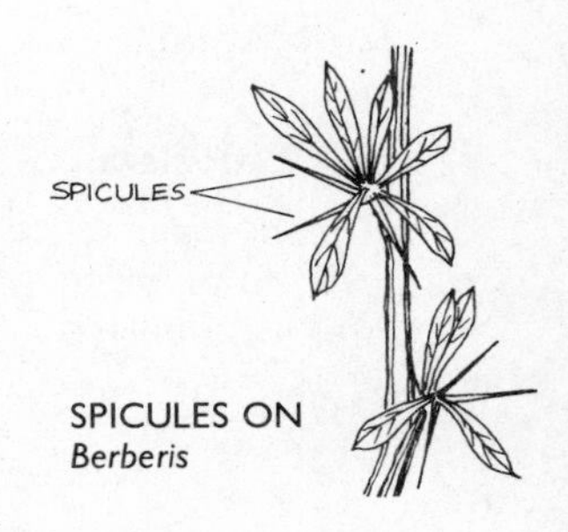

SPICULES ON *Berberis*

Spike

(1) An inflorescence in which racemose stalkless (or semi-stalkless) flowers are produced on a long axis.
(2) An ear of corn.

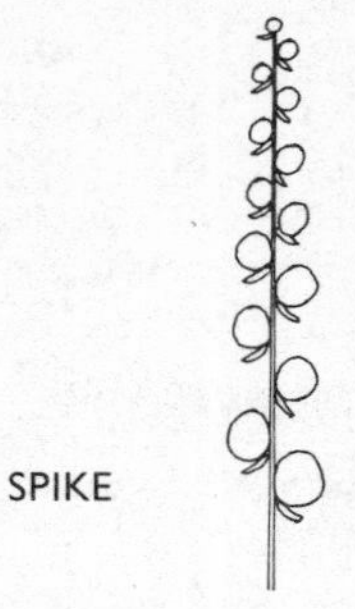
SPIKE

Spikelet

(1) A small or secondary spike, itself forming part of a cluster.
(2) The flower of grasses, usually one or more flowers with basal bracts.

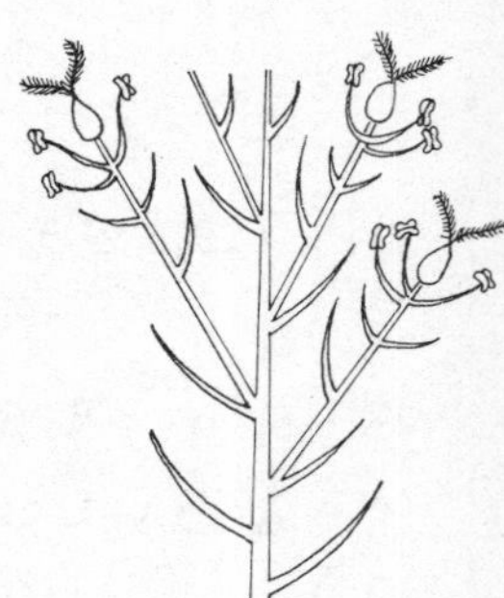
A GRASS SPIKELET

Spiking

A method of aerating lawns by pushing a garden fork into the ground up to its full depth, rocking it, and repeating at 6 in (15 cm) intervals until the entire area has been covered. A mechanical device fitted with rotating spikes is available for the same purpose but it is unable to penetrate to the same depth and is therefore more often used for ensuring that dressings or fertilizers reach the roots. The use of hollow-tine forks is an elaborate way of doing the same job.

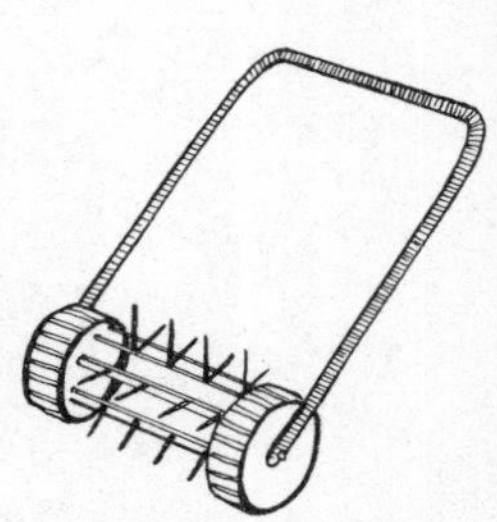
MECHANICAL SPIKER

Spiling

A method of producing exhibition quality root crops such as carrot and parsnip in poor ground. An iron bar is pushed into the soil to a depth of 18–24 in (45–60 cm), then rotated until a conical hole is excavated, which is filled with good sifted soil or potting compost. The seed or seedling is then planted in this soil and will grow down into the cone of good material.

Spine

A stiff, sharply pointed prickle or thorn, usually an outgrowth on a leaf or stem, as on holly.

Spinescent

(1) Having spines; spiny.
(2) Terminating in a spine or thorn.

Spinney

A thicket; a small wood or wooded area.

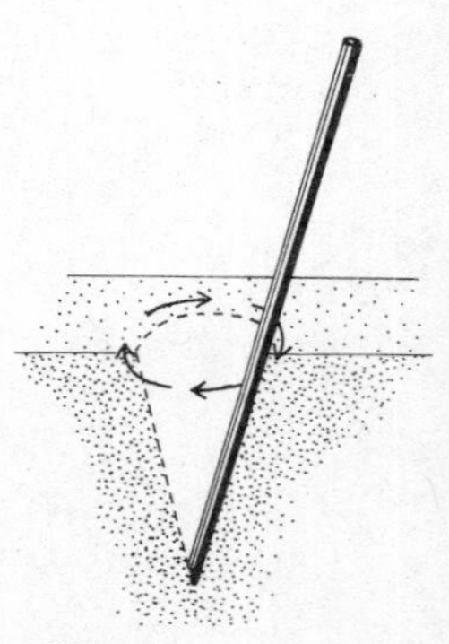
SPILING

Spinose, Spinous

Sharp-pointed; having spines or slender thorns.

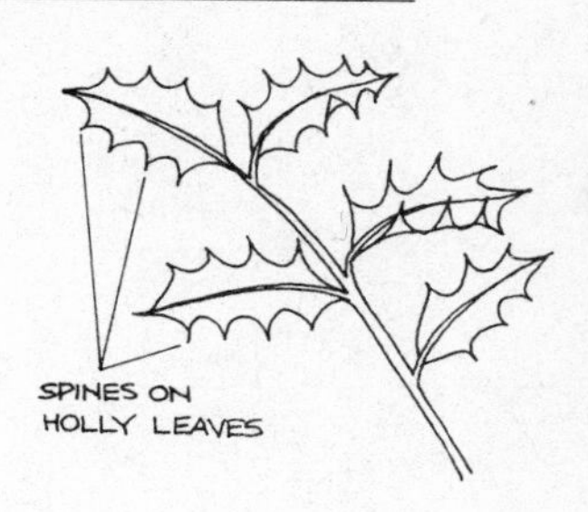

Spinule

A small spine or thorn.

Spinulose

Having small spines.

Spiny

Prickly; thorny; having an abundance of spines.

Spire

(1) A long slender stalk or stem.
(2) A tapering or conical plant, such as certain trees, including *Chamaecyparis lawsoniana* 'Spek', and flower spikes of *Lupinus*.

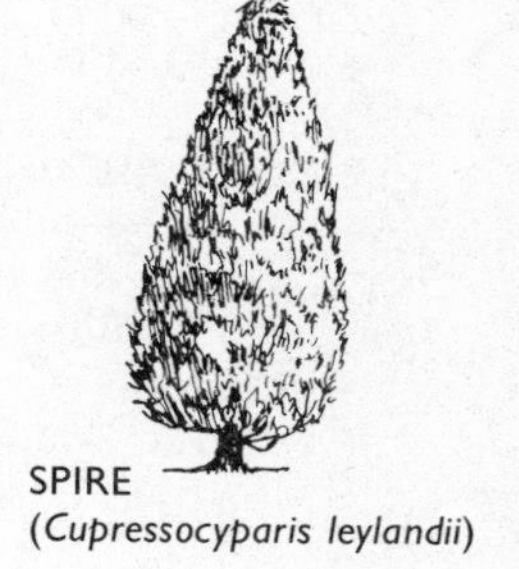

SPIRE
(*Cupressocyparis leylandii*)

Spired

(1) Having a crown with several growing points.
(2) Having parts arranged in spirals.

Spit

A spade's or fork's depth of soil, usually about 10 in (25 cm). In earlier times used as a verb – to dig or plant with a spade or fork.

Spittlebug

The name used in USA for a froghopper or cuckoo-spit insect.

SPITTLEBUG OR
FROGHOPPER
(*Philaenus spumarius*)

Splice Grafting

A simple method of joining a scion and stock which are equal in girth. *See* Grafting.

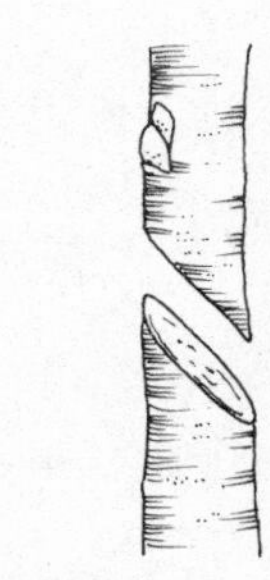

SPLICE GRAFT

Split Stone

A condition of stone fruits, especially plums, in which the exocarp splits to reveal the kernel or seed.

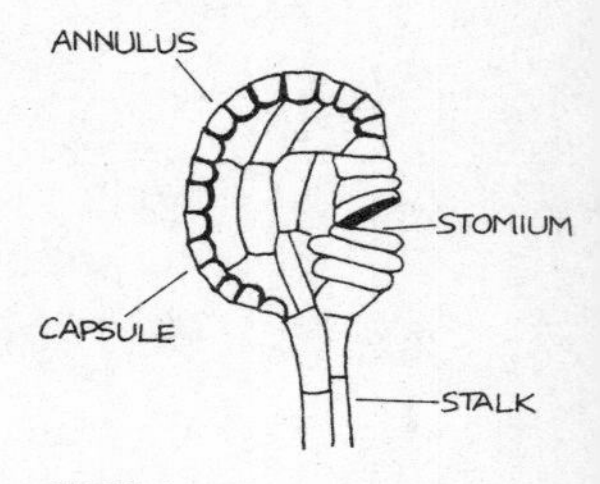

SPORANGIUM OF FERN

Spoon

The narrow rolled petals of some types of chrysanthemum which expand at the perimeter end into a spoon-shape, so that the plant is referred to as a spoon-chrysanthemum.

Sporangium

A spore-case or sac in which spores are produced.

Spore

A simple, minute reproductive body, usually associated with ferns, fungi, mosses, etc. Functionally equivalent to the seed in higher plants.

Sporophore

The aerial spore-bearing structure of certain fungi, such as the cap of a mushroom or toadstool.

SPOROPHORE (*Pleurotus ostreatus*)

Sport

A mutation, whether induced or accidental. A plant that varies spontaneously from the normal, either in leaf- or flower-form.

Sporule

A small spore.

Spot Treatment

Attacking weeds with a weedkiller individually rather than by broadcast application.

Sprawling

Low, straggling mass or cluster; an uncontrolled plant often used for ground-cover.

Spray

(1) Florist's term for a number of flowers on a single stem.

(2) To eject a stream of fine droplets of water or diluted chemicals from a syringe or rose, usually under pressure.

Sprayer

Any device used for applying insecticides, chemicals or plain water to plant foliage. The old type of syringe has now been superseded by smaller hand-held plastic models, or by the larger units incorporating a pump-action pressurized tank and lance.

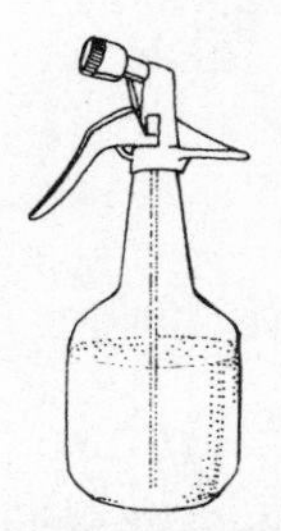
HAND SPRAYER

Spreader

(1) Saponin or similar soapy substances added to chemicals to provide better coverage of the foliage when spraying insecticides.
(2) Dry sand mixed with seeds to give more even distribution.
(3) A mechanical device used to spread fertilizer, seeds, etc. evenly over a lawn.

Spreading

Extending outward or horizontally.

Sprig

Small branch, twig or spray, often ornamental and having berries or flowers, but not necessarily so; e.g. a sprig of mint.

Springbok

Trade name for a type of rake with flexible sprung tines, used for raking moss out of lawns, etc: *see also* Rake.

Springtail

One of about 250 species of the order *Collembola* that is able to jump when disturbed, by means of a forked 'spring' that is normally kept folded under the body. They feed on living and dead vegetation and can be a pest in greenhouses. Control is by an insecticide applied either as a dust or soil drench.

SPRINGTAIL
(*Orchesella villosa*)

Springwood

Inner part of the annual rings in tree trunks, consisting of thin-walled water-conducting tissue: *cf.* Sapwood, Summerwood.

Sprinkle Bar

See Dribble bar.

Sprinkler

Any automatic device for applying water to garden or lawn in the form of a fine spray.

Sprit

Small shoot or sprout. Seed potatoes, for instance, will be showing sprits when ready for planting. 'Chit' is used in some parts of the country for the same thing.

Sprout

(1) To begin to grow; to throw out a shoot or shoots.
(2) A new shoot on a plant; the tender side-shoots produced on some brassicas, e.g. Brussels sprouts.
(3) The initial growths on potato tubers emanating from the 'eyes'.

SPUD (CAN HAVE A WALKING-STICK HANDLE OR A KNOB)

Spud

(1) A narrow digging or weeding implement; a slim-bladed spade.
(2) A slang word for potato, ('tettie' in Devonshire).

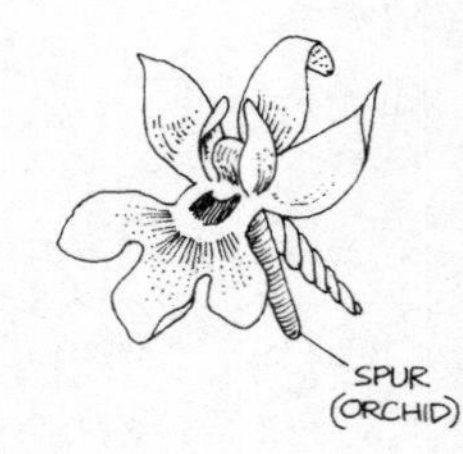
SPUR (ORCHID)

Spur

(1) A large lateral root or branch of a root.
(2) A tubular pouch at the base of a petal, usually containing nectar, as in some orchids.
(3) A short lateral twig or branchlet of a fruit tree upon which the flowers and fruit are produced.

Spur Back

To trim back the side-shoots of a tree or bush to within two or three buds of the main shoot.

Squama, *pl.* Squamae

A scale or scale-like structure.

Squamate

Having or producing scales.

SPURS ON A CORDON FRUIT TREE

Squamella

A small scale.

Squamellate

Consisting of or covered in small scales.

Squamose

Composed of scales or covered in scales.

SQUAMOSE BUDS

Squarrose

Rough with projecting or deflexed scales or bracts.

Staddle

(1) A sapling or young tree left standing when those around it have been felled.
(2) A stump left for coppice.
(3) A support for a stack of hay, etc.

Stag-headed

Trees with dead ends of branches projecting through the leafy crown in the manner of stags' horns. Usually a sign of senility but may be caused by disease.

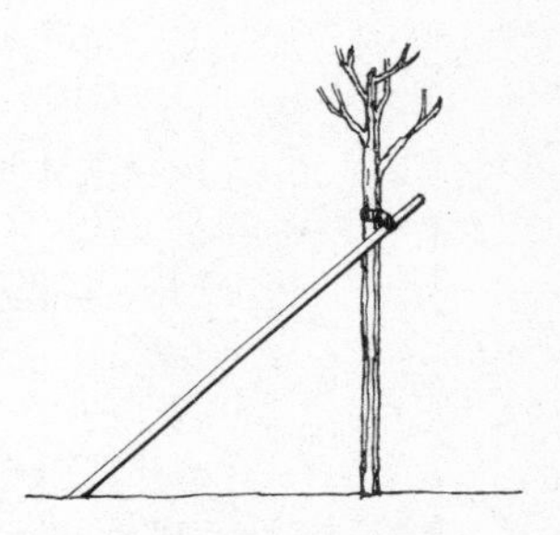
STAKING A YOUNG TREE

Staging

Greenhouse benching, or temporary benching at a horticultural show on which the exhibits are arranged.

Staking

Supporting young plants by securing them to a cane or stake to prevent movement.

Stalk

The common term for the stem supporting a flower or leaf. The correct term for a leaf-stalk is petiole and for a flower-stalk a pedicel; a stem can produce buds, a stalk cannot.

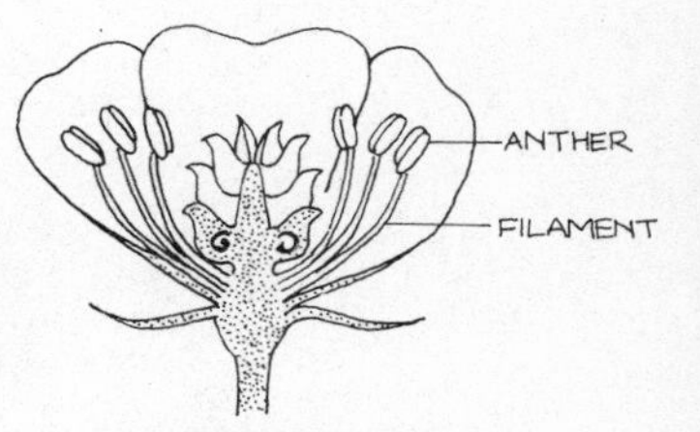

STAMENS OF BUTTERCUP (*Ranunculus repens*)

Stamen

The male pollen-producing part of a flower consisting usually of anther and filament.

Staminal Column

A column formed by the fusion of the bases of filaments.

Staminate

Male, having stamens but no pistils.

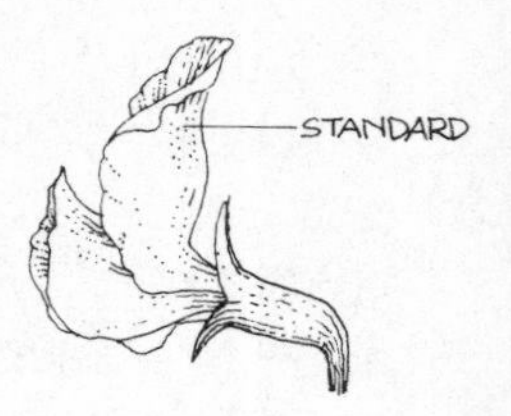

SWEET PEA

Staminode

A sterile stamen.

Standard

(1) The uppermost petal of a papilionaceous flower, such as pea, bean, gorse, etc.
(2) Any tree, shrub or plant grown on a single leg or stem.
(3) The upright-growing petals of the iris flower as opposed to the drooping or spreading falls.

STANDARD SHRUB

Standing Crop

The total amount of living material in a specified population such as a forest where the standing crop can vary from season to season.

Starting

Bringing a subject out of dormancy by planting in moist peat or soil and applying slight warmth, e.g. to bulbs.

Statocytes

Starch-containing, gravity-sensitive cells in the root-tips of plants.

Steeping

(1) Soaking corms or seeds with hard coats before planting or sowing to expedite germination, e.g. sweet pea seeds.
(2) Suspending a sack of animal manure or soot in a barrel of water to produce liquid manure.

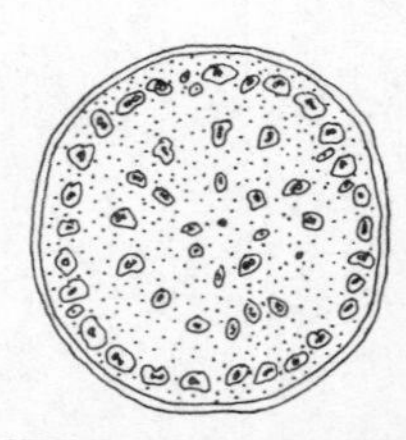
STELE OF MAIZE STEM (TRANSVERSE SECTION)

Stele

Axial cylinder containing the vascular bundles in stems and roots of vascular plants.

Stellate, Stellar

Star-like or star-shaped, as on the flat branched hairs on leaves and stems of Cruciferae.

Stellulate, Stellular

Like a very small star; can apply to both leaves and flowers.

STELLATE FLOWERS OF *Stellaria holostea*

Stem

(1) The leaf- or flower-bearing axis of a plant. Sometimes erroneously called a stalk.
(2) The main part of a plant or tree above ground.

Stem and Bulb Eelworm

A group of eelworms noted for attacking many types of bulb including onions, narcissi, bluebells, etc. and the stems of some plants. Control in greenhouses is by chemical or steam sterilization of the soil; in open ground there is no cure, but rotation of crops may help.

Stem Rooting

Of plants which develop roots above ground from their stems or branches, including ivies, Virginia creeper, etc. Some lilies produce underground roots from the stems above the neck of the bulb, so require deep planting.

Stem Rot

A general term applied to any disease which attacks the stems of plants. Sometimes called foot rot, damping off, black rot, etc.

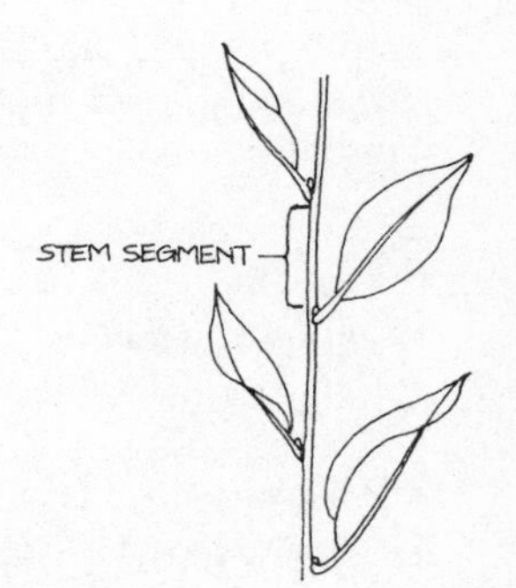

Stem Segment

The portion of a stem between two nodes.

Sterile

(1) Unfruitful, barren, not producing fruit, seeds or spores. Many hybrid plants are sterile, as are some varieties of grape, bananas, etc.
(2) Sand, vermiculite, perlite, etc. are sterile until contaminated by contact with disease organisms (principally bacteria and fungi).

Sterilization

Totally sterile soil is of little use. Partial sterilization is what is meant when a gardener refers to 'sterilization' in which the temperature of the soil is raised sufficiently to destroy all weed seeds, insects and fungi, but most of the beneficial bacteria are able to survive. Sterilization by chemical means is widely practised commercially with similar results.

Sterol

One of a class of complex solid alcohols found in animals and plants.

Stigma

The part of the pistil which receives the pollen; when receptive it is often slightly sticky.

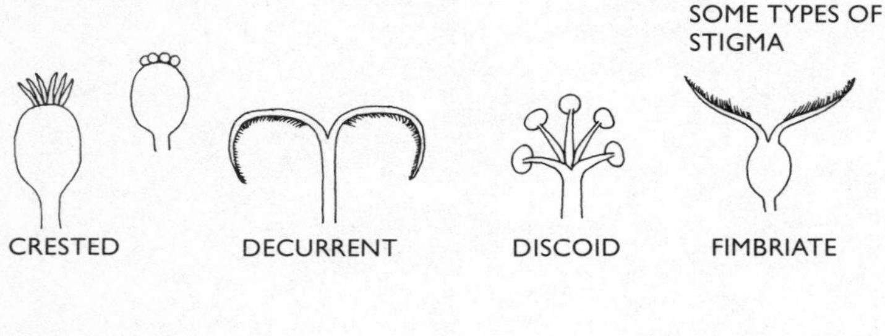

SOME TYPES OF STIGMA

Sting

Stiff, sharp-pointed, hair-like projection on some plants, emitting an irritating fluid which can produce a rash or inflammation when touched or brushed against, e.g. stinging nettle.

STINGING HAIR OF NETTLE (SECTION)

Stipe

Usually the stalk of a fungal fruit-body; a fern leaf; a stalk or pappus.

Stipitate

Having a stalk or stipe.

Stipulate

Having stipules.

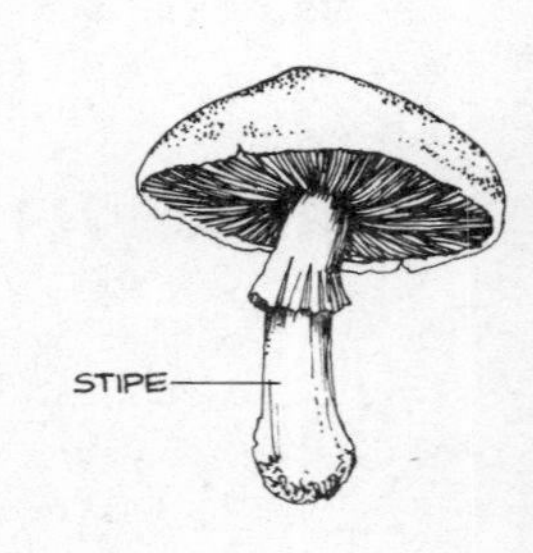

Stipule

Leaf-like organ usually found in pairs at the base of a petiole or the node on a stem.

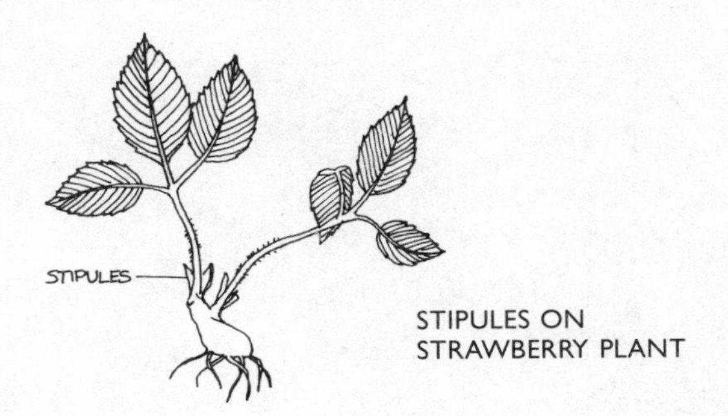

STIPULES ON STRAWBERRY PLANT

Stock

(1) The plant into which a graft or scion is inserted.
(2) A rhizome.
(3) The stem or trunk of a tree.
(4) A stump left after a tree has been felled.

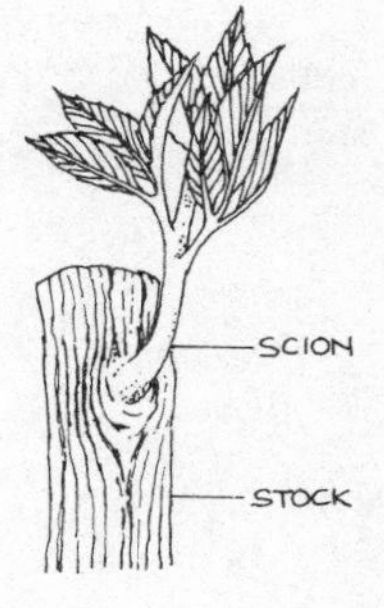

Stock Plant

A mother plant from which cuttings or scions will be taken for propagation purposes.

Stockade

An enclosure formed basically of upright stakes or planks.

Stockholm Tar

A dressing for the wounds of trees obtained from pine trees, and popular with gardeners until the beginning of the twentieth century when it was proved to be ineffective in preventing infection.

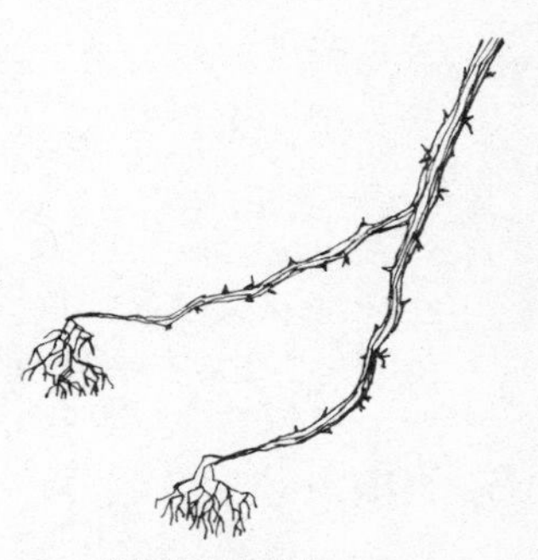

STOLON OF GOOSEBERRY

Stolon

A shoot that arches or runs along the ground, taking root and producing a new plant at the tip. Distinct from a runner which roots at the nodes along the length of the stem.

Stoloniferous

Producing stolons.

Stoma, *pl.* Stomata

A minute, mouth-like opening in the epidermis of a leaf or stem by which gases are exchanged. Usually more numerous on the undersurface of most leaves: *cf.* Pore.

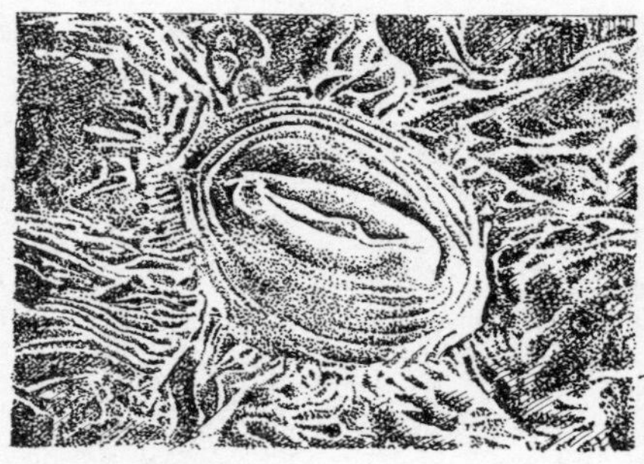

STOMA CLOSED

STOMA OPEN

STOMA ON UNDERSIDE OF PELARGONIUM LEAF × 3000

Stone

The hard, woody, protective case containing the kernel of a drupe such as plum, peach, etc. A pyrene.

Stone Pit

A virus disease of pears resulting in fruits having small, gritty spots in the flesh.

Stook

See Shock.

Stool

(1) A tree stump or an old root from which new shoots will grow, such as a chrysanthemum stool.
(2) A clump of basal shoots.

Stooling

(1) Cutting down to ground level to induce new growth.
(2) Developing several stems from the base of a plant.

Stopping

The removal of the growing tip from a plant to force the production of laterals or side-shoots. Often employed in the growing of chrysanthemums, dahlias, carnations, etc.

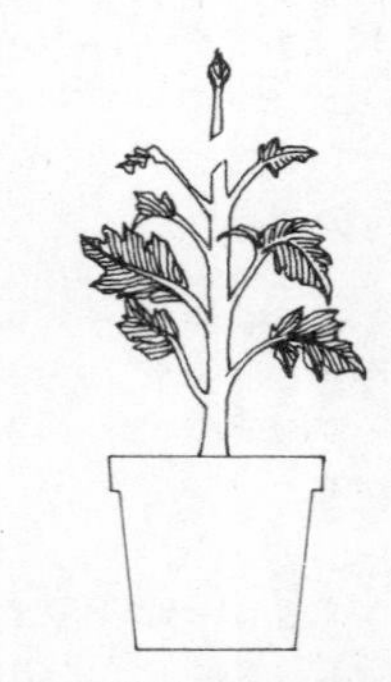

STOPPING A PLANT

Storage Organs

Bulbs, corms, tubers and rhizomes are all organs which store food and fluids during the dormant period. There are also plants which store liquids above ground including many succulents and cacti.

Stove House

A type of south-facing greenhouse very popular in Victorian times, usually built against a wall to ensure minimum heat loss. It was used mainly for raising tropical subjects needing a temperature of 70°F (21°C) or more, and was normally equipped with heating pipes fed by a hot-water boiler.

Stove Plant

A plant which needs to be housed permanently in a heated greenhouse with a temperature of not less than 70°F (21°C) and high humidity.

Strain

Line of descent; selection by which the best varieties are obtained and maintained. A gardener who consistently takes first prize for his carrots, for instance, could be said to have 'a good strain' of carrot seed.

Strangle

(1) To restrict the development of a plant by garotting, as in the case of unfruitful trees which can often be brought into production by this method.
(2) To so constrain a host plant as to prevent further growth, as is done by some woody climbing plants such as wistaria.

Strap-shaped

Of a narrow leaf that is at least three times as long as wide, and with parallel or near-parallel sides.

Stratification

The layering of certain hard seeds in coarse sand and encouraging germination by exposing them to winter conditions or refrigeration. Used mainly with very hard seeds such as those from trees which grow naturally in snowy climates, e.g. the Norway spruce (*Picea abies*).

Straw

The dried cut stems of any cereal plant, especially after the seed-head has been removed. It may be composted and dug into the soil or chopped and applied as a mulch. Bunches of it wrapped around the trunk and large branches of fruit trees can be used as traps for a variety of crawling insects. Straw from corn which has been sprayed with a selective weedkiller may be harmful to plants and should be avoided.

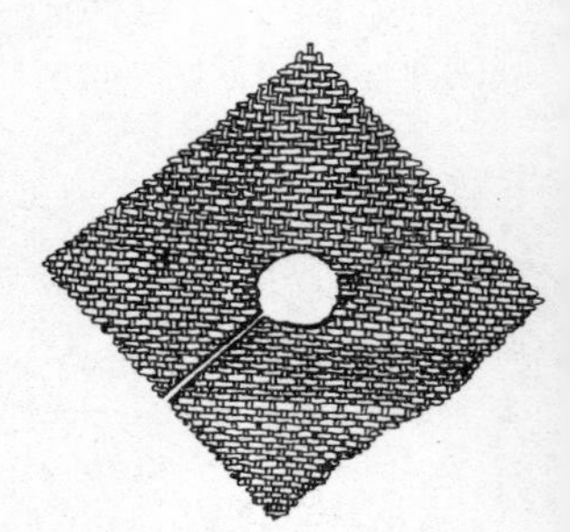
STRAW MAT

Straw Mats

Straw woven into mats about 12 in (30 cm) square with a hole in the middle were widely used in earlier times to protect strawberries from slugs and mud splashes. Similar results can be obtained by the use of black plastic sheeting or by loose straw tucked beneath the foliage.

Strawberry Pot

Originally a clay pot, about 30 in (75 cm) tall with holes at intervals in the sides into which strawberry crowns can be planted. Wooden barrels are equally effective but these are becoming difficult to find now that beer is transported in metal containers, Plastic substitutes are available, and the latest idea is to grow the plants in a plastic bag about 30 in (75 cm) long by 12 in (30 cm) wide suspended from a wooden or metal hanger, and the plantlets inserted into slits cut in the sides. This method is widely used in places such as the Channel Islands and can easily treble the yield from a greenhouse.

STRAWBERRY BARREL

Striate

Having fine longitudinal lines or ridges, often parallel; usually applied to leaves.

Strict

Stiff and straight; upright.

Strig

A cluster of fruits such as currants, complete with the main stem and smaller stalks. The stripping of the fruit from the stalks is termed strigging.

STRIG OF BLACKCURRANTS

Strigose

Having stiff, appressed hairs or bristles.

Strike

To root a cutting, or to prepare a cutting in such a way as to encourage it to root.

Strip Cropping

(1) Cutting the entire crop in one go without selecting mature from immature plants. This is normal commercial practice today with lettuce, cabbage, etc. which are hybrid varieties all specially bred to mature together, but in the past the crop would be 'picked over' several times before being finally cleared.
(2) A method of protected cultivation whereby the plants are arranged in lines with paths between the cloches to enable them to be moved from one crop to the next with the minimum of labour.

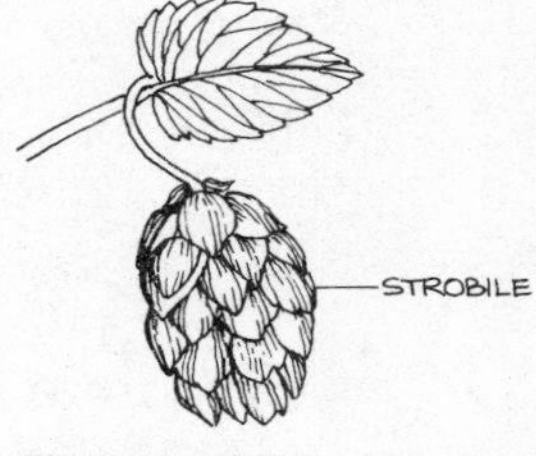

FEMALE HOP FLOWER (*Humulus lupulus*)

Strobile, Strobilus

(1) A cone-like structure, a scaly spike of female flowers as in the hop (*Humulus lupulus*).
(2) A pine-cone or similar inflorescence consisting of imbricated scales.

Stroma

Tissue in which fructification organs of certain fungi are embedded.

Struma

Cushion-like dilatation of a plant's organ.

Strychnine

A crystaline alkaloid obtained from plants of the genus *Strychnos*, especially the *nux-vomica*. It is colourless and extremely poisonous.

Stub

(1) The stump of a tree, or the base of a branch left projecting after the branch itself has been removed.
(2) To stub out a tree is to dig it out completely, including the root.

TREE STUB OR STUMP

Stubble

The short lower ends of grain-stalks left in the ground after harvest.

Stump

(1) The base of a tree left in the ground after felling. *Same as* Stub.
(2) The stem of a plant after the head has been removed, e.g. cabbage.

BONSAI STUNTED GROWTH

Stunt

To check the growth of a plant by chemical or physical means, e.g. bonsai.

Stylar

Having a style.

Style

That part of the reproductive system of a flower between the ovary and the stigma.

Stylopodium

The fleshy swelling at the base of the style in plants of the carrot family (Umbelliferae).

Sub-shrub

A suffrutescent or suffruticose perennial with hard woody stems, such as the hardy fuchsia (*F. magellanica*). Often the soft top growth dies back in winter but in the spring new young shoots are produced. The word is sometimes erroneously applied to a low-growing shrub such as sage (*Salvia officinalis*). *Same as* Undershrub and Suffrutex.

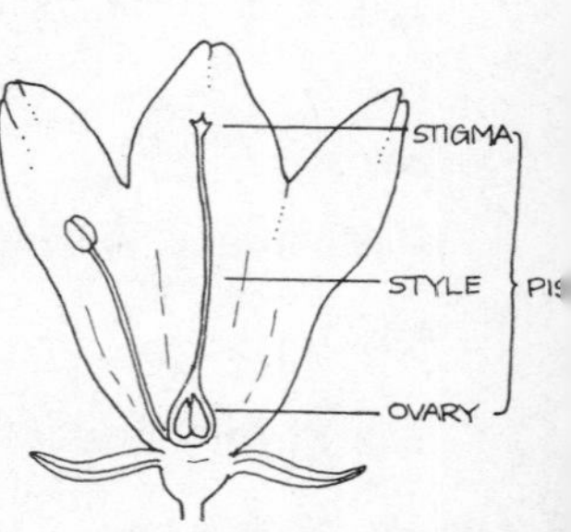

DIAGRAM SHOWING POSITION OF THE STYLE

Subalpine

A plant indigenous to mountain regions just below the alpine zone.

Subapical

Below the apex.

Subcutaneous

Correctly, below the skin, but used horticulturally as below the surface (of a leaf or stem).

Suber

Cork.

Suberin

A mixture of waxy substances present in the thickened cell-walls of many trees; the chemical basis of cork.

Suberose

Suberose trees with corky bark, such as the cork oak (*Quercus suber*), are grown extensively in Spain and Portugal. Every ten years or so the outer bark is stripped from the tree trunks, revealing the inner bark. Some trees are more than 500 years old and have been stripped 50 times.

Suberous

Corky-textured; cork-like.

Subglobose

Not quite spherical.

Subherbaceous

Herbaceous but becoming more woody as it matures.

Subirrigation

Providing water below the soil surface, naturally in a moraine, artificially in some methods of soilless cultivation.

Submerged

Under water; growing below the surface.

Suborder

Subdivision of order, next below order in classification.

Subsoil

The soil immediately beneath the topsoil layer, usually one or two spits down. The subsoil is usually infertile or mainly so.

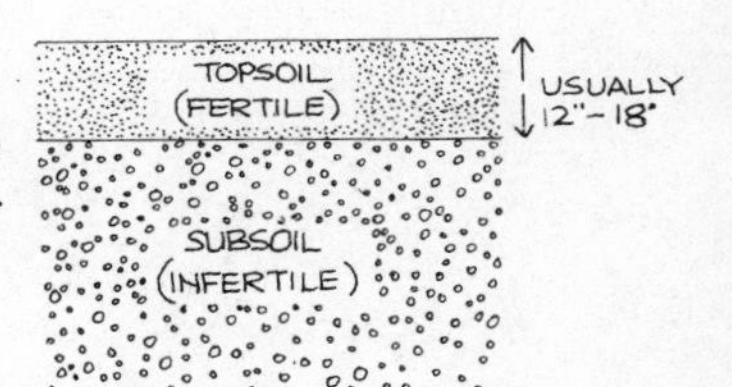

Subspecies

A major subdivision of a species, between species and variety.

Substrate, Substratum

(1) The material in which a plant grows.
(2) A layer of soil below the topsoil.

Subtending

Located close to but below a flower; usually appertaining to a bract, especially when this is prominent.

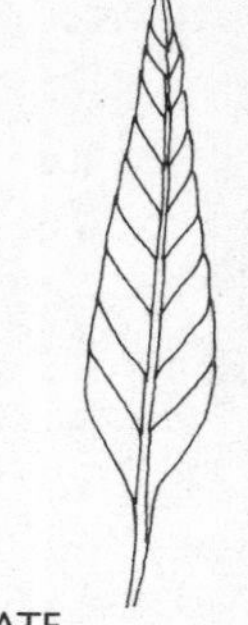

SUBULATE

Subterranean

Under or below the surface of the ground: *cf.* Epiterranean.

Subtropical

Plants originating in subtropical areas of the world, usually grown in heated greenhouses but some may be planted outside in favoured parts of Britain including the south coast of Devon and the west of Scotland.

Subulate

Awl-shaped, tapering from the base to a sharp point, usually applied to leaves.

LEAF SUCCULENT
(*Aloe variegata*)

Succession

(1) The sequence of crops in a garden or plot of land to secure the maximum yield.
(2) The natural sequence of development as the vegetation in an area matures.

Successional

A system of sowing vegetable seed at intervals of two or three weeks to provide a continuous supply for the kitchen. An essential part of a Victorian head-gardener's planning.

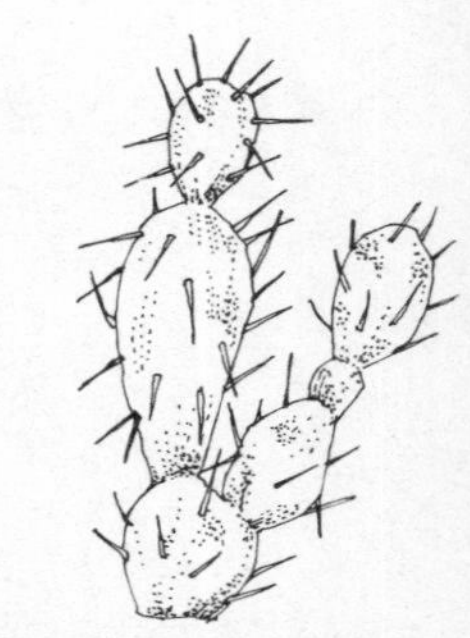

STEM SUCCULENT
(*Opuntia vulgaris*)

Succulent

Thick, fleshy. Usually applied to plants with swollen leaves or stems, enabling them to survive arid conditions, e.g. cacti.

Sucker

(1) A shoot arising from underground near the base of the parent plant and often directly from the rootstock. Also called a Turion.

(2) Any type of sap-sucking insect including aphids, apple-suckers, pear-suckers, etc.

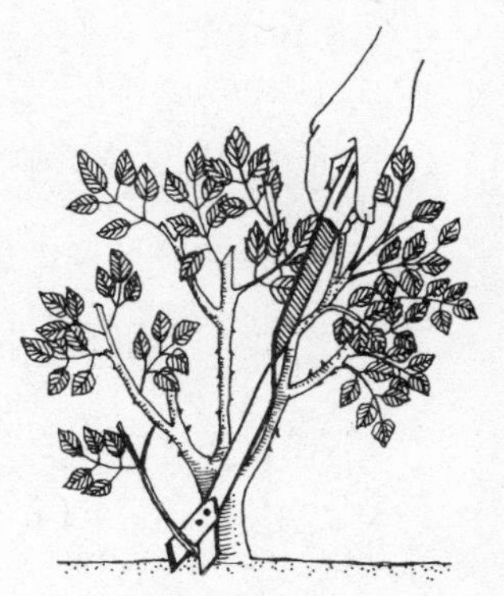
SUCKERING IRON

Suckering Iron

A metal tool specially designed for removing suckers from the rootstocks of roses and other bushes.

Suffrutescent

Slightly shrubby perennials having stems woody at the base but softer higher up. *See* Sub-shrub.

Suffrutex

See Sub-shrub.

Suffruticose

Short and shrubby, and having woody stems: *see also* Sub-shrub.

SULCATE LEAF OF WOOD GARLIC (*Allium ursinum*)

Sulcate

Of leaves that are grooved with longitudinal parallel (or near-parallel) furrows, e.g. bluebell (*Endymion non-scriptus*).

Sulfur

See Sulphur.

Sulphate of Ammonia, SULFATE OF AMMONIA (USA)

A fast-acting nitrogenous fertilizer for use with broad-leaved subjects such as cabbage. The chemical must not touch the foliage or the leaves will be scorched. Sulphate of ammonia is a main ingredient of lawn sand which not only feeds the grass but destroys broadleaved weeds in the lawn.

Sulphate of Iron, SULFATE OF IRON (USA)

Iron is one of the minerals required by plants in very small amounts. It can be applied as a chelate to the soil, or used as a foliar spray when diluted.

Sulphate of Magnesium,
SULFATE OF MAGNESIUM (USA)

An essential trace element usually and easily available as 'Epsom Salt'. It can be incorporated in the soil or used as a foliar spray when dissolved and diluted.

Sulphate of Potash, SULFATE OF POTASH (USA)

Potash is one of the three main plant foods (the K of NPK formulae) essential for the production of fruit and flowers. It can be applied as a top-dressing or watered around established plants.

Sulphur, SULFUR (USA)

One of the trace elements needed by plants in minute quantities. It is also used as a dust to control airborne fungi such as botrytis.

Sulphur-shy, SULFATE-SHY (USA)

Of certain varieties of fruit trees and bushes which can be adversely affected by spraying with a lime-sulphur wash.

Summer Deciduous

Of a plant that naturally loses its leaves in summer.

Summerwood

(1) Outer part of the annual rings in a tree trunk consisting of thick-walled water-conducting tissue.
(2) Growth made by trees and shrubs between June and August – mainly applied to fruit trees: *cf.* Sapwood, Springwood.

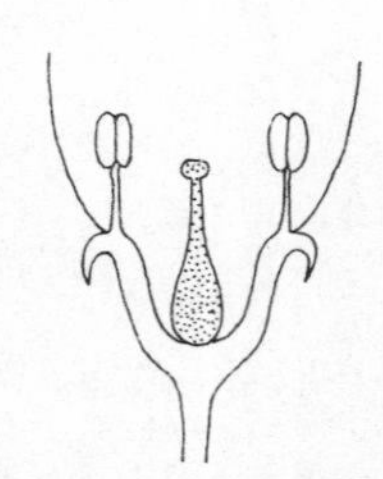
SUPERIOR OVARY

Superficial

On or very close to the surface.

Superior

Upper; above; raised. Of an ovary which occurs above the sepals, petals and stamens: *see also* Inferior.

Superphosphate of Lime

A fertilizer made by treating phosphate-containing rock with sulphuric acid. Often simply called superphosphate.

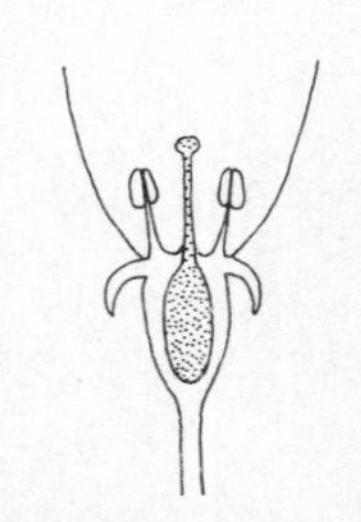
INFERIOR OVARY

Supine

Prostrate; lying flat.

Supple

Pliable and flexible; of a twig or stem that is easily bent.

Support

Any stake or other device used to shore up a plant. Many subjects need supporting if they are to give of their best, e.g. chrysanthemums and runner beans. Young trees must be supported by strong stakes or the wind will rock them and they will die.

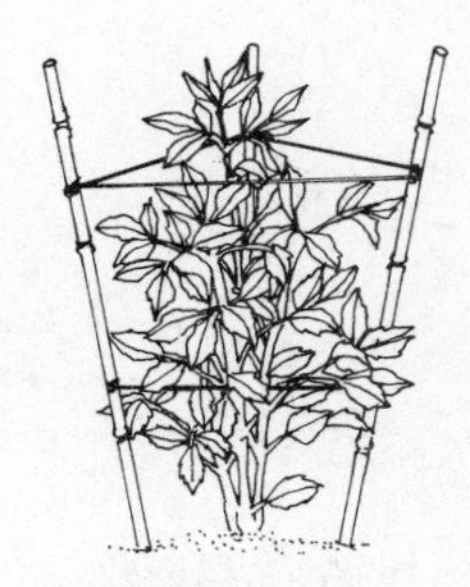

A CHOICE PLANT SUPPORTED BY A FRAME

Suprafoliar

Carried above the leaf or foliage.

Surface Tension

The property of a liquid that results in it behaving as though its surface is covered in an invisible elastic skin. The surface tension of water is responsible for the formation of drops and bubbles, the rise of water in a capillary tube and the wetting of surfaces. Capillarity in a plant is essential for transporting water and nutrients within the plant against gravity.

Surfactant

A detergent added to water (or other liquid) before spraying to reduce its surface tension and so increase its wetting properties.

Suspension

The mixing of finely powdered insoluble fungicides or pesticides with water or other liquid. If stirred frequently, the powder is suspended evenly thoughout the water and can then be discharged through a sprayer.

Sussex Trug

A type of flat basket used for carrying small tools, plants, cut flowers, etc. Originally made in Sussex from strips of interwoven wood, they can now be obtained in plastic.

SUSSEX TRUG

Suture

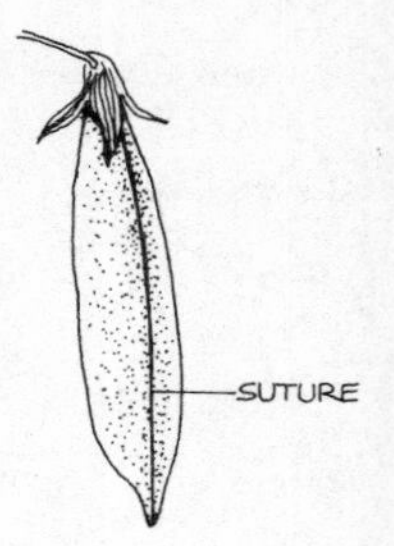

A seam or join, especially of carpels; a line along which dehiscence occurs.

Swag

See Festoon.

Swamp

Area of wet marshy land, usually too wet for cultivation but often supporting some natural plant growth such as rushes (Juncaceae).

Swap Hook

A sickle.

Swarm Spore

Type of spore produced by certain fungi which has limited powers of mobility in water films in soil or plant surfaces.

Swath, Swathe

(1) Row of corn, etc. as it falls where reaped and left to dry on the ground.
(2) Width of grass or corn cut by the mower.
(3) The extent of the sweep of a scythe.

Sweal

To scorch or burn off, as heather, gorse, etc.

Switch

Slender flexible shoot or cane cut from a tree and often used to control animals.

Switch Plant

A plant with long slim green shoots and few or no leaves, the shoots performing the same function as leaves. Many brooms (*Cytisus*) are switch plants.

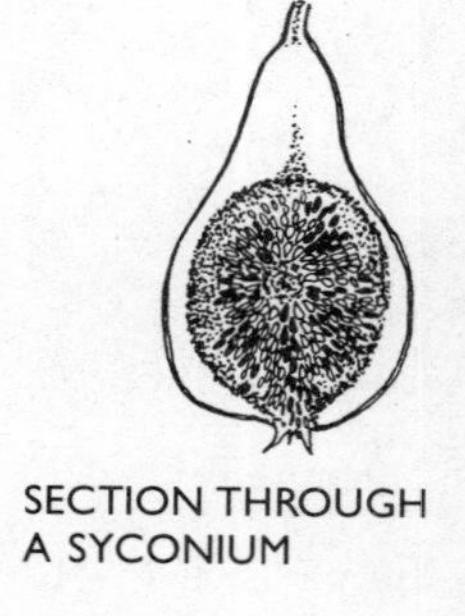

SECTION THROUGH A SYCONIUM

Syconium

A multiple fruit in which the true fruits (pips) are contained in a hollow open-ended receptacle e.g. the fig.

Sylvan

Wooded; abounding in woods; groups of trees. A sylvan glade is one surrounded by numerous trees.

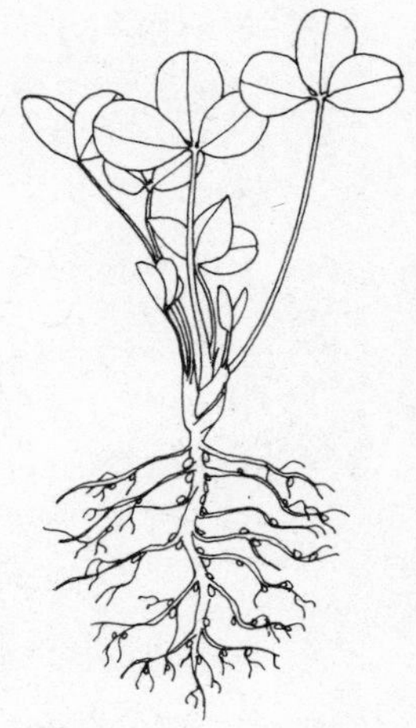

SYMBIOSIS NODULES OF NITROGEN-FIXING BACTERIA ON CLOVER ROOTS

Symbiosis

The association of two different plants which live attached to each other and contribute to each other's support. Different to parasitism in which one organism preys upon another. *Same as* Commensalism and Consortism.

Sympatric

Groups of similar organisms which although theoretically capable of interbreeding do not do so because of different flowering times or other preventative reasons.

Sympetalous

See Gamopetalous.

Symphilids

Soil-borne pests which attack the roots of many subjects including tomatoes, especially if grown in unsterilized soil. In appearance they resemble tiny white centipedes about $\frac{1}{8}$ in (3 mm) long.

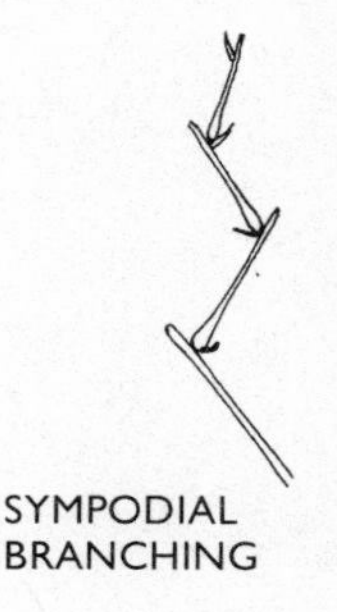

SYMPODIAL BRANCHING

Sympodial

Having a stem composed of several branches, each seeming to continue the parent branch, as in some orchids and vines: *cf.* Monopodial.

Synandrium

A mass of united stamens.

Synangium

A cluster of united sporangia.

Syncarp

A compound fruit formed from two or more carpels, or when separate pistils within one flower are united or partially so, e.g. the pineapple (*Ananas*).

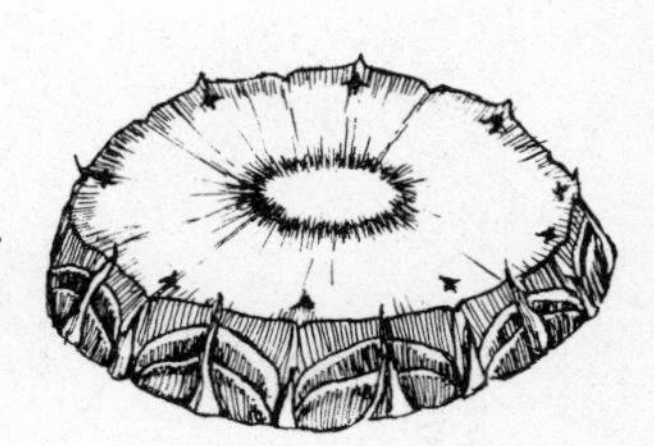

SYNCARP (PINEAPPLE)

Syncarpous

Of ovaries with two or more fused carpels.

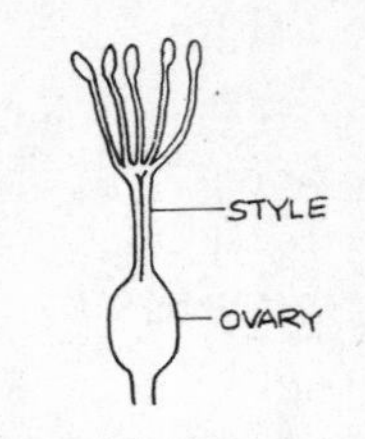

SYNCARPOUS

Synoecious

Having male and female flowers in the same flowerhead, e.g.: narcissi.

Synonym

A systematic name to which another is preferred; or an earlier name now replaced; or in some cases a name in common use but misapplied.

Synsepalous

Having sepals united.

Systemic

A type of insecticide which is absorbed into the structure of the plant and circulates in the sap, thus destroying only sap-sucking insects: *cf.* Contact.

T

Tactic Movement

The movement of a cell in response to an external stimulus, such as light. *Same as* Taxis.

Tamping

Firming; gently compacting.

Tang

(1) A prong, or tine, of a garden fork.
(2) The tapering end of a file or trowel which is secured in a handle.

Tannin

One of a group of organic chemicals often present in leaves, unripe fruit, etc. It is thought that the unpleasant taste discourages grazing by animals and attacks by birds.

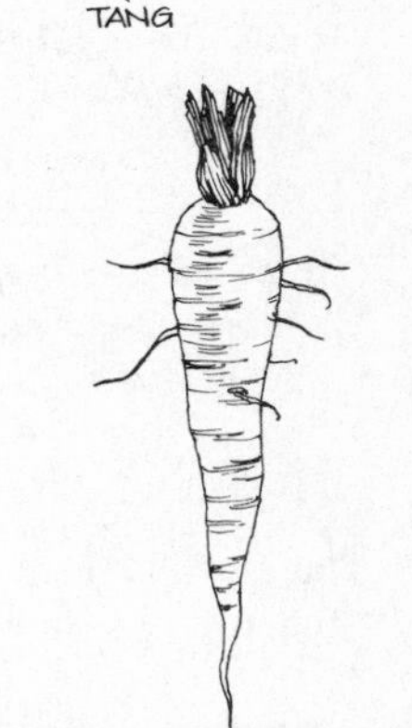

TAP ROOT (PARSNIP)

Tap Root

A strong main root descending vertically with little or no lateral growth. Well-grown carrots and parsnips develop tap roots but in some species (e.g. conifers) the tap root can be as much as 6–8 ft (1.8–2.4 m) long.

Tapering

Becoming gradually smaller toward one end.

Tapetum

A layer of nutritive tissue mainly found in the reproductive organs of plants.

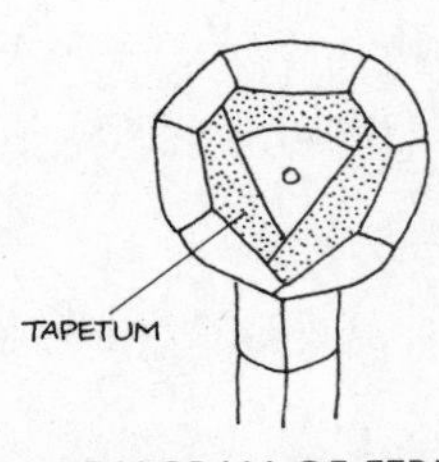

DIAGRAM OF FERN SPORANGIUM

Tapis-vert

Ornamental geometrical garden layout, usually executed partially or entirely in dwarf box.

Tar Disc

Impregnated disc of tarred felt or thick paper about 3 in (7.5 cm) in diameter with a radial slit and central hole. Widely used in earlier times for the control of cabbage root fly, now largely superseded by chemical insecticides.

TAR DISC

Tar Oil Wash

Otherwise known as winter wash because it can be applied only during the dormant period of deciduous trees or the foliage will be severely scorched. Used to control a variety of pests such as aphids, scale insects, caterpillars, etc.

Tare

Types of vetch; the biblical name for darnel (*Lolium temulentum*), a species of rare rye-grass.

Tarsonemid Mites

Tiny sap-sucking, almost invisible creatures (Tetranychidae) which are particularly harmful in greenhouses and on houseplants.

Tassel

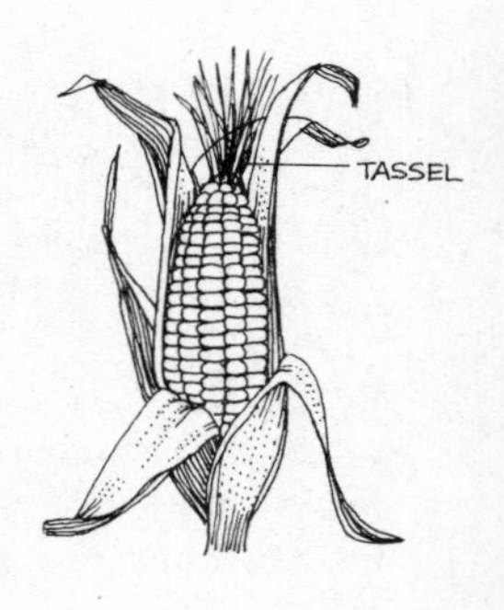

The inflorescence at the top of a stalk of Indian corn (maize) and some other plants.

Taxis

See Tactic movement.

Taxon, *pl.* Taxa

A biological category or group.

Taxonomy

The general principles of classification.

Ted

To turn over hay or grass and spread out to dry before gathering.

Tedder

An implement for turning over hay or grass, usually towed behind a tractor or horse.

Teleutospore, Teliospore

A type of spore developed by rusts to survive the winter. Also known as 'winter spores', they normally require a period of dormancy before becoming active: *cf.* Uredospore.

Tender

A vague term used to describe a plant that will not withstand frost.

Tendril

The thread-like coiling organ of a plant enabling it to climb or grasp a support. May be part or all of a stem, leaf or petiole.

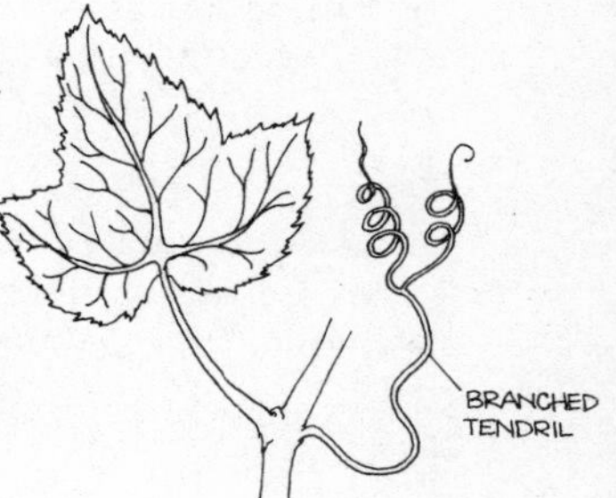

Tent Caterpillars

Otherwise known as web caterpillars. A group of larvae which live communally inside a web that covers a shoot of a plant on which they feed.

Tentacle

Sensitive, flexible hair or filament.

Tenuous

Slender; thin; delicate.

Tepal

A flower segment not clearly distinguishable as being either a petal or a sepal, e.g. crocus and tulip.

Teratological

Abnormal or malformed growth; distorted.

Terete

Smooth and cylindrical, usually circular in cross-section but sometimes tapering.

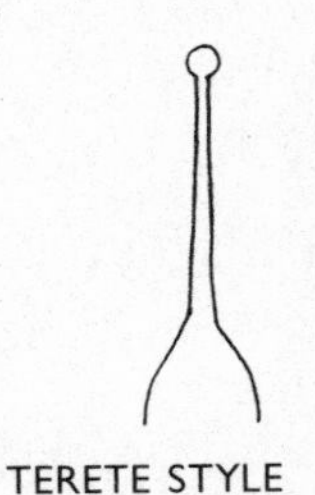
TERETE STYLE

Terminal Point

The bud or shoot at the apex of a plant; the tip or growing point.

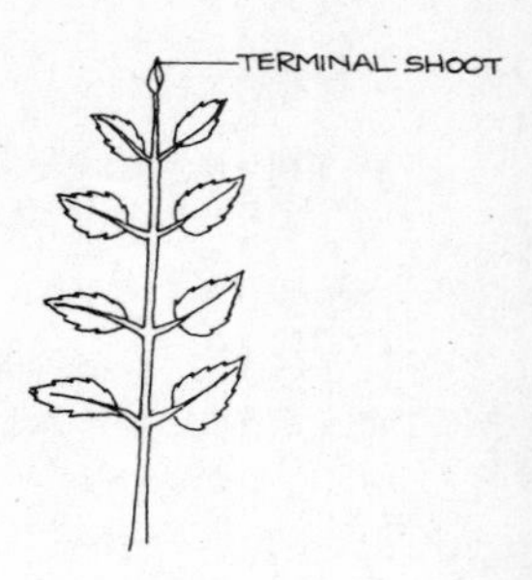

Ternate

Having three leaflets; grouped in threes or a leaf divided into three parts, as in laburnum.

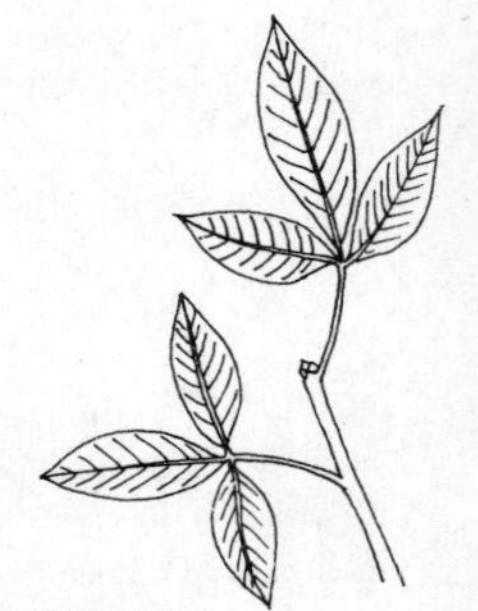
TERNATE
(*Laburnum anagyroides*)

Terrarium

A completely enclosed glass vessel in which to grow certain plants, the moisture transpired by the plants being returned to the soil by condensation: *cf.* Wardian case.

TERRARIUM

Terrestrial

(1) Living or growing in the soil.
(2) A land plant as opposed to an aquatic or epiphyte.

Tessellate

Mosaic-patterned or chequered leaves or petals, or having a marbled appearance as in the flowers of the snake's head fritillary (*Fritillaria meleagris*).

TESSELLATE
(*Fritillaria meleagris*)

Testa

A hard shell or seed coat.

Testaceous

Of a brownish-red colour; brick red.

Testiculate

Applies to certain orchids which grow from double tubers shaped like testicles.

Tetrad

A group of four, such as a group of four pollen grains or spores.

Tetradynamous

Having four long stamens in pairs and two short, as the Cruciferae.

TETRADYNAMOUS STAMENS

Tetragonous

Having four angles and concave faces.

Tetrahedral

Pyramid-shaped, with a base and three sides – four surfaces in all.

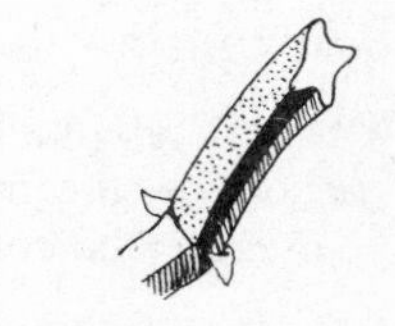

TETRAGONOUS STEM OF *Pelargonium tetragonum*

Tetramerous

Having four parts, or with parts in fours. Sometimes written 4-merous in technical journals.

Tetrandrous

Having four stamens.

Tetraploid

Having four basic sets of chromosomes instead of the more usual two (diploid).

Tetraspore

Spores formed in groups of four in red seaweeds.

Thalamus

The receptacle of a flower.

Thallophyte

A member of the lowest main division of the vegetable kingdom (Thallophyta) including fungi, bacteria, etc. A plant whose body is a thallus. This is an out-of-date term as these plants are not now considered to be closely related; it includes seaweeds.

Thallus

A plant body not differentiated into leaf, stem and root, such as algae, bacteria, fungi, etc.

THALLUS OF LIVERWORT (*Pellia epiphylla*)

Thatch

Reeds or straw arranged to form a waterproof covering for houses, barns, etc. or for the protection of root crops in winter storage clamps.

Theca

(1) A sheath, case or sac; a spore case.
(2) One half of an anther containing two pollen sacs.

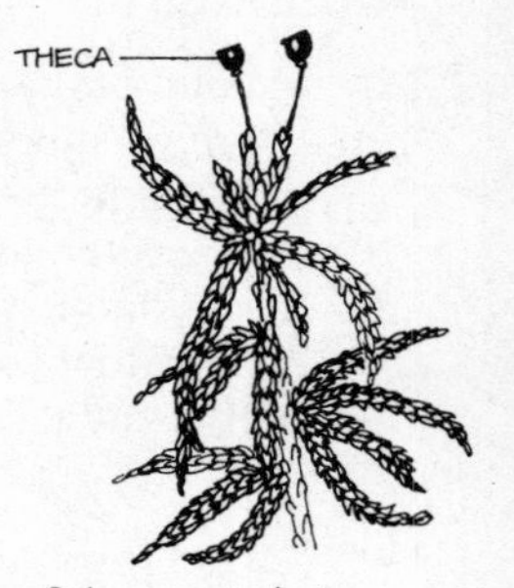

Sphagnum palustre

Thermophile

A type of bacteria needing a high temperature for development.

Thermotropism

The ability of plants to turn towards (positive thermotropism) or away from (negative thermotropism) a source of heat such as the rays of the sun.

Thicket

(1) An area of dense undergrowth, usually with some shrubs and small trees.
(2) An overgrown area of scrubland which has reverted to the wild.

Thigmotropism

The response of certain organs of some plants to physical contact, as the tendrils of sweet pea which twine around a support when contact has been made.

Thimble

The smallest size of flowerpot in the days of clay pots, measuring 2 in (5 cm) wide at the rim and 2 in (5 cm) deep. Much smaller plastic pots can nowadays be obtained.

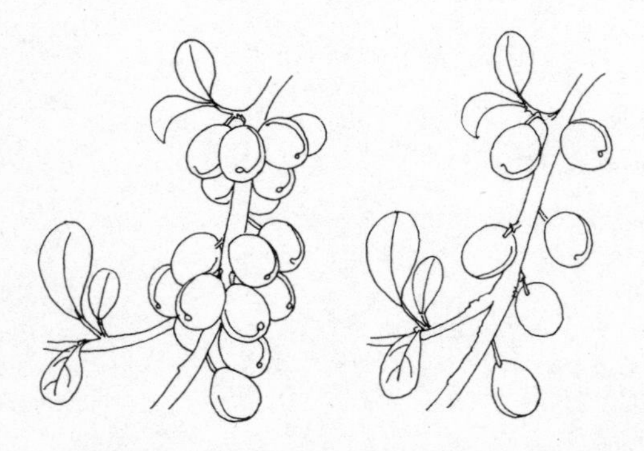

PLUMS, BEFORE AND AFTER THINNING

Thinning

(1) Reduction of the number of plants in beds or boxes to allow the remainder room for development. The same principle is applied to orchards and forests where alternate trees are removed.
(2) The reduction in the crop of fruitlets at an early stage of development so that fewer larger fruits are eventually produced, e.g. grapes.

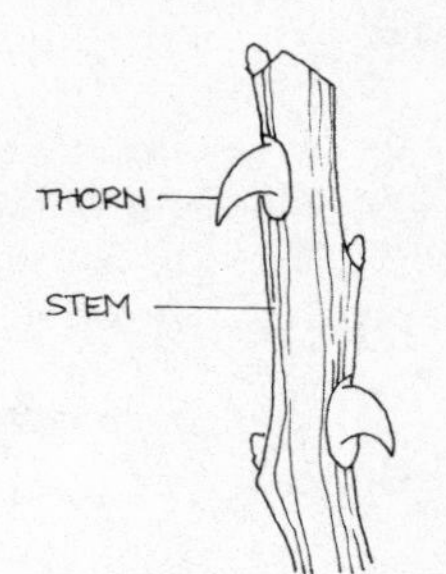

THORN ON A ROSE STEM

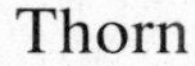

Thorn

Correctly a sharp, spine-like extension from a stem or other part of a plant. Some authorities differentiate between thorns and prickles – thorns being carried on stems (as in the rose) and prickles on leaves (as in holly).

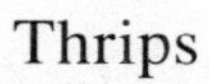

Thrips

A tiny sap-sucking insect of the order Thysanoptera with four hair-fringed wings. Many are injurious to plants. Otherwise known as thunderbugs or thunderflies because they are more troublesome in hot weather. Thrips is both singular and plural.

THRIPS

Throat

The tubular opening of a calyx or perianth.

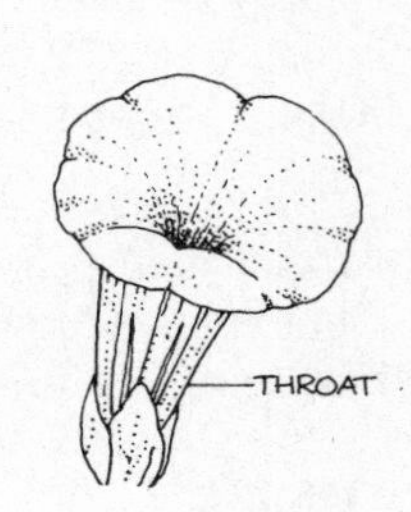

Thrum-eyed

A short-styled inflorescence with the stamens in the throat of the corolla as in primulas: *cf.* Pin-eyed.

Thumb

The second-smallest size of flowerpot in the days of clay pots. It measured 2$\frac{1}{2}$ in (6 cm) wide at the rim and 2$\frac{1}{2}$ in (6 cm) deep.

Thyrsus, Thyrse

A dense panicle, especially one whose lateral branches are cymose as in the horse-chestnut.

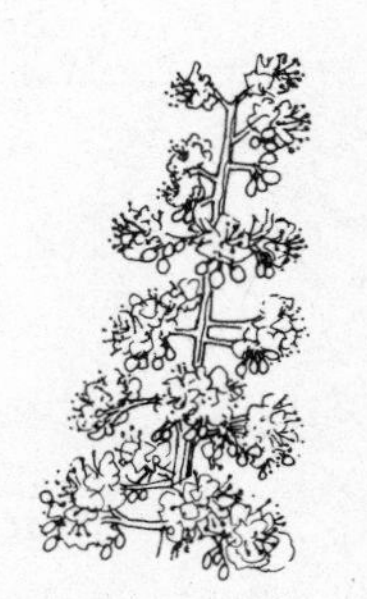
THYRSUS OF HORSE-CHESTNUT (*Aesculus hippocastanum*)

Tillage

The results of tilling; the condition of the soil after tilling.

Tiller

(1) A sapling; a sucker from the base of a stem.
(2) A side-shoot arising from the base of a plant, particularly applied to grasses.
(3) One who tills the soil in order to prepare it for sowing.

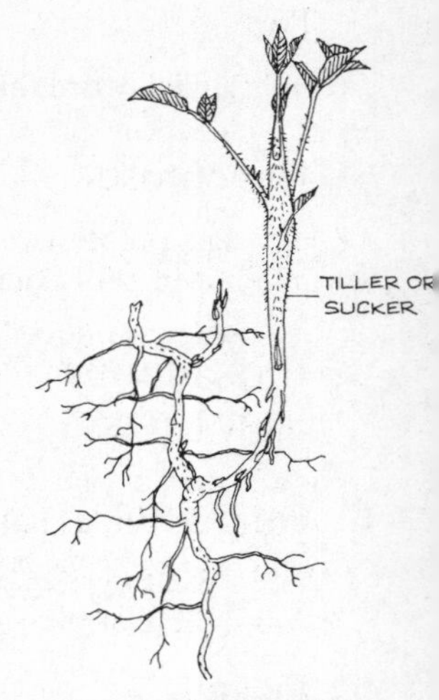

TILLER OF RASPBERRY

Tillering

The production of dense growth from the base of a cut-down tree.

Tilling

The act of working or cultivating the soil, whether manually or using an implement such as a plough.

Tilth

The fine crumbly surface layer of soil produced by tilling.

Tine

A prong or spike of a garden tool such as a fork or rake. *Same as* Tang.

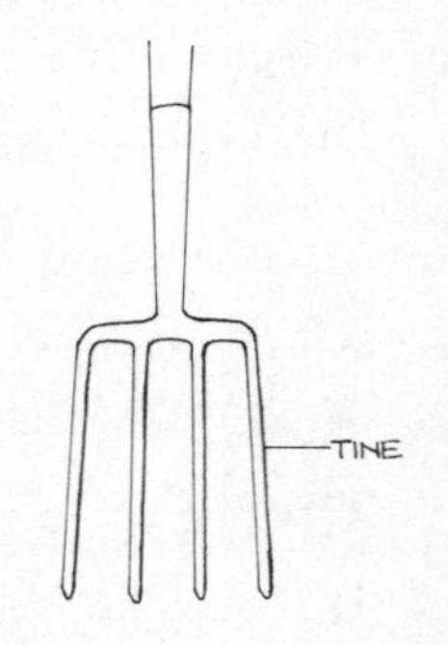

Tip-bearer

A fruit tree which produces fruit buds mainly on the tips of the young shoots, e.g. the apple Worcester Pearmain. Such varieties are unsuitable for training as cordons, espaliers, etc.

Tissue

The material of a plant body; an aggregate of similar cells.

Tissue Culture

The growing of detached pieces of plant tissue in nutritive fluids.

Toadstool

A general term covering a number of fungi with fruiting bodies similar to mushrooms. Many are edible but some are toxic – a few fatally so, including the death cap (*Amanita phalloides*).

Tod

Old English word for a bushy mass of plants, especially ivy.

Tomentose

Having a cottony matted pubescence; furry; densely covered in short fine hairs.

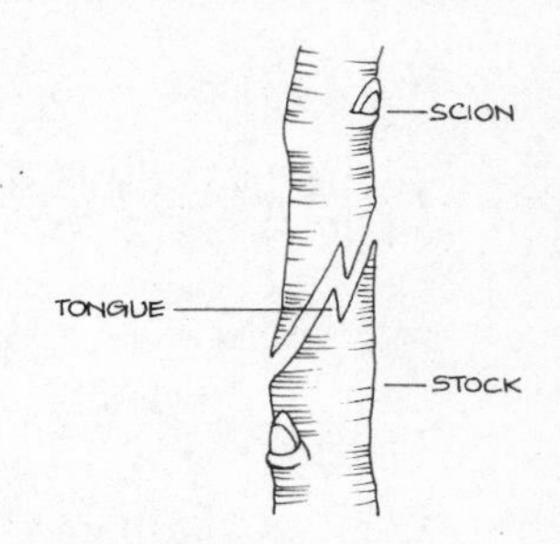

TONGUE GRAFTING

Tomentulose

Finely or slightly tomentose.

Tongue Grafting

Whip grafting in which a wedge-shaped scion is inserted into a cleft in the stock. *See* Grafting.

Tooth

A marginal lobe on a leaf or petal, usually small, often pointed or angular, occasionally rounded.

Toothed

Serrated in a number of ways, as applied to the margins of many leaves; one with regular teeth is called dentate, when the teeth are angled it is serrate, with small teeth serrulate, and with rounded teeth crenate.

Top

To remove the top or growing point of a plant to encourage development of side-shoots.

Top Dressing

(1) Scattering fertilizer on the surface of the soil and allowing it to be incorporated naturally without digging-in.
(2) Removing the top inch or so of soil from pot plants and replacing with fresh compost.

TOPIARY

Topiary

The ancient art of training and clipping hedges and trees into fancy shapes. Box and yew are the most used for this purpose, some complicated specimens being trained on wire frames.

Topping

Removing the growing tip of a plant to force the production of laterals (stopping), also removing the apex of broad bean plants to control black fly infestation.

Topsoil

The spit of soil nearest the surface down to a depth of 10–12 in (25–30 cm) below which lies the subsoil.

Tortrix Moth

Small moth from whose eggs leaf-rolling caterpillars are produced.

Tortuous

Having many twists and turns; contorted, as the stems of the twisted willow *Salix tortuosa*.

Torulose

Having small bulges or swellings at intervals.

Torus

(1) The receptacle of a flower.
(2) The disc-shaped structure in vessel segments, part of the water-conducting tissue of plants.

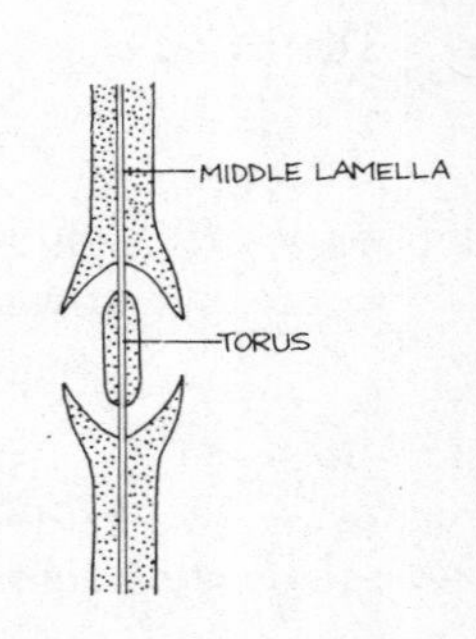

Total Weedkiller

A weedkiller that will destroy any plant with which it comes into contact, e.g. ammonium sulphamate (AMS). Not to be confused with ammonium sulphate, which is a chemical fertilizer.

Toxaphene

Chlorinated camphene used as an insecticide.

Toxic

Poisonous.

Toxin

A specific poison of organic origin.

Trace Elements

Chemical substances essential in very small quantities to plant life, the main ones being boron, copper, iron, manganese, molybdenum and zinc.

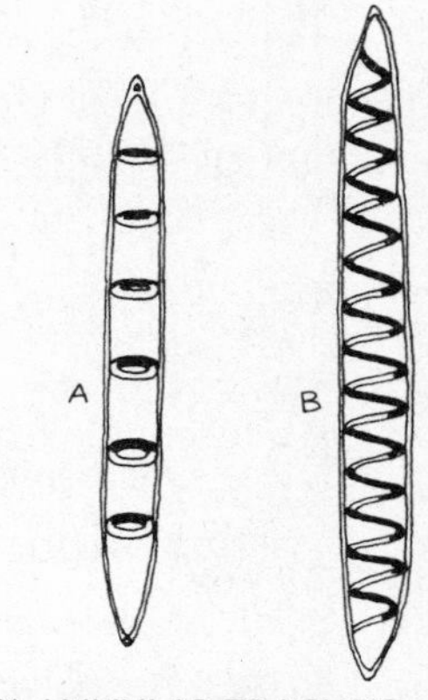

(A) ANNULAR TRACHEID
(B) SPIRAL TRACHED

Trachea

A duct or vessel, a fluid-conducting tube in xylem.

Tracheid

A long tube-like, but closed, cell in xylem.

Trailing

Of plants with long stems which would naturally extend along the ground, rooting on the way, but which are often trained to grow upward or downward in special containers. Trailing plants are unable to climb without assistance.

TRAILING PLANT
(*Sedum morganianum*)

Translocation

The natural process by which water and nutrients travel through the vascular system of a plant.

Transpiration

The continual evaporation of water from leaf and stem surfaces. The rate of transpiration is affected by several factors such as temperature, light, humidity, etc. It is controlled to an extent by the guard cells of the stomata.

Transplant

To transfer a plant from one place and establish it in another.

Trapeziform

Four-sided, with only one pair of sides parallel.

Tread

The upper horizontal edge of the blade of a digging spade or fork upon which the foot is placed to force the tool into the ground. It is sometimes widened or flattened to provide better support for the boot.

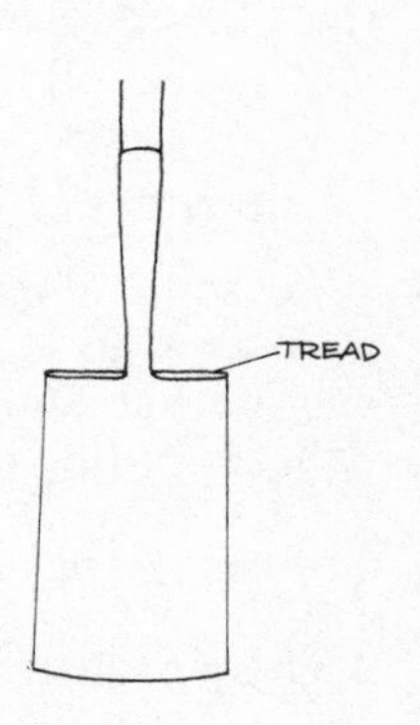

Tread-in

To press into the soil with the foot.

Treading

Firming the soil with the feet, usually before sowing seed.

TREADING

Tree

A large perennial plant with a single woody stem and elevated crown of branches usually at some distance from the ground.

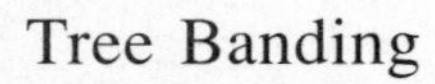

Tree Banding

The fixing of grease bands around the bole or main branches of a tree to trap crawling ascending insects. Sticky bands serve the same purpose as do bunches of hay or straw tied around the tree trunk, but these must be inspected frequently and the trapped pests destroyed.

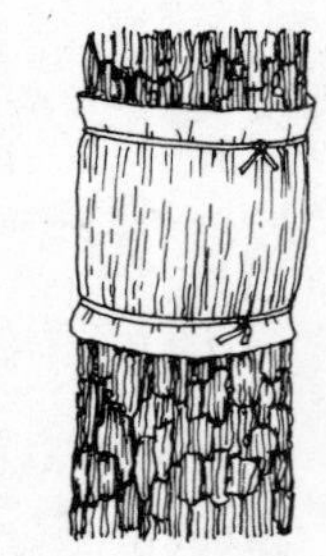
GREASE BANDING

Trellis, Trelliswork

Structure of intersecting narrow strips with open square spaces in between, used as a support for plants or as a screen. Can be made of wood or plastic.

Trembling Bog

Swampland over water or soft mud, shaking at every footstep; also known as a quaking bog.

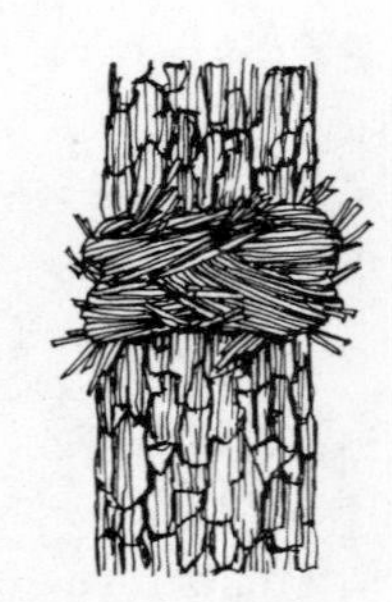
STRAW BANDING

Trench

(1) Long narrow slit cut out of the ground.
(2) A method of digging or ploughing so as to bring the lower soil to the surface.

Trenching

A method of digging to a depth of nearly 30 in (75 cm) by removing the soil in three spits whilst retaining the various strata at their original levels: *cf.* Full trenching, Mock trenching, Ridging.

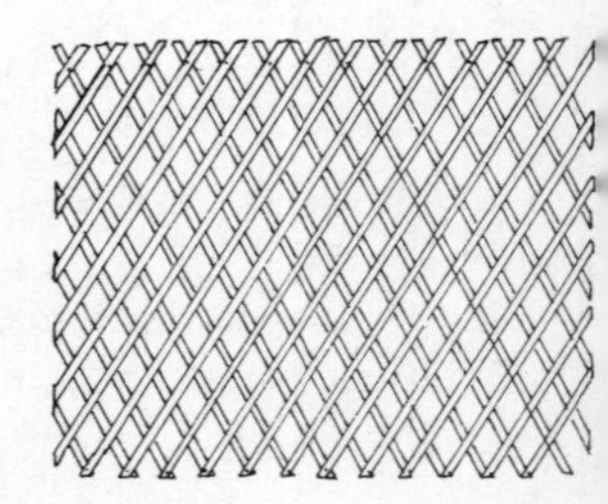
TRELLIS

Triad

A group of three.

Triadelphous

Having stamens united by the filaments into three bundles.

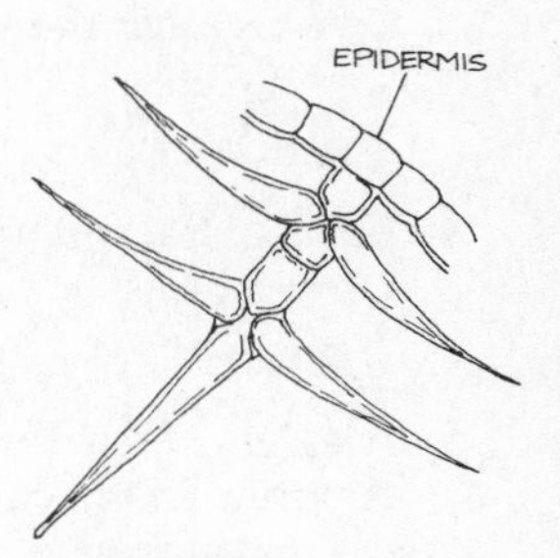

TRICHOME ON LEAF OF COMMON MULLEIN (*Verbascum thapsus*)

Tribe

A term, rarely used today except by taxonomists, for a group of plants ranking below subfamily and above genus, with names usually ending in -eae such as Bambuseae and Paniceae, which are tribes of grasses.

Trichome

A plant-hair or bristle growing from the epidermis, distinct from an emergence which grows from the inner tissues.

Trichotomous

Divided or forking in threes.

Trickle Irrigation

A system of automatic watering and feeding whereby the fluids trickle or drip slowly over a long period into the compost in which the plant is growing. *Same as* Drip-feeding; *cf.* Spaghetti watering.

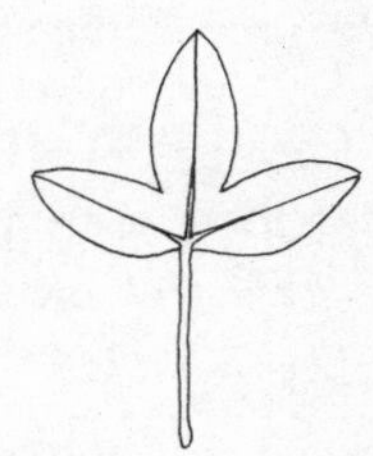
TRIFID

Tricolpate

Having three grooves, as in some pollen grains.

Trifid

A leaf cleft into three parts but not to the base.

Trifoliate

Having three leaves, or leaves in groups of three.

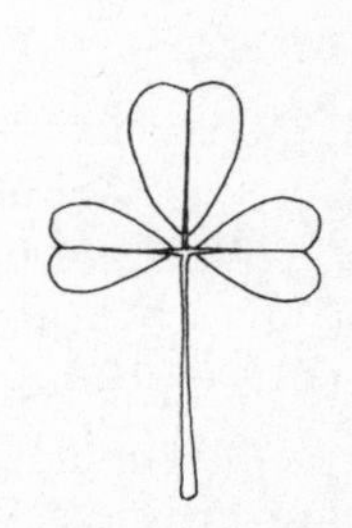
TRIFOLIATE

Trifoliolate

A leaf composed of three leaflets, sometimes called a trefoil.

Trigonal

Triangular, having three sides and angles, not necessarily equal.

Trijugate

Having three pairs of pinnate leaves.

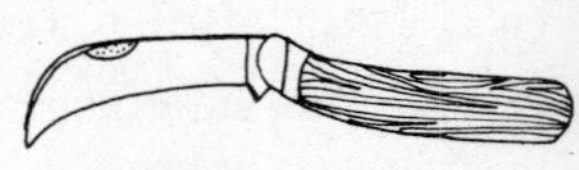
TRIMMING KNIFE

Trimerous

Having three parts, or parts in groups of three. Often written 3-merous in technical journals.

Trimming Knife

A pruning knife, particularly one with a folding blade.

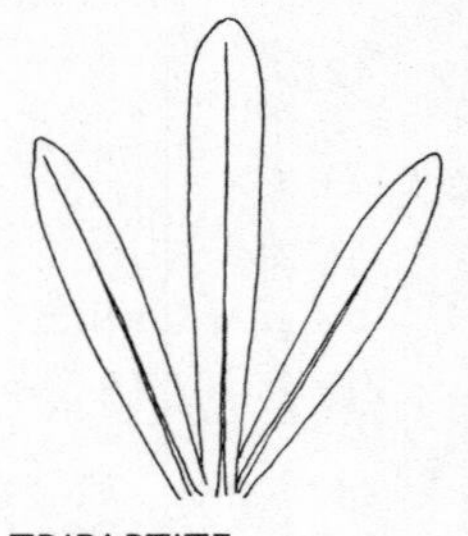
TRIPARTITE

Trimorphic

Occurring in three forms in the same species.

Tripartite

Divided into three parts or types; leaves, sepals, petals, etc. having three lobes.

Tripinnate

Applied to a bipinnate leaf with leaflets further divided pinnately.

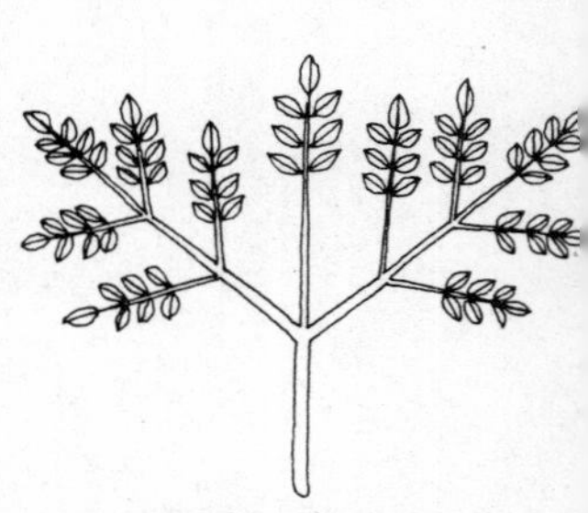
TRIPINNATE

Triploid

Having three basic chromosome sets rather than the usual two. Triploid plants are usually sterile.

Triquetrous

Triangular; a three-sided stem with concave faces in cross-section.

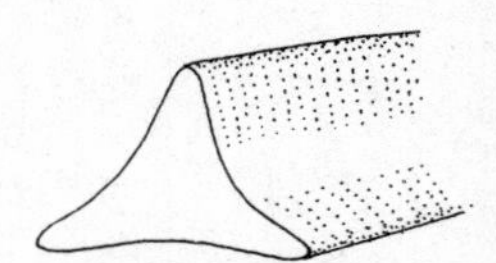
SECTION OF A TRIQUETROUS STEM

Tropical

Of plants that originate in the tropics and need an optimum temperature of 70°–80°F (21°–27°C).

Tropism

The turning of an organism or part of one (such as a leaf) in response to an external stimulus, e.g. light.

Trowel

A gardener's hand-tool with short concave blade. There are several types and designs.

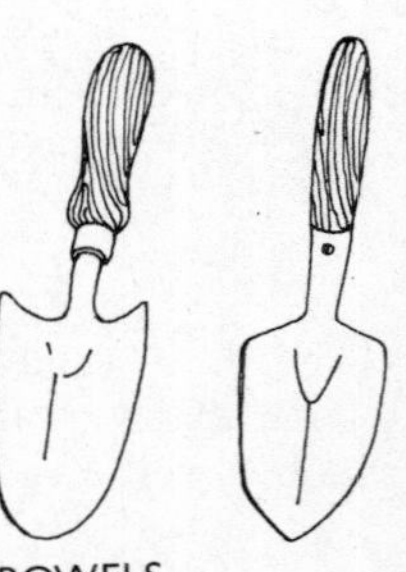
TROWELS

True

Seedlings retaining the characteristics of the parent. Homozygous.

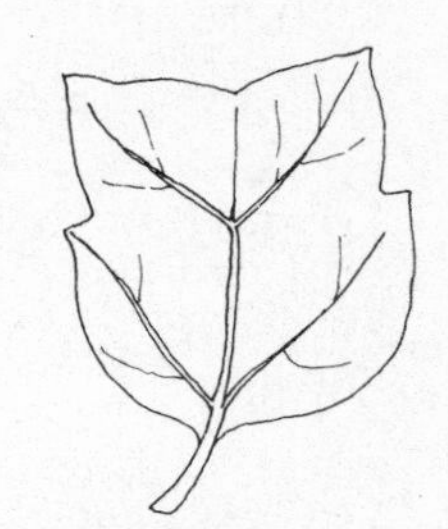

TRUNCATE

True Leaves

The first leaves of the plant following the cotyledons or seed leaves.

Trug

Shallow garden basket, formerly made of wood strips, now often a plastic imitation. *Also called* a Sussex trug.

Trumpet

A loosely applied term to describe flowers with trumpet-like coronas, as in narcissi.

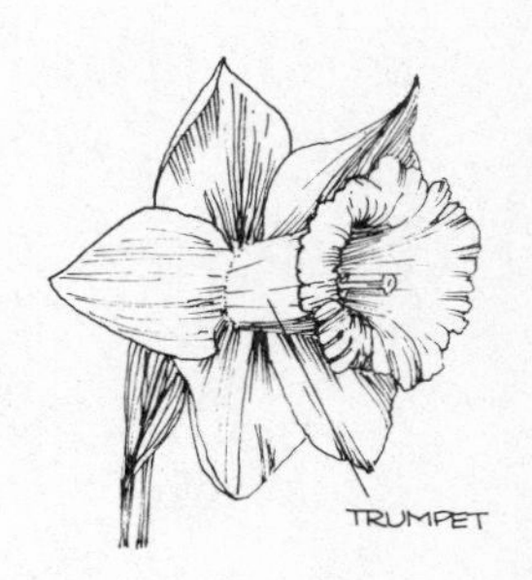

Truncate

Of a leaf, appearing as if cut off squarely or nearly so at the tip.

Trunk

Main stem of a tree as distinct from roots and branches.

Trunk-slitting

Scoring vertically down the trunk of a tree with a sharp knife in order to slit the bark and thus release the pressure to allow the trunk to expand.

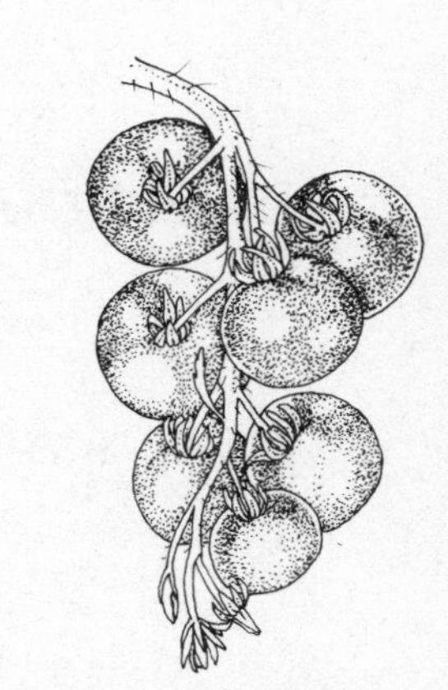

TRUSS OF TOMATOES

Truss

A cluster of fruits or flowers growing on one stem.

Tube

The long tubular section occasionally occurring between the base and perianth of a flower.

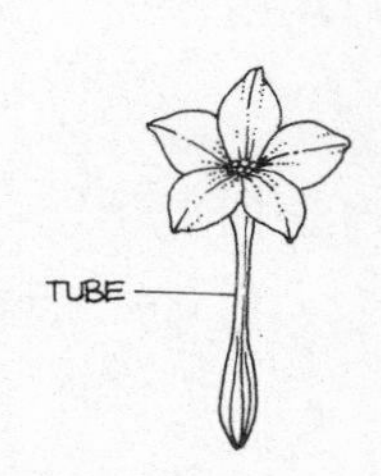

Tuber

The usually underground storage organ of certain plants such as the potato and dahlia, from which fresh growth will develop in due course.

Tubercle

(1) A small tuber or similar type of root-growth.
(2) A small protuberance upon a stem or leaf. On a cactus with tubercles, such as *Mammillaria*, the areoles are usually on the tubercles.

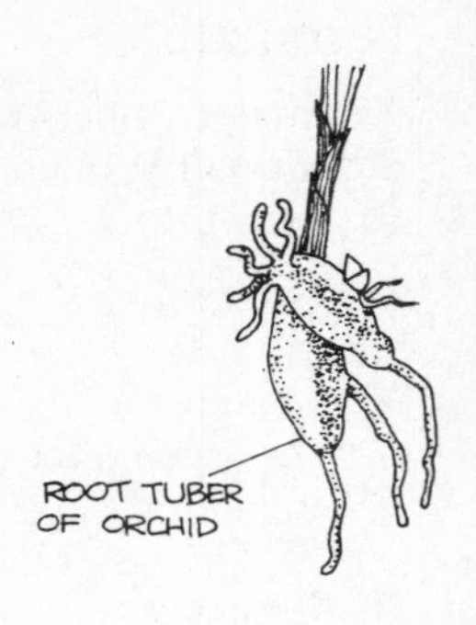

Tuberous

(1) Having or producing tubers.
(2) Tuber-like.

Tubule

Minute tubular structure in a plant.

Tufa

Correctly, a porous deposit of calcium carbonate found around mineral springs, but now widely employed to describe a man-made porous rock-like substance manufactured from peat and cement, often used in the construction of artificial rock gardens, etc.

Tufted

Having many short crowded branches arising at or near ground level; not spreading, densely clustered.

Tumid, Tumescent

Swollen or enlarged.

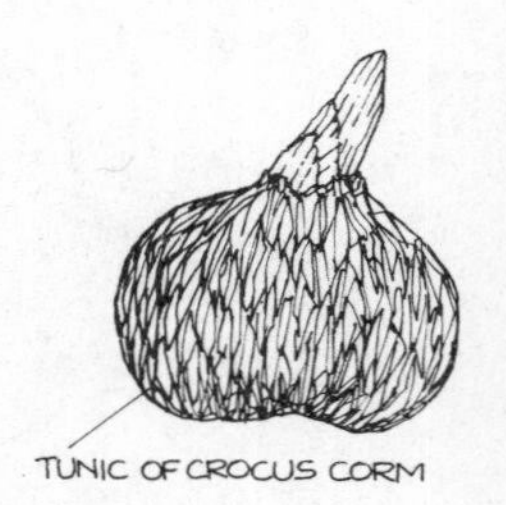

Tump

A hillock or clump; a mound, usually grass-covered.

Tunic

The loose flaky outer skin or membrane around a corm or bulb. All corms, but not all bulbs, have tunics.

Tunicate

Having several concentric layers, as in onions.

TUNICATE

Turbary

(1) Area of land given over to the digging of turf or peat.
(2) The right to extract peat from another's land.

Turbinate

(1) Shaped like a top or inverted cone.
(2) Spirally coiled.
(3) Scroll-like.

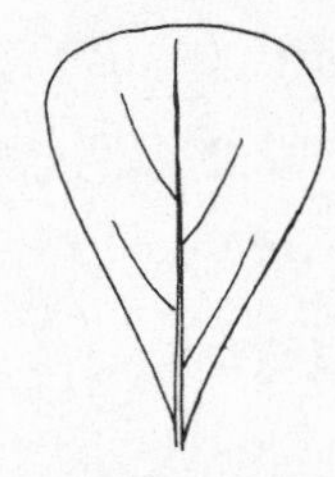

TURBINATE LEAF

Turf

(1) An area of land matted with the roots of grass.
(2) A rectangular piece of grass with matted roots for replanting elsewhere; a sod in USA.
(3) A slab or block of peat dug for fuel, mainly in Ireland.

Turf Seat

A seat covered in turf set on to a raised bank, popular in Britain in the sixteenth century.

Turfing Iron, SOD-LIFTER (USA)

A tool used for lifting turves after the vertical cuts have been made with an edging tool.

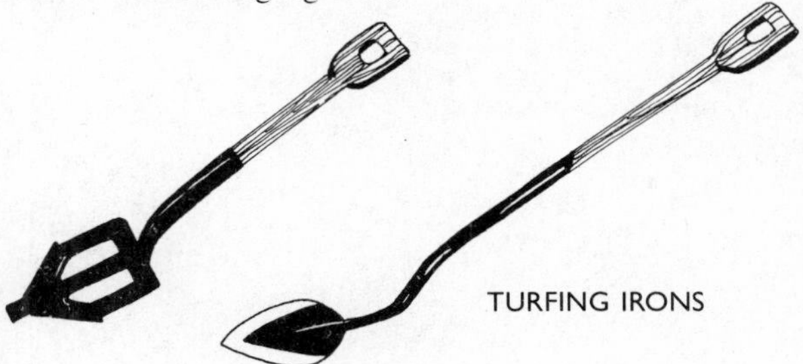

TURFING IRONS

Turgid

(1) Swollen beyond the natural size; distended with fluid contents; inflated.
(2) The balance of osmotic pressure and elasticity of the cell-wall when plants are fully charged with water.
(3) Not wilting. Wilting is the result of a decrease in turgidity.

Turgor

The condition of being stiff or rigid, due to the uptake of water into living tissues.

Turion

(1) An underground bud on a rhizome from which a new stem will arise.

(2) A young shoot or sucker.
(3) An overwintering bud on certain water-plants.

TURION OF ASPARAGUS

Turk's Cap

A flower that resembles an ancient Turk's cap, e.g. the Pyrenean lily (*Lilium pyrenaicum*) in which the petals are severely reflexed.

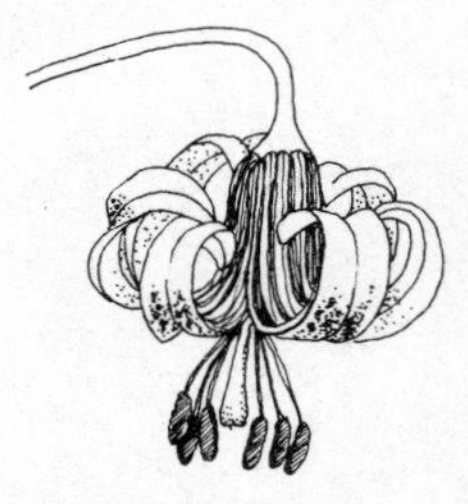
TURK'S CAP LILY (*Lilium pyrenaicum*)

Turnsole

A plant whose flowers are supposed to turn to follow the sun, such as heliotrope (*Heliotropium arborescens*).

Tuskar, Tusker

A peat-spade.

Tussock

Small hillock of grass, sedge, etc.; a tuft or clump, usually grass-covered.

Twig

(1) A small shoot or branch; a side-growth.
(2) A water-divining rod.

Twin

Growing in pairs.

Twiner

A plant which climbs and twines around another for support, such as honeysuckle (*Lonicera* species).

TWINER

Twitch

Twitch-grass or quitch-grass. *Also called* Couch-grass.

Twitten

A narrow path between two walls or hedges. Also known as a ture in some rural areas.

Tylosis

An ingrowth from a neighbouring cell through a pit into a vessel.

DIAGRAM OF TYLOSES IN A VESSEL

U

U Cordon, Double-U Cordon

Methods of training fruit trees and bushes, more often used with soft fruits such as red and white currants: *see also* Cordon.

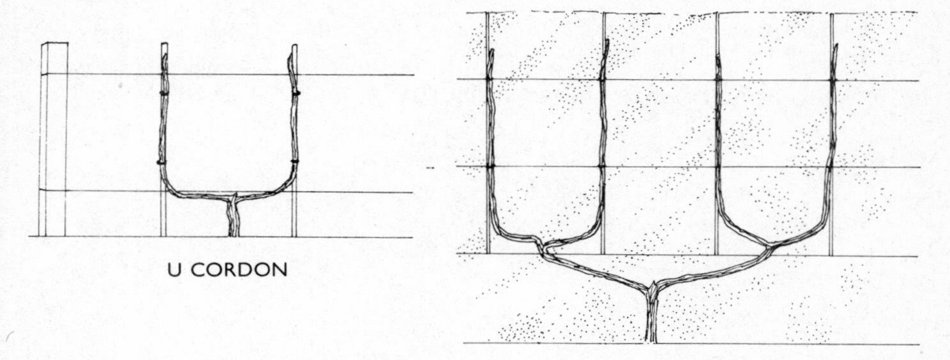

DOUBLE-U CORDON

U CORDON

Uliginous

Growing in muddy or swampy areas, a bog plant.

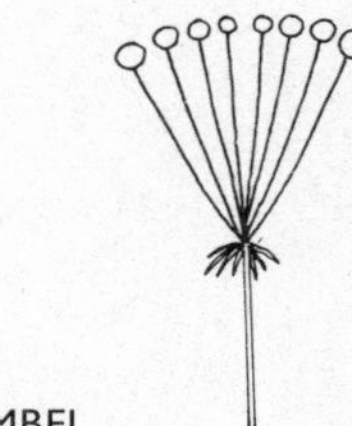

UMBEL

Ulmin

A gummy substance which exudes from certain trees, particularly the elm (*Ulmus procera*).

Ultramicroscopic

Too small to be seen with an ordinary microscope.

Umbel

A flowerhead in which the individual flower stems spring from a common point, often flat-topped or nearly so, as in cow parsley (*Anthriscus sylvestris*).

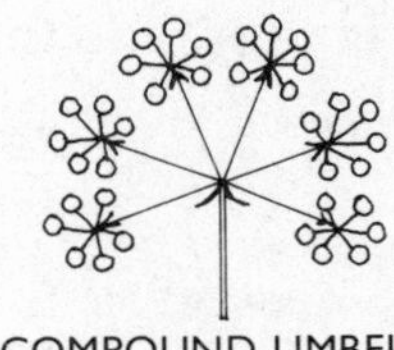

COMPOUND UMBEL

Umbellate

Resembling an umbel or carried in an umbel; umbrella-like.

Umbellet

A secondary umbel in a compound umbel.

Umbelliferous

(1) Any plant having an umbel or umbels.
(2) A member of the family Umbelliferae.

Umbelliform

Resembling an umbel.

Umbellule

A partial umbel.

Umbilicate

Having a depression, usually central; navel-like.

UMBILICATE FUNGUS
(*Russula aeruginea*)

Umbo

A knob or projection, occasionally found on the caps of some fungi.

Umbonate

Having a central boss.

Unarmed

Lacking thorns or prickles.

Uncinate

Hooked at the end, or having a hook; hook-shaped.

Uncus

A hook or hook-like appendage.

Undergrowth

Growth of shrubs and plants under trees, often dense or matted.

Undershrub

See Sub-shrub.

Understorey

A ground cover of shrubs often found underneath the tree canopy in tropical rainforest and deciduous woodland.

Underwood

Small trees or shrubs growing beneath larger trees. Brushwood.

Undulate

Wavy: of leaves or petals with wavy surfaces or edges, usually applied to those that are wavy up and down; those that are wavy in and out are termed sinuate.

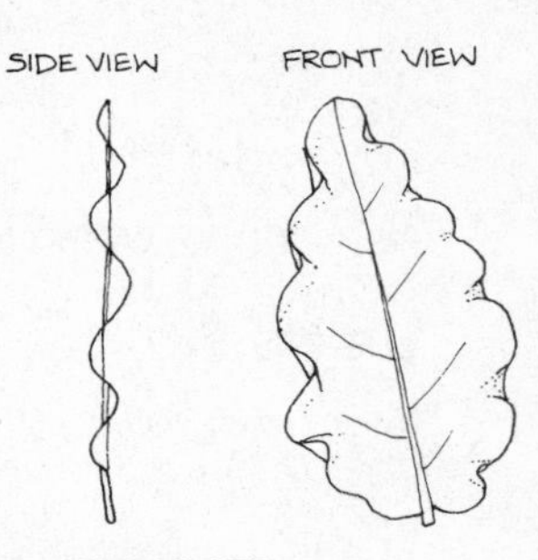

UNDULATE

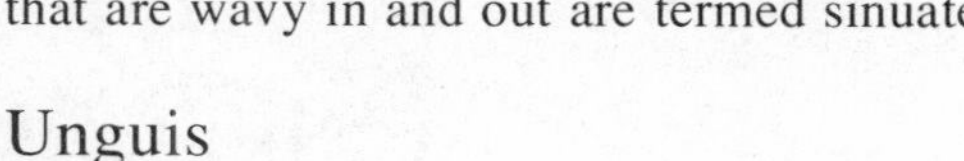

Unguis

The claw of a petal.

Unicellular

Tissues, organs or organisms consisting of a single cell, e.g. bacteria.

Unifoliate

Having a single leaflet but compound in structure.

Unilateral

One-sided.

Unilocular

Of ovaries with a single cell or chamber.

Union

The joint where a scion has been grafted on to the rootstock, usually indicated by a slight swelling.

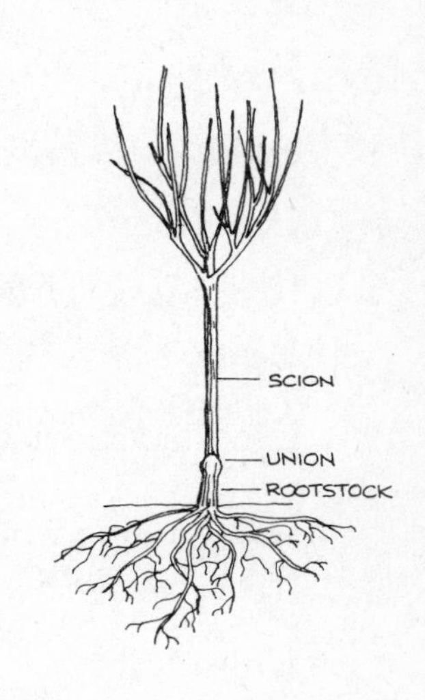

Uniseriate

Arranged in a single row or layer.

Unisexual

Flowers of one sex only. Each flower is of a single sex, but flowers of different sexes may be borne on the same plant (monoecious) or different plants (dioecious).

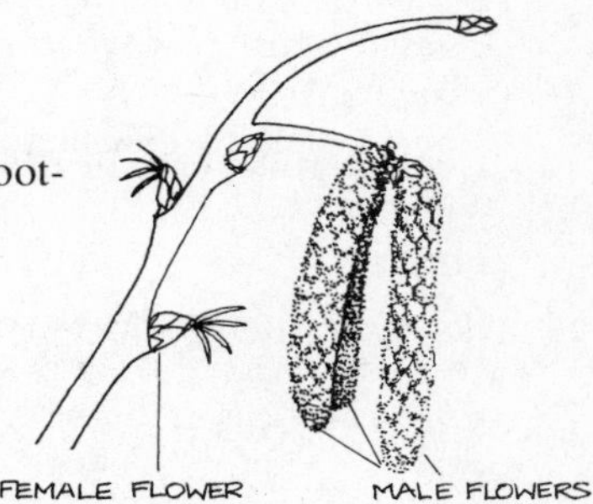

UNISEXUAL FLOWERS OF THE COMMON HAZEL (*Corylus avellana*)

Upland

Area of high ground, a stretch of hilly country.

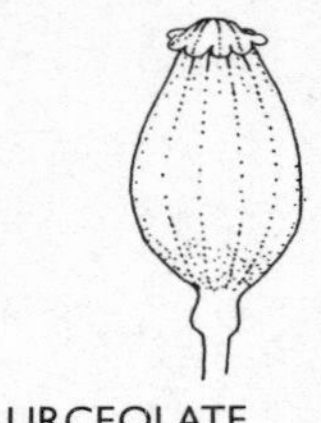

URCEOLATE

Upright

Habit of a plant with vertical or semi-vertical main branches or stems.

Urceolate

Pitcher-shaped.

Urea

An organic chemical derived from animal urine and used as a high-nitrogen fertilizer, either as a drench or a foliar feed when correctly diluted.

UREDOSPORES

Uredo

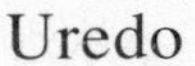

Rust in plants; rust-fungus in its summer stage.

Uredosori, Uredia

Clusters of uredospores.

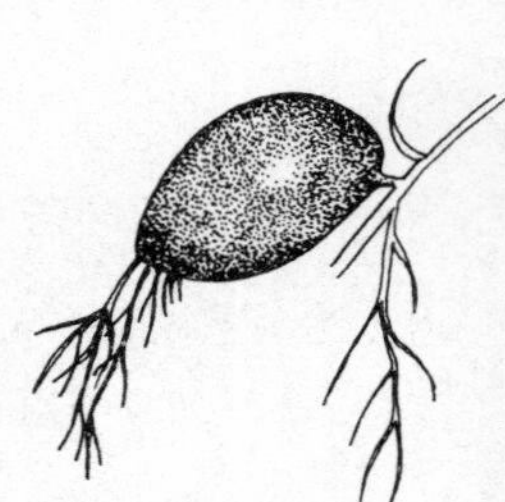

UTRICLE ON BLADDERWORK: BLADDER ENLARGED (*Utricularia vulgaris*)

Uredospore, Urediospore

The rust spore most familiar to gardeners, responsible for the rapid proliferation of rusts during the summer months, which is why they are often called 'summer spores': *cf.* Teleutospore.

Urine

Diluted animal urine is a fast-acting fertilizer, high in nitrogen with some potash. Poured over the compost heap, it will accelerate decomposition.

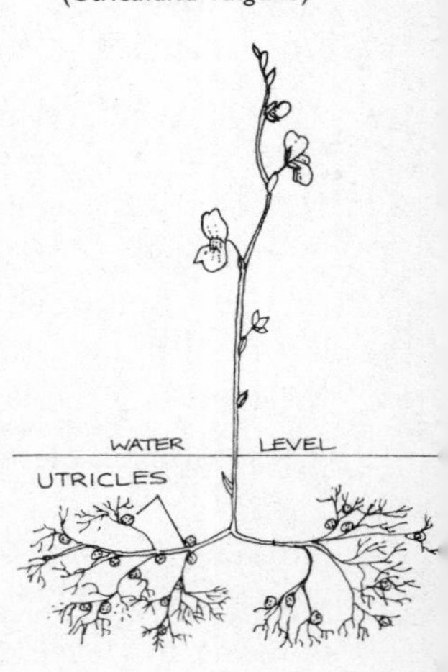

Utricle

(1) A bladder or bladder-like envelope of certain fruits, usually enclosing a single seed.
(2) A bladder on bladderwort, which contains air to give the plant buoyancy.

Uva

A grape-like berry formed from a superior ovary.

V

Vacillate

To sway to and fro; unsteady; in need of support.

Vacuolation

The formation of vacuoles during the development of living cells.

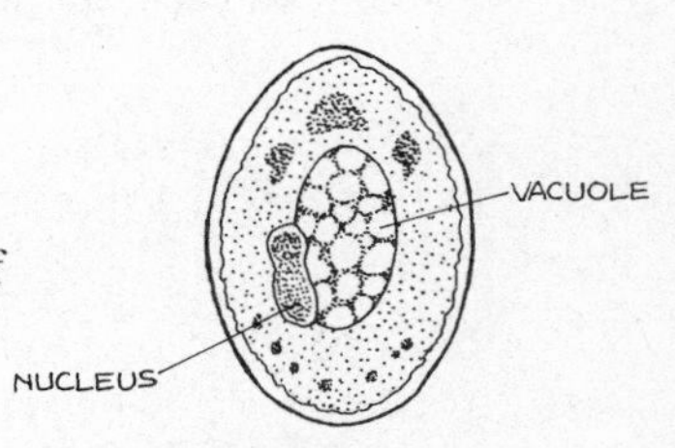

VACUOLE IN A YEAST CELL

Vacuole

Space within protoplasm usually containing a fluid or air.

Vagina

A sheath; a sheathing leaf-base.

Vaginate

Sheathed; having a sheath.

VAGINATE OR SHEATHED

Vaginule

A little sheath, especially one surrounding the base of a moss seta.

Vallecula

A groove or furrow.

Vallum

A wall or bank of sods or earth thrown up from a ditch.

Valvate

Meeting at the edges without overlapping; often applied to petals and leaves in the bud stage.

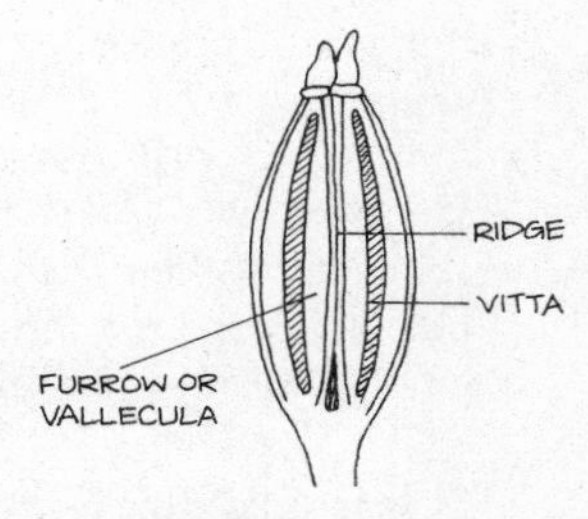

DIAGRAM OF THE FRUIT OF FENNEL (*Foeniculum vulgare*)

VALVATE AESTIVATION

Valvate Implicative

Of petals in the bud stage which do not overlap but are slightly entwined or tangled.

Valvate Involute

Of petals in the bud stage which do not overlap but curl inwards at the edges.

VALVATE INVOLUTE AESTIVATION

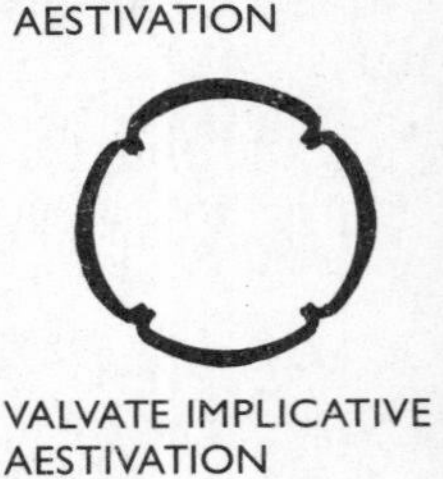

VALVATE IMPLICATIVE AESTIVATION

Valve

(1) A flap over a pore.
(2) One part of a dehiscent pod, capsule, etc.
(3) The lower bract in grasses. *Same as* Lemma.

Valvule

The upper bract in grasses: *see also* Palea.

Vaporizer

A heated container containing a vaporizing chemical which releases pesticides into the greenhouse atmosphere: usually electrical but not always so.

Varicoloured

Diversified in colour, same as variegated.

Variegated

The diversification of colouring in the foliage of plants, often the result of a reduction in the chlorophyll content. Frequently green with cream, white or silver, but sometimes with other colours including reds and yellows.

Variety

A group of plants having distinctive features but not sufficiently distinct to be classed as a species. 'Cultivar' is a compound word derived from CULTIvated VARiety, now officially and widely used. The recommended usage now is variety for natural varieties, and cultivar for those of cultivated origin; the former may have latinized names and the latter vernacular names.

Variform

Of various forms.

Vascular Bundles

Groups of strands of conducting tissue in the higher plants, composed of xylem, cambium and phloem; the means by which liquids are circulated within the plant.

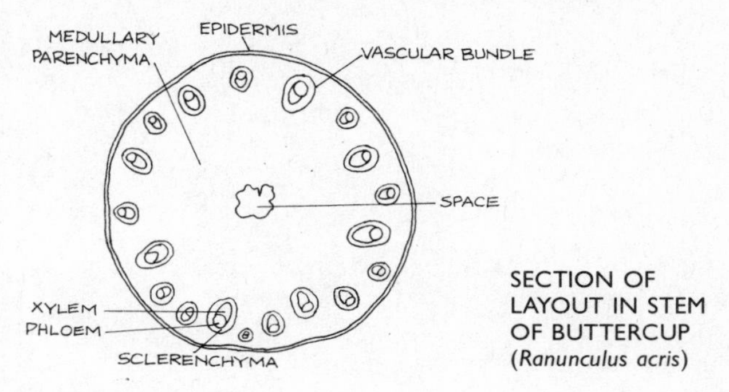

SECTION OF LAYOUT IN STEM OF BUTTERCUP (*Ranunculus acris*)

Vascular Cryptogams

Ferns and similar pteridophytes.

Vascular Plants

Pteridophyta and spermatophyta; all plants possessing organized vascular tissue.

Vascular Strands

The vessels by which water and nutrients travel from one part of a plant to another. They are grouped into vascular bundles.

Vase

A fruit tree with branches trained in the form of a 'U'. Otherwise known as goblet- or cup-shaped.

Vector

A transmitter of disease or infection such as an aphid.

Vegetative Organs

The leaves, stems and roots of a plant, excluding flowering parts.

Vegetative Propagation

The method of propagation using pieces of living plants (cuttings) as opposed to seed. In vegetative propagation the new plant will carry all the characteristics of the parent. All plants vegetatively produced from one parent are called a clone.

Veil

The ring of tissue which in the early stages of development of a mushroom or similar fungus connected the stipe (stalk) with the cap (pileus), and which ruptures as the fungus matures: *cf.* Velum.

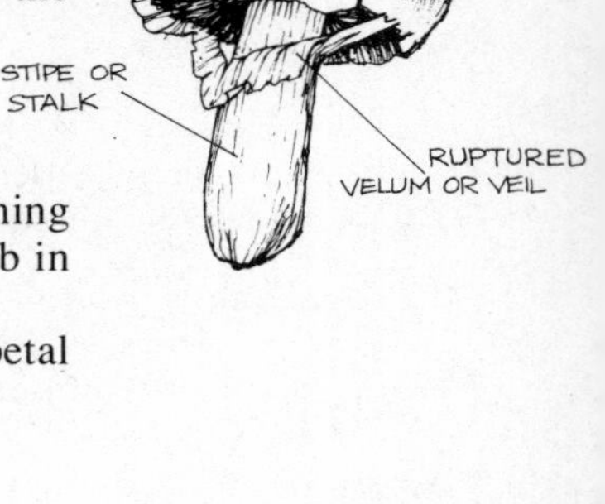

Vein

(1) Any one of the visible conducting and strengthening tissues of a leaf. A vascular bundle forming a rib in a leaf: *cf.* Nervure.
(2) A streak or minor variegation in the colour of a petal or leaf.

Velamen

A multilayered corky covering of dead cells on some aerial roots, as in some orchids, enabling them to absorb moisture from the atmosphere.

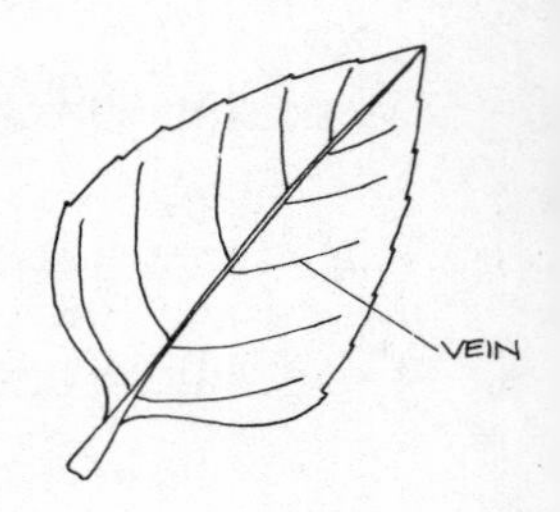

Velum

A membrane, especially the membrane joining the stalk of certain fungi to the rim of the cap, as in mushrooms, toadstools, etc: *cf.* Veil.

Velutinous

Velvety; having a velvet-like covering of fine hairs.

Venation

Veining or the arrangement of veins in a leaf or petal.

Vent

(1) A ventilation opening in the sides and roof of a greenhouse.
(2) To expose to air.

Ventral

(1) The upper surface of a leaf.
(2) The front or inner surface of an organ.
(3) Facing toward the axis.

Ventricose

Distended or inflated on one side, or all round at the base. More pronounced than gibbous.

Verdure

Green vegetation or the fresh green colour of young healthy vegetation.

Verge

(1) The edge of a lawn or the edge of a turf path, especially beside a border or flowerbed; a boundary.
(2) To slope or incline downward.

Vermicular

Worm-eaten or of worm-eaten appearance.

Vernal

Appearing in the spring.

Vernalize

To advance the germination of seeds and bulbs by submitting them to special treatment before planting. Hyacinths are vernalized (prepared) to force them into flower for Christmas.

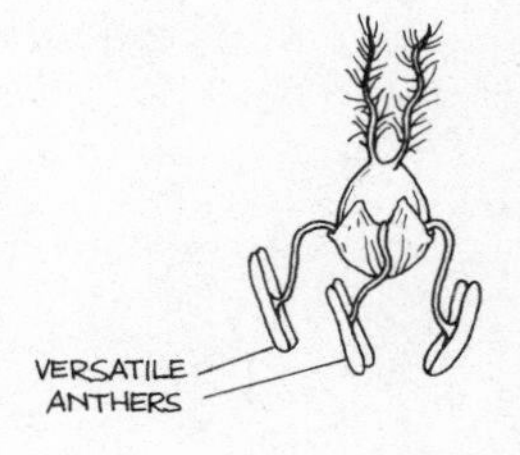

Vernation

The arrangement of the young leaves in the leaf bud.

Verrucose

Warty, covered with warts or similar protrusions.

Verruculose

Covered with or having small warts.

Versatile

An anther attached by the middle of the back. Dangling; capable of free movement.

Verticil

A whorl; a circle of three or more petals or leaves arranged around an axis.

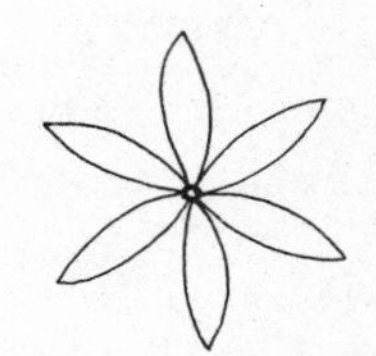

DIAGRAM OF A VERTICIL OF SIX LEAVES

Verticillate

Grouped in whorls, or appearing to be so.

Vesicle

A small bladder or cavity filled with fluid or air, as in some seaweeds. Vesicles also occur within the cytoplasm of living cells and form part of the Golgi apparatus.

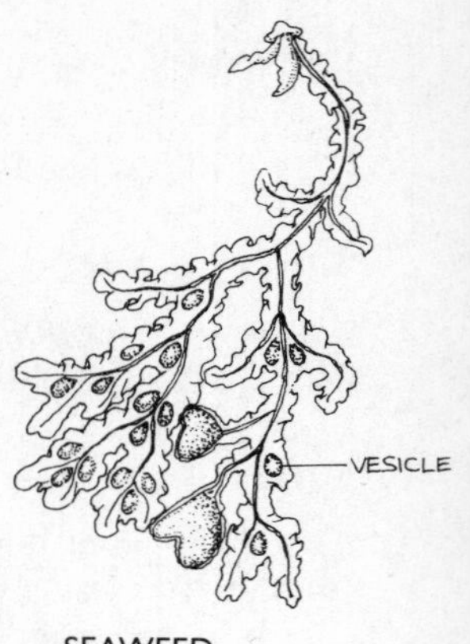

SEAWEED

Vespertine

Types of flower that bloom in the evening, e.g. evening primrose (*Oenothera acaulis* and *O. biennis*).

Vessel

A conducting tube for water and nutrients in plants.

Vestigial

Imperfectly developed; usually a functionless structure which was fully developed and functional in ancestors.

VESTIGIAL STAMINODE

FIGWORT
(*Scrophularia macrantha*)

Vestiture

Any surface covering such as hairs, scales, etc.

Vexillum

The erect uppermost petal on sweet peas, etc., otherwise known as a 'standard'.

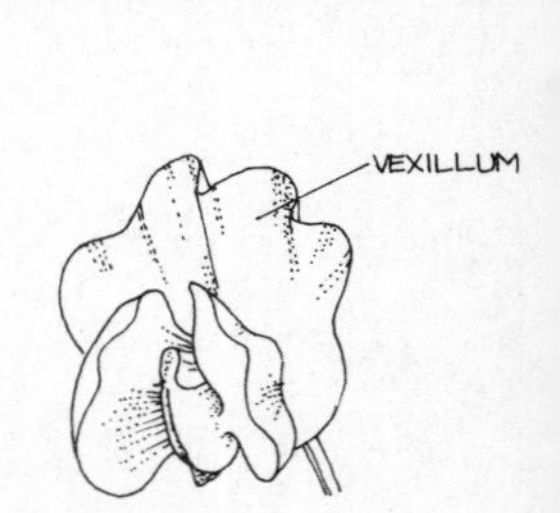

SWEET PEA
(*Lathyrus odoratus*)

Viable

Able to live; of seeds that have the ability to germinate.

Villous

Covered with long soft shaggy hairs, not matted.

Vimineous

Producing long flexible shoots or pliable stems, as those used in wickerwork and basket-making.

Vine

(1) A grape-vine in Britain and Europe.
(2) Any climbing plant in USA.

Vine Eye

A wedge-shaped perforated iron nail for driving into a brick wall to support wires up which plants can be trained.

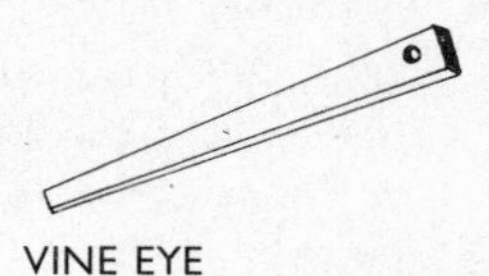

VINE EYE

Vinery

Glasshouse or similar structure, usually heated, for the cultivation of grapes.

Vineyard

An area of land planted with grape vines, usually on a commercial scale, for wine-making.

Virescence

The green tinge sometimes present in petals (especially white ones) which would not normally contain any green pigment or chlorophyll.

Virgate

(1) Rod-like; straight and slender; twiggy.
(2) An ancient land measure, about 30 acres.

Virgin Soil

Soil which has not yet been cultivated.

Virgulate

Shaped like a rod or wand.

Virulent

Highly poisonous or malignant; often describes a fast-spreading pathogen that causes serious disease.

Virus

A microscopic pathogenic agent capable of rapid reproduction within the cells of a plant. There are numerous viruses which attack plants and can destroy or debilitate them, and others which are considered beneficial. Virus organisms are too small to be visible with an ordinary microscope, and are usually transmitted from plant to plant by sap-sucking insects.

Viscid, Viscous

(1) Sticky; semi-fluid.
(2) A leaf-surface covered with a sticky solution, such as the oak-leaf pelargonium (*P. quercifolium*).

Vista

A view, usually through a narrow opening such as an avenue of trees, etc.

Viticulture

Cultivation of vines, usually for the purpose of commercial wine-making.

Vitta, *pl.* Vittae

(1) One of several oil-tubes in the pericarp of fruits of most umbelliferous plants: *see also* Vallecula.
(2) A stripe of colour.

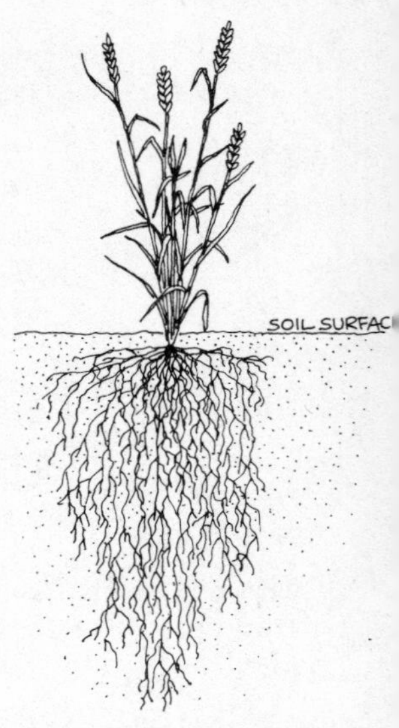

ROOT SYSTEM OF WHEAT SHOWING MASSED VIVERS

Viver

A rootlet or root-fibre.

Viviparous

The strict meaning is 'producing live young', such as the aphid or greenfly. Also used to describe seeds that germinate while still contained in the parent plant, or plantlets produced on the leaves or stems of the parent, such as *Bryophyllum daigremontianum*.

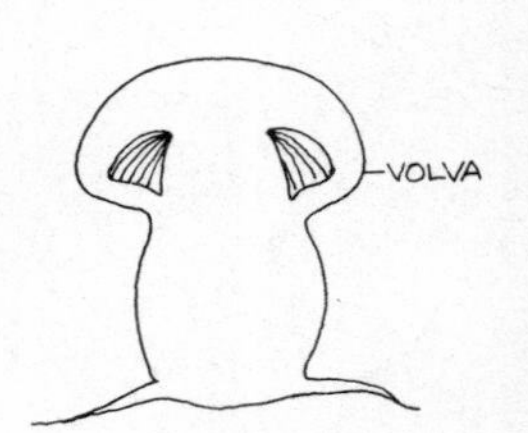

Volute

Twining or twisting; spiralling.

Volva

The membrane protecting many fungi during the early stages of growth.

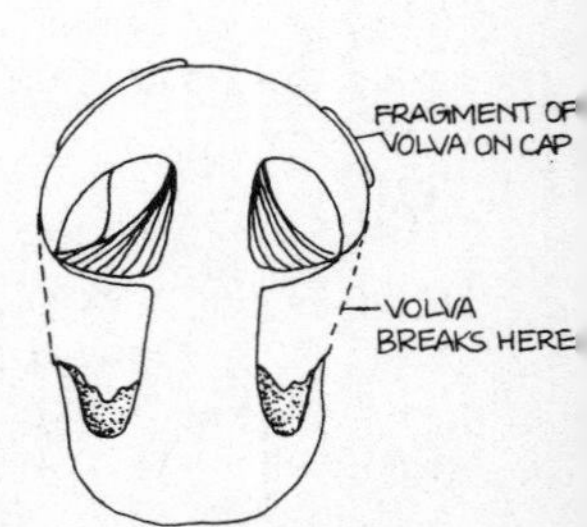

DIAGRAM OF THE GROWTH OF A MUSHROOM (*Agaricus campestris*)

Vraic

A type of seaweed popular in the Channel Islands for manure and fuel.

Vulvate

Oval.

W

Wain

A wagon for conveying agricultural produce such as hay and roots. A wainwright is one who makes wagons.

WAIN

Wall Nail

A chisel-shaped nail with a soft lead tang attached to the head. Used for driving into the mortar between bricks so that the lead tang can be twisted around wires or the stems of plants as a support.

WALL NAIL

Wand

A supple thin stick or branch, a flexible twig; a young shoot of willow used in basket-making.

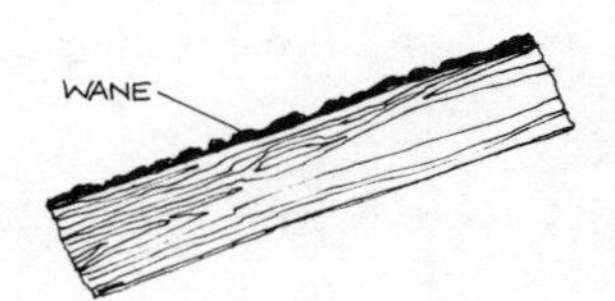

Wane

The edge of a plank of wood still carrying the bark, or a plank of wood with a defective edge.

Wardian Case

A totally enclosed glass container in which certain plants needing humid conditions such as ferns can be raised. Otherwise known as fern cases. Originally dome-shaped and invented in 1830 by Dr Nathaniel Bagshaw Ward (1791–1868): *cf.* Terrarium.

WARDIAN CASE

Warren

Formerly an enclosed protected area for the breeding of game; now an area of land where rabbits live.

Wart

A rounded protuberance on the surface of a plant.

Wasps

Any of a large number of insects of the order Hymenoptera, including the common wasp *Vespa vulgaris*, which is often considered a pest in summer when attracted to ripe fruit and similar sweet substances. In spring and early summer they are very beneficial since they feed their offspring on a variety of other insects.

COMMON WASP
(*Vespa vulgaris*)

Water Bud

An underwater bud growing upward to form a new stem on a waterplant such as frogbit (*Hydrocharis*).

Wattle

Interlaced thin branches formed into fencing and screens: *cf.* Hurdle.

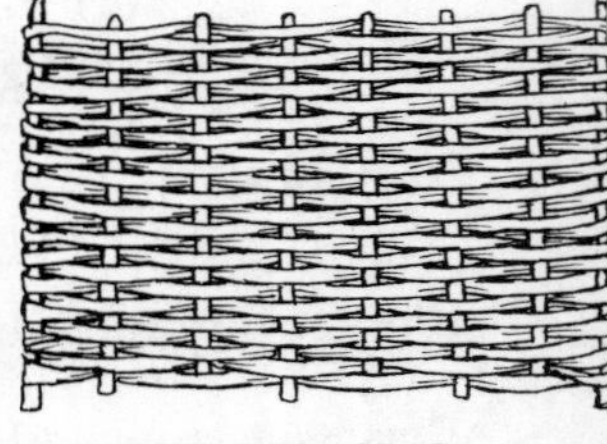
WATTLE FENCING

Wax

The natural material produced by some plants as a form of protection or to reduce evaporation. It appears in several forms such as the bloom on grapes and plums, a powdery or floury coating on some leaves, or as scales on stems, etc.

Weald

Formerly an area of fairly dense woodland, now a tract of open or slightly wooded country particularly between the North and South Downs of England.

Weed

(1) A plant not valued for beauty or use, regarded as hindering or suppressing the growth of more valued plants.
(2) To clear ground of weeds.

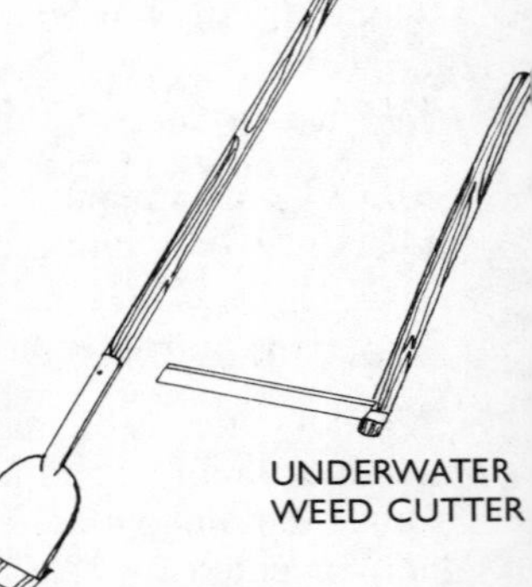

Weed Cutter

A tool with a narrow horizontal blade attached to a long handle, used to sever weeds just below ground level. A different design is used to cut weeds in rivers and lakes.

Weeding Fork

A small hand-fork.

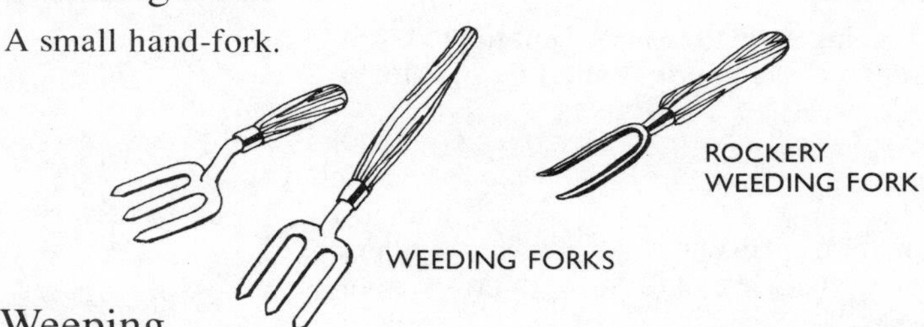

Weeping

A tree or shrub of pendulous habit, either natural as in some willows (*Salix*) or induced as in weeping ornamental cherries such as *Prunus serrulata*.

CHEAL'S WEEPING CHERRY
(*Prunus serrulata*)

Weevil

(1) A member of the beetle family Curculionidae, usually of small size with elongated head. Most larvae and many beetles cause much damage, particularly to fruit, nuts, bark of trees, etc.
(2) Any insect injurious to stored grain.

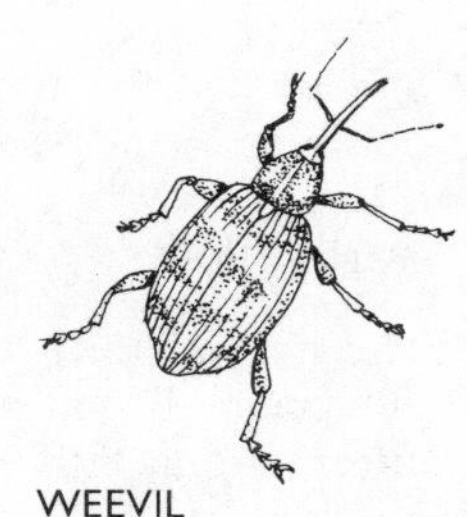
WEEVIL
(*Curculio nucum*)

Wettable Powder

A finely ground chemical fungicide or pesticide which although not soluble can be suspended in water and applied through a spray if frequently stirred.

Wetting Agent

A spreader such as saponin or soft soap which when added to a spraying solution allows the surfaces of a plant to be covered with a fine mist rather than larger droplets.

Whale Hide

A trade name for bituminized pulpboard made into temporary pots for container-grown plants. No connection with aquatic mammals.

Whin

Another name for gorse.

Whip

(1) A long twig or slender, pliant thin branch.
(2) A young tree consisting of a single erect stem with no side-branches.

Whip-and-Tongue Graft

A graft formed by making a tongue-shaped cut in the scion (graft) which fits into a corresponding notch in the stock. *See* Grafting.

Whorl

A circle of three or more flowers or leaves at one node, arranged as the spokes of a wheel and springing from stem or axis at the same level.

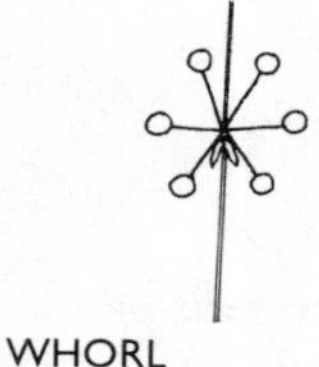
WHORL

Wicker

A small flexible osier or twig, used for making wicker-work or basketry.

Wicket

A small gate often adjacent to a larger one.

Wigwam

Three or more canes or poles arranged in a conical shape and tied at the top to form a pyramid as a support for climbing plants such as sweet peas, runner beans, etc.

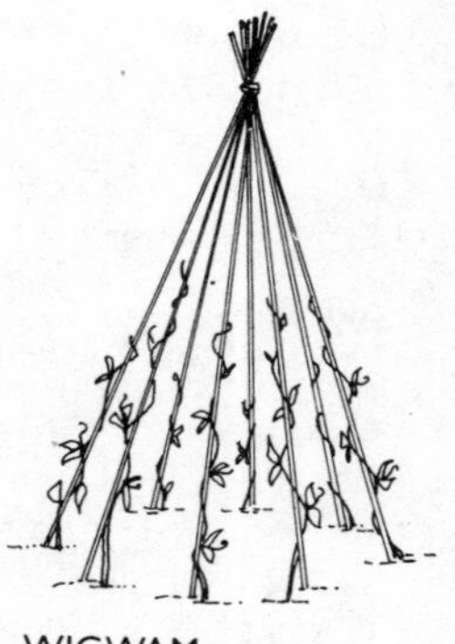
WIGWAM

Wild Type

The natural form of a species, as it occurs in the wild, unchanged by cultivation or selective breeding. Wild type alleles are usually dominant.

Wilt

(1) To fade; droop; become limp.
(2) A general term for several unrelated diseases which cause plants to collapse, including *Verticillium* and *Fusarium*.

Wilting

A condition of plants when short of water, or when low humidity (usually in a greenhouse in hot weather) results in the leaves transpiring liquids faster than the root-system can replace it.

Win

To dry (grass, etc.) by exposure to the wind.

Wind Pollination

The transfer of male pollen to female flowers (or parts of flowers) by the action of wind or air currents, as in hazel and many other catkin-bearing plants.

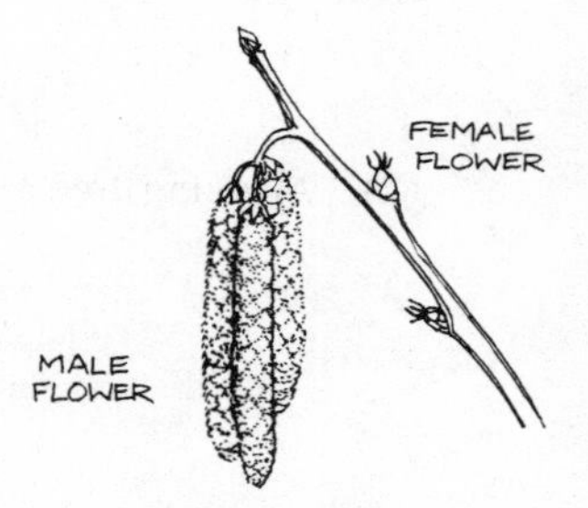

COMMON HAZEL (*Carylus avellana*) IS AN EXAMPLE OF A WIND POLLINATED PLANT

Windbreak

A group or row of plants, trees or material fixed to stakes, specifically arranged to break the force of a strong wind. A permeable windbreak (or windscreen) will give protection for a distance of about 10 ft (3 m) for each foot (30 cm) of height; whereas a solid wall or fence is effective for about twice its height but creates violent turbulence beyond.

WINDBREAK

Windfall

Fruit which drops to the ground naturally, either when ripe or diseased.

Windlestraw

A thin dry stalk of grass; any of the long-stalked species of grass.

Wing

(1) One of the pair of lateral petals of papilionaceous flowers such as the sweet pea.
(2) A dry flat appendage of a seed to assist its distribution by wind or air currents, such as sycamore.

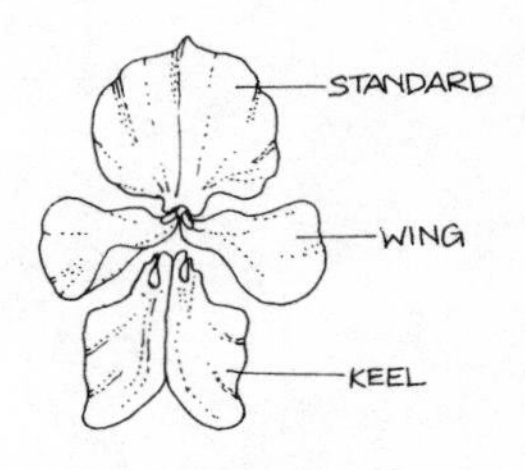

DIAGRAM OF SWEET PEA FLOWER

Winnow

To separate the chaff from the grain by wind action; to sift or separate by wind action.

Winter Crop

Any vegetable which can be harvested during the winter months such as Brussels sprouts, spinach beet, savoy cabbage, etc.

Winter Fallow

Agricultural term for land left vacant during the winter months, not planted.

Winter Spores

The special type of spore produced by fungi in autumn which are adapted to withstanding winter conditions.

WINTER SPORES OF WHEAT RUST (TELEUTOSORUS)

Winter Wash

See Dormant oil, Tar oil wash.

Wire Rake

Same as lawn rake or springbok rake; one made of fanned out sprung wires bent at the ends.

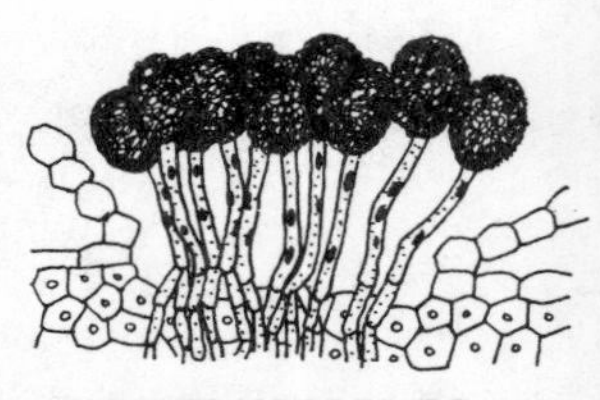

SUMMER SPORES OF WHEAT RUST (URDEDOSORUS)

Wireworms

The larvae of click beetles (*Agriotes lineatus*) which live in the soil. The grubs are serious pests and attack the roots of many crops including tubers, bulbs and corms. They are particularly prevalent in old pasture and in soils which have been recently reclaimed from grassland.

WIREWORM

Wisp

(1) A small bundle of hay, straw, etc. as used in the making of corn-dollies, etc.
(2) A small hand-broom or brush, originally made from a small bundle of straw.

Witch's Broom

(1) A tuft of underdeveloped branches produced usually on woody plants and trees; if cuttings are taken from them they will remain dwarf. Often caused by fungal attack.
(2) Bundle of twigs tied to a rustic handle used for sweeping leaves, etc. from lawns. A besom.

WITCH'S BROOM OR BESOM

Withe

A flexible twig usually of willow; a handle for a basket or tool made from a number of withes twisted together.

Withwind

A climbing plant such as honeysuckle or bindweed.

Withy

Flexible shoot of willow or osier grown specially for use in the manufacture of baskets and for tying bundles of canes or rods.

Wold

Formerly a wooded area, now usually upland country.

Wood

(1) A group of trees growing fairly close together but not so densely as a forest and not so extensive in area.
(2) The hard part of the substance of trees and shrubs, xylem.

Wood Ash

The substance obtained by burning wood and similar materials which contains potash, phosphate, iron, magnesium and manganese in varying amounts. It is very soluble in water, so should be added to the soil immediately or stored in a dry place.

Woodlice

Arthropods which inhabit dark moist situations such as decaying vegetation, bark, etc. They will sometimes eat seedlings but, having soft mouthparts, in general present no threat to the gardener.

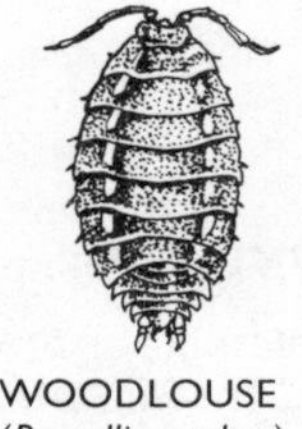

WOODLOUSE
(*Porcellio scaber*)

Woody

Of a plant with lignified stems and branches, usually somewhat lacking in foliage: *cf.* Soft-stemmed.

Woolly

Having long, soft matted hairs; pubescent; downy.

Woolly Aphid

A type of aphid which secretes and surrounds itself with a waxy, cottony substance which is resistant to water. They attack several kinds of tree, including pear and apple. Also called American blight.

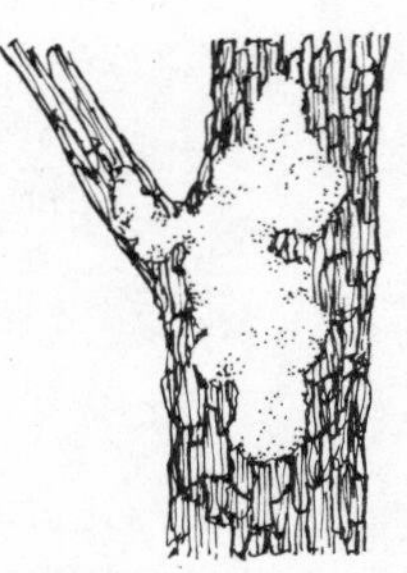

WOOLLY APHID
(*Eriosoma lanigerum*)
FORMS WOOLLY CLUSTERS ON APPLE TREES IN SPRING

Worm

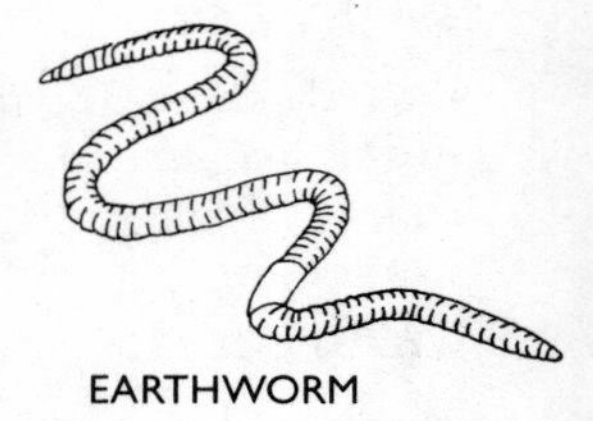
EARTHWORM

(1) Slender burrowing elongate invertebrate animal of the genus *Lumbricus* with reddish-brown body. The gardener's greatest asset, known to the Victorians as 'nature's ploughman'.
(2) The general (erroneous) name for a small maggot or grub, especially in USA.

Wound Dressings

See Sealing agent.

Wrack

Seaweed cast ashore or growing where it is exposed by the tide; any of the Fucaceae, the bladder-wrack family of seaweeds.

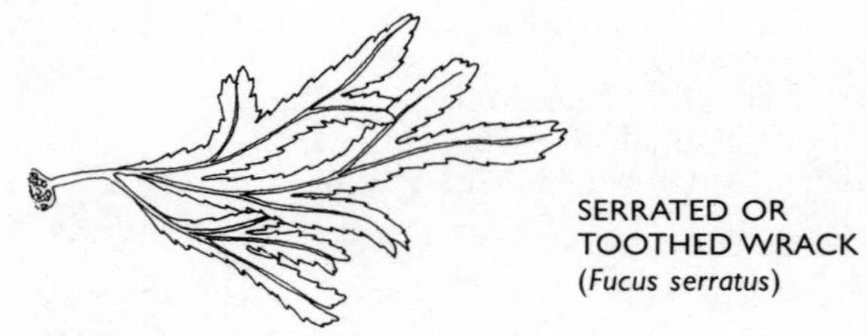
SERRATED OR TOOTHED WRACK (*Fucus serratus*)

Wrenching

A method of root-pruning carried out in commercial nurseries to make later transplanting easier, or to prevent the development of a tap-root.

X

Xanthein

The yellow colouring matter present in some plants, e.g. *Xanthoria parietina*, the lichen often seen on walls and roots.

Xanthic

Yellow, or the yellow pigment (xanthin) in flowers, often combined with carotene (carotin) in yellow flowers and fruit.

Xanthophyll

The dark brown or yellow compound in plants forming the colouring-matter of autumn foliage.

Xenogamy

Cross-fertilization.

Xenograft

A graft from a member of a different species.

Xeromorphic

Plants, or parts of plants, protected against excessive loss of moisture by hairs, thick cuticles, or similar structural characteristics.

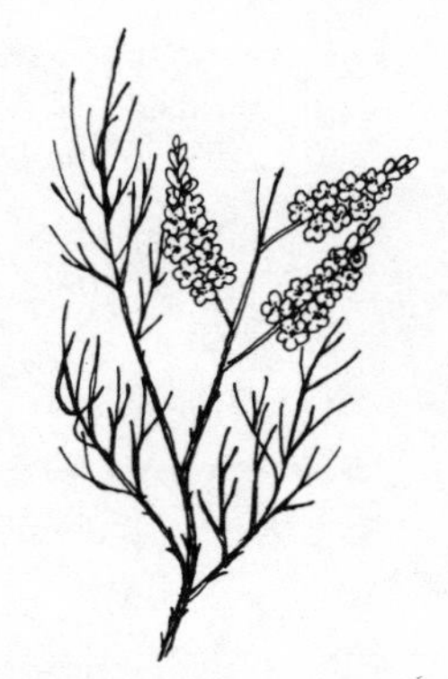

XEROPHYTE
(*Tamarix anglica*)

Xerophyte

A plant adapted to inhabit places where fresh water is scanty, or where water uptake is difficult owing to an excess of salts, e.g. on a seashore. Tamarisk (*Tamarix anglica*) is an example: *cf*. Hydrophyte, Mesophyte.

Xerophytic

A plant capable of survival under conditions of extreme drought, or one adapted to arid conditions, e.g. cacti.

Xiphophyllus

A type of plant having sword-shaped leaves, e.g. iris.

XIPHOPHYLLUS
LEAVES OF IRIS

Xylem

Woody tissue in plants usually consisting of vessels, fibres, etc., all with lignified walls, for the conducting of aqueous solutions and for support.

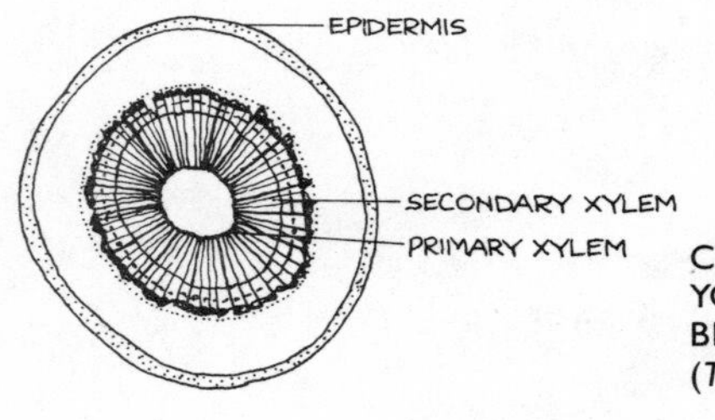

CROSS SECTION OF A YOUNG BRANCH OF LIME (*Tilia cordata*)

XYLOCARP
(COCONUT)

Xylocarp

A hard, woody fruit such as the coconut.

Xylogenous

Of organisms which grow on wood; some fungi are xylogenous.

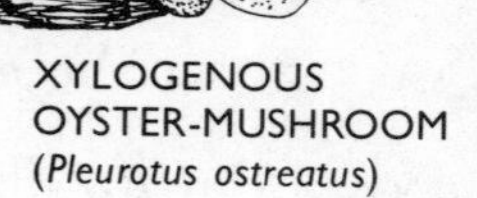

XYLOGENOUS
OYSTER-MUSHROOM
(*Pleurotus ostreatus*)

Xylology

The study of the composition and structure of wood.

Xylophagous

Insects which feed on, or bore into wood, e.g. the death-watch beetle.

THE XYLOPHAGOUS
DEATH-WATCH
BEETLE
(*Xestobium rufovillosum*)

Xylose

A pentose (sugar) found in many plants. Also known as wood-sugar.

Xystus

A tree-planted walk, an avenue between trees.

Y

Yard

(1) A straight thin branch.
(2) USA term for a garden, especially one at the rear of the house; backyard.
(3) Unit of length equal to 3 feet (90 cm).

Yarpha, Yarfa

A peat-bog; clayey or sandy peat; peaty soil in Shetland.

Yeast

Unicellular fungi which convert sugars to alcohol and carbon dioxide; certain species are of importance in baking and brewing.

Z

Zap

To hit, strike or kill swiftly, as in the swatting of a fly. Probably of USA origination.

Zebra

Any plant or tree having stripes reminiscent of a zebra's. The hard and striped wood of *Guaiacum* is called zebra-wood.

Zinc

One of the minerals needed in small quantities for optimum plant growth. Its availability decreases as the pH rises above 6, and this is particularly noticeable on light soils.

Zone Lines

Thin black or dark brown lines in decayed wood, usually caused by fungi.

Zoning

(1) Leaf-markings, usually circular and darker than the main leaf area but can be lighter or differently coloured.
(2) Climatic zones detailed in USA and other large countries to help assess the hardiness of plants in a particular part of the country.

LEAF-ZONING ON PELARGONIUM

Zoophyte

Any plant thought to resemble an animal.

Zoospore

A spermatozoid, an asexual reproductive cell. A spore occurring in certain fungi, algae, etc. having power of locomotion. Some rusts are reproduced by zoospores.

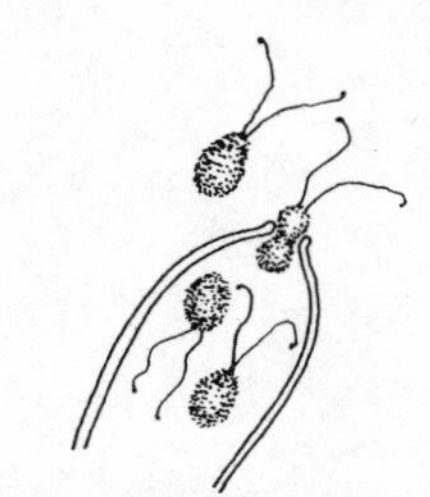

ZOOSPORES BEING RELEASED

Zygomorphic

A flower divisible into two equal parts in one plane only, such as antirrhinum, pea, etc. *cf.* Actinomorphic, Radial.

Zygomycetes

A group of fungi, moulds, etc. which produce zygospores.

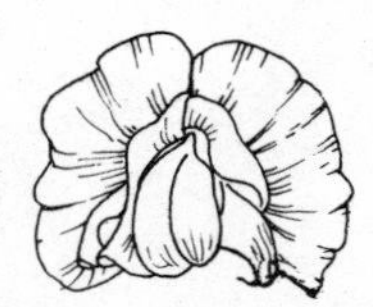

ZYGOMORPHIC FLOWER OF SWEET PEA (*Lathyrus odoratus*)

Zygophyte

A plant which reproduces by means of zygospores.

Zygospore

A spore produced by the union of buds from two adjacent hyphae; the method by which some fungi and algae multiply.

Zygote

A fertilized female gamete.

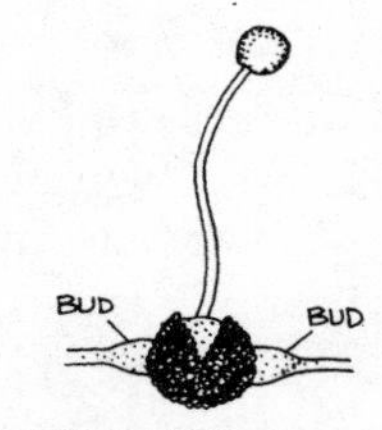

GERMINATING ZYGOSPORE

Zymase

Any group of enzymes inducing fermentation.

Zymolysis

The action of enzymes.

APPENDIX

Leaf Shapes

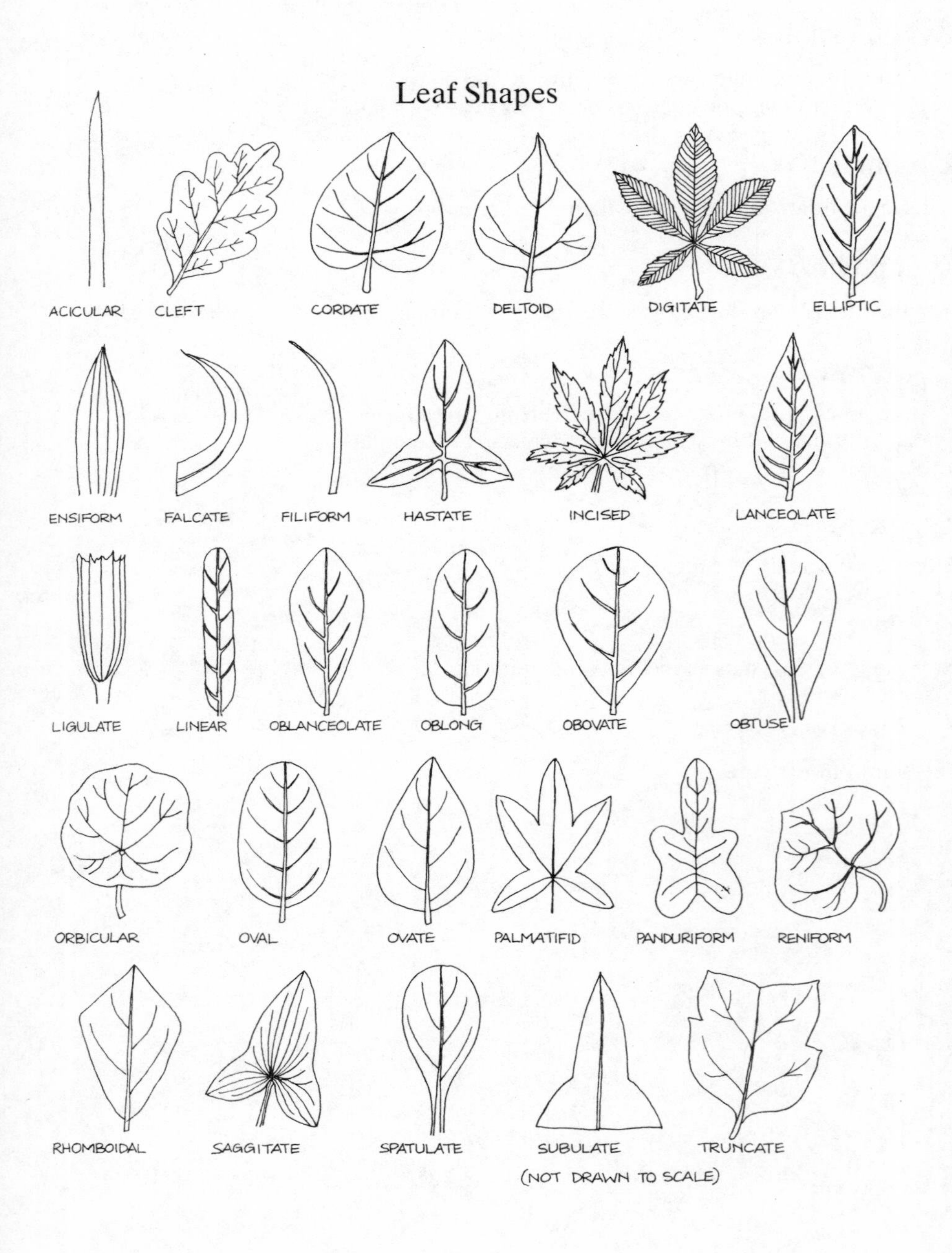

Leaf Margins

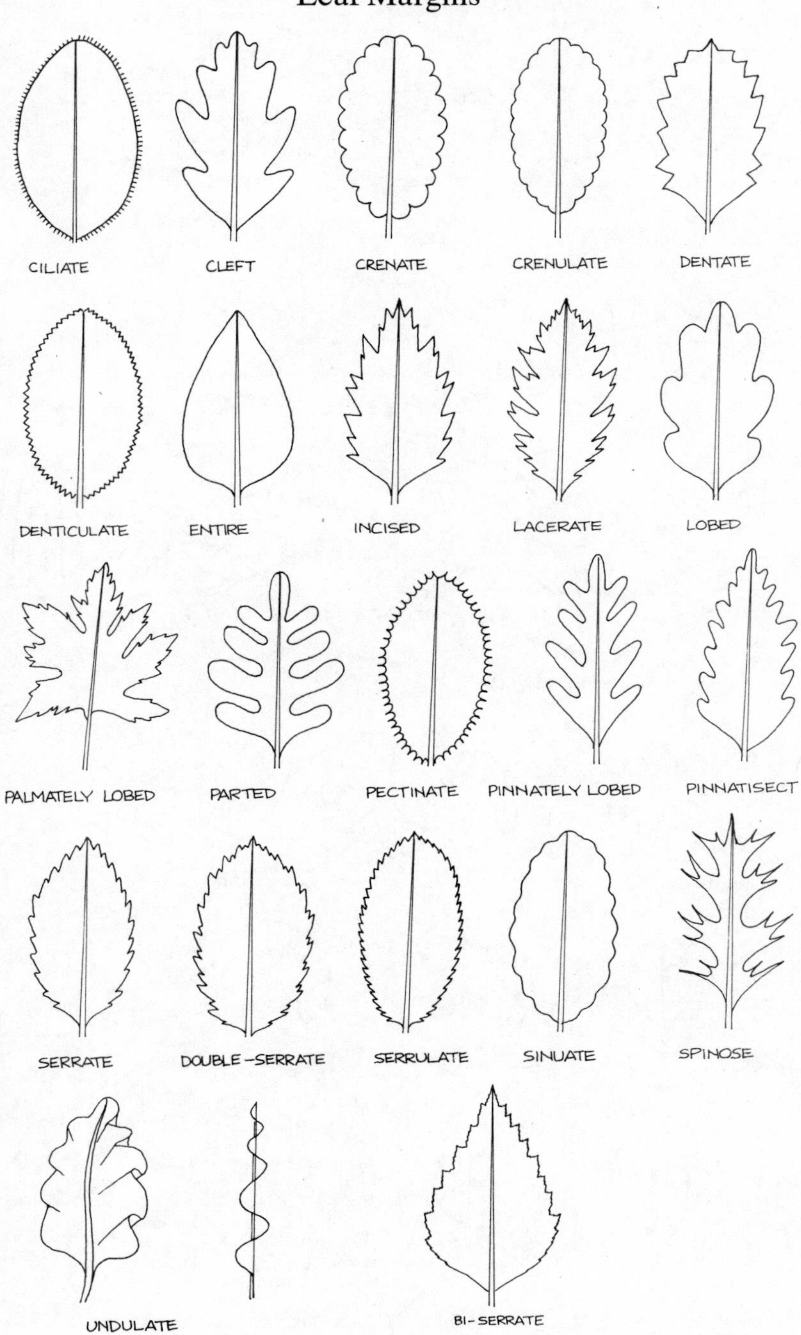

Leaf Tips

ACUMINATE

ACUTE

APICULATE

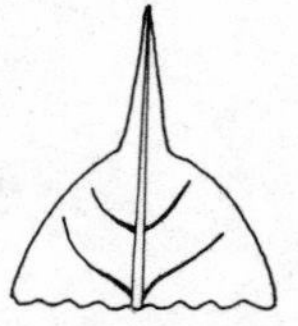
ARISTATE

CAUDATE

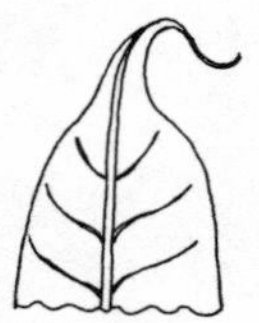
CIRRHOSE

CLEFT

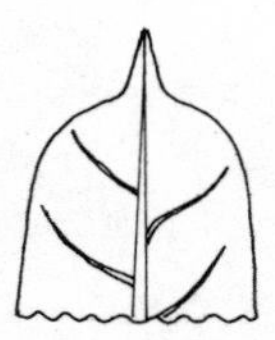
CUSPIDATE

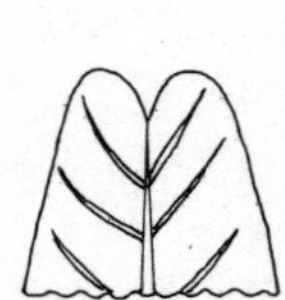
EMARGINATE

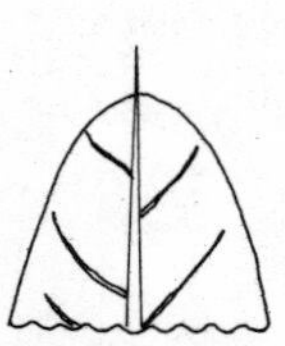
MUCRONATE

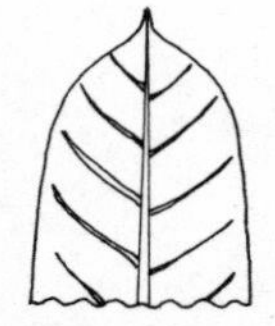
MUCRONULATE

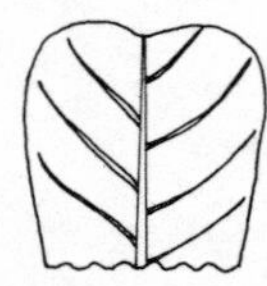
OBCORDATE

OBTUSE

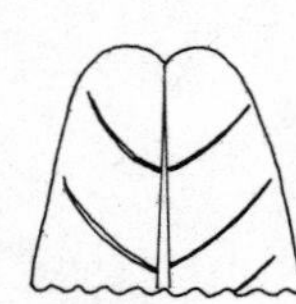
RETUSE

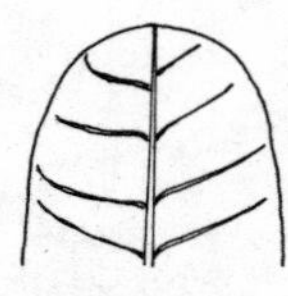
ROUNDED

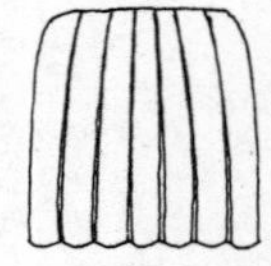
TRUNCATE

Leaf Bases

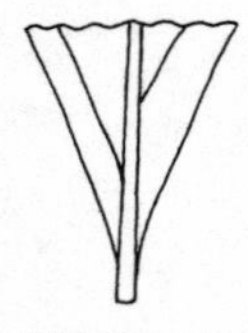
ATTENUATE

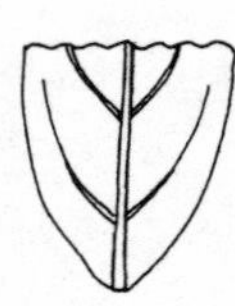
ACUTE

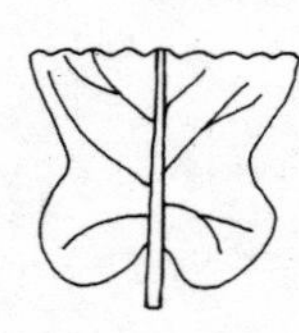
AURICULATE

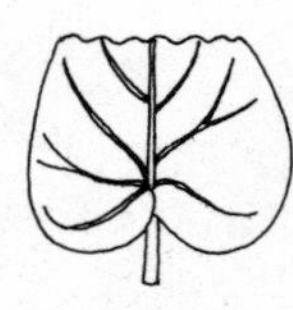
CORDATE

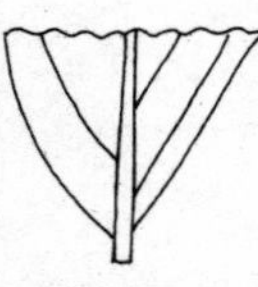
CUNEATE

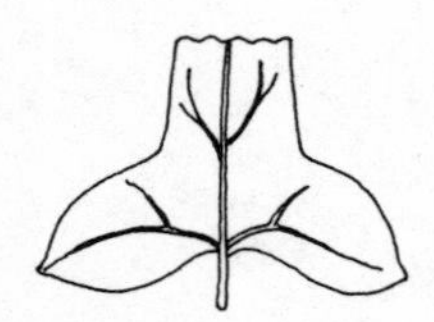
HASTATE

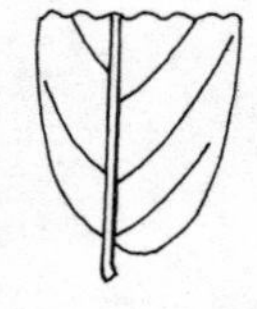
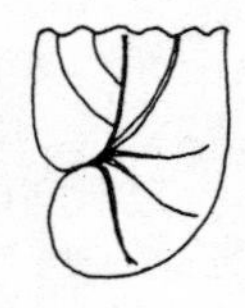
OBLIQUE

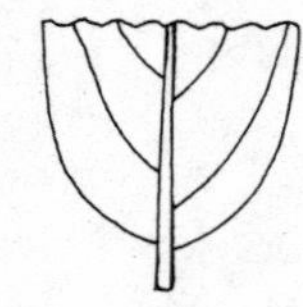
OBTUSE

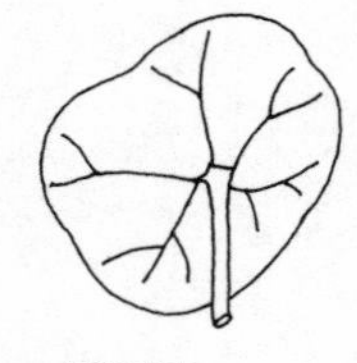
PELTATE

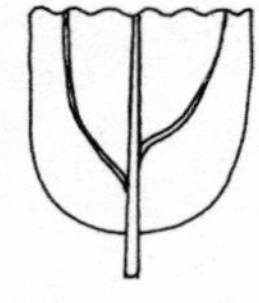
ROUNDED

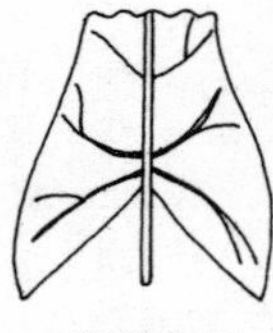
SAGITTATE

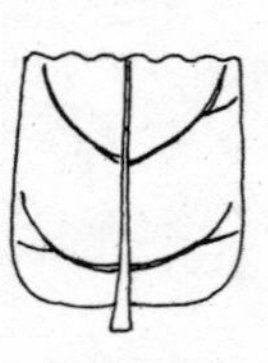
TRUNCATE

Compound Leaves

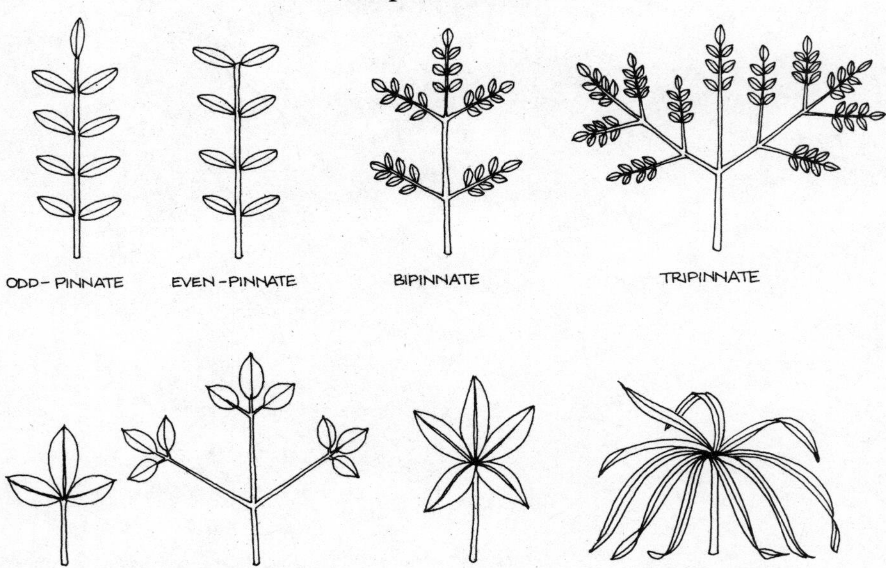

Leaf Lobing

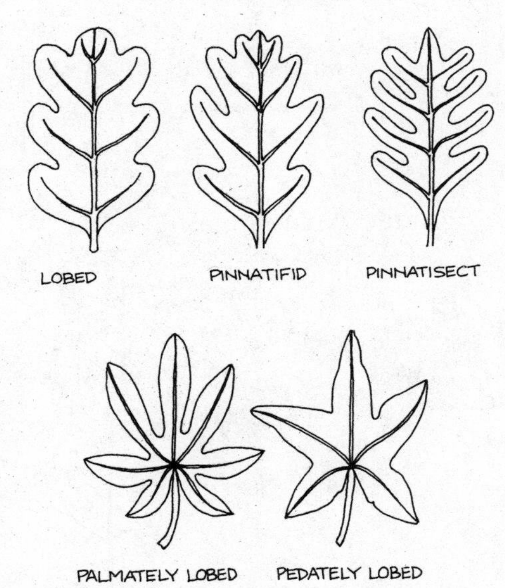

Attachment to Stem

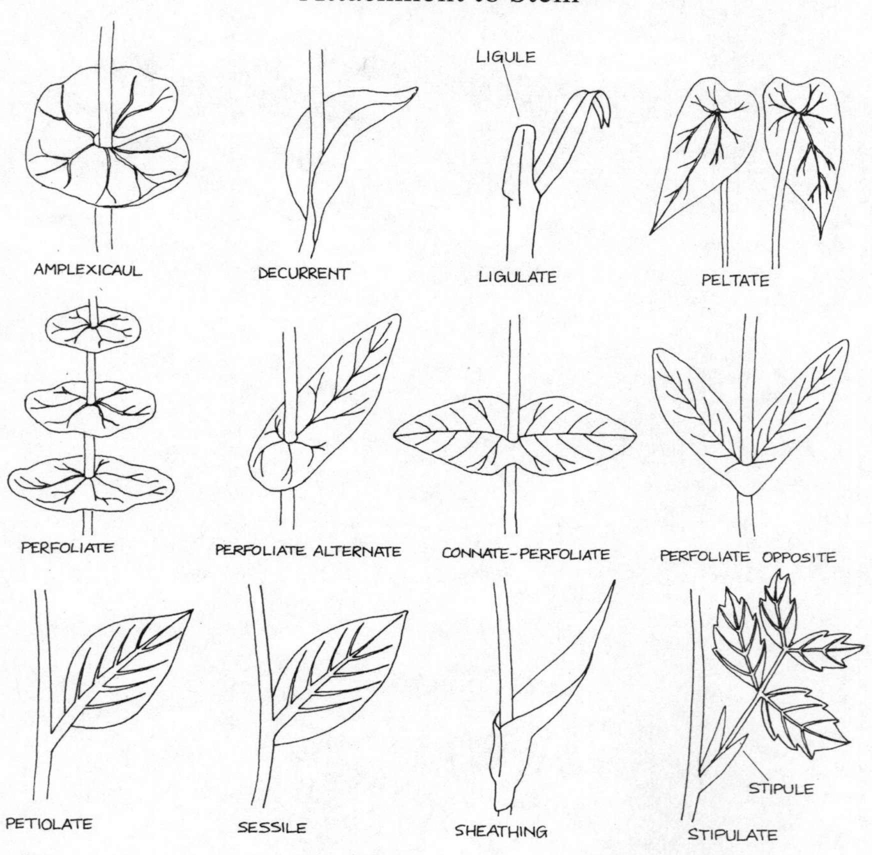

Foliage Arrangement

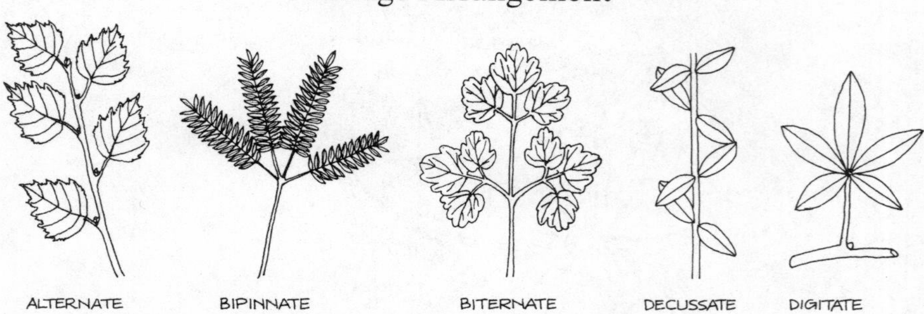

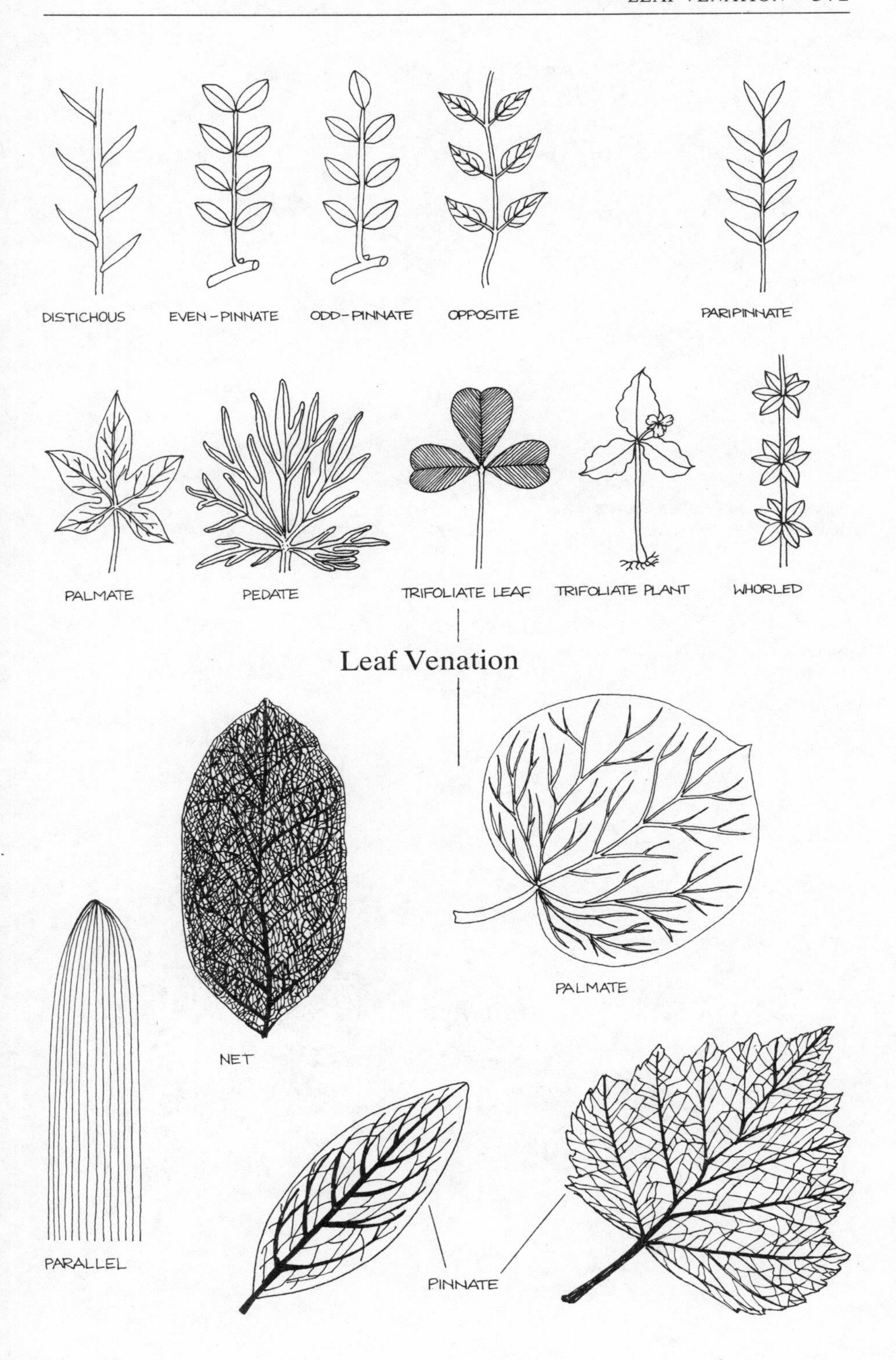

Leaf Venation

Forms of Ovule

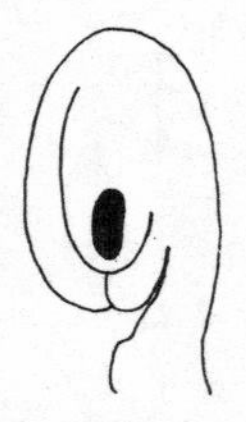

ANATROPOUS

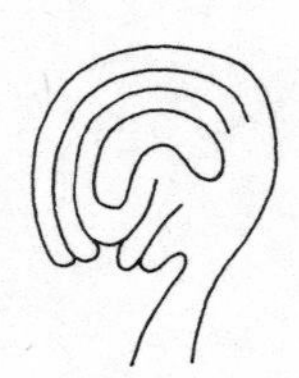

AMPHITROPOUS

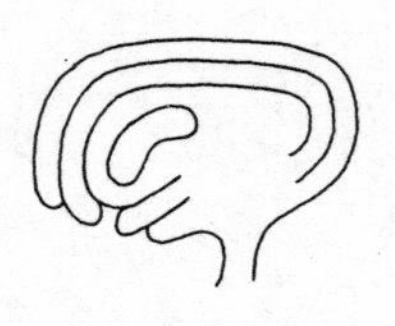

CAMPYLOTROPOUS

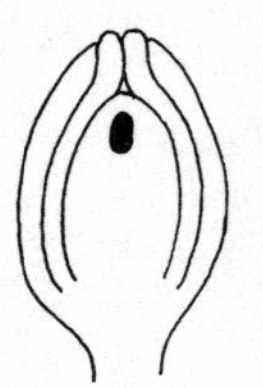

ORTHOTROPOUS

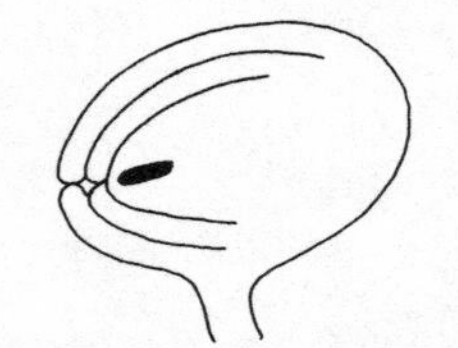

INTERMEDIATE WITH INTEGUMENTS CONTRIBUTING TO MICROPYLE

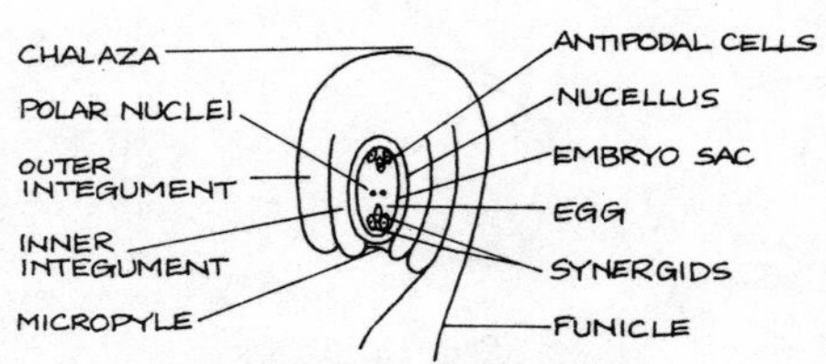

BASIC STRUCTURE OF AN OVULE

Floral Diagrams

GENERAL SECTION OF A FLOWER BUD

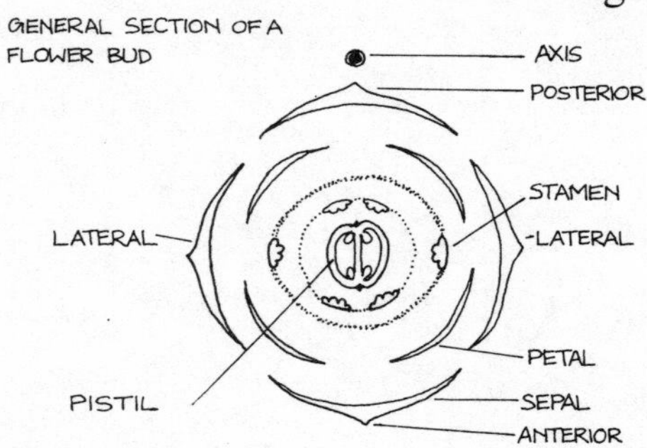

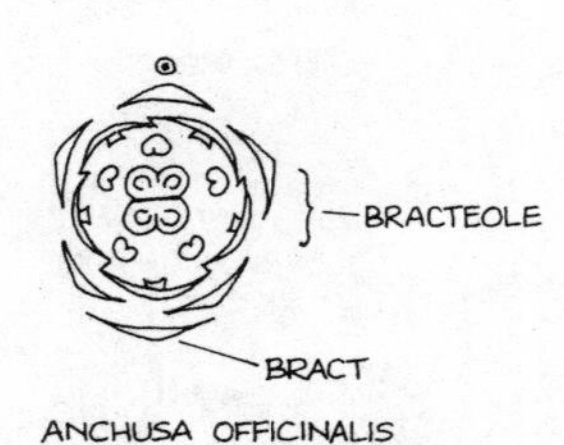

ANCHUSA OFFICINALIS

TYPICAL CRUCIFIER. PETALS SHOWN IN BLACK, AND THE FOUR CARPELS SLIGHTLY SEPARATED

BUTTERCUP (RANUNCULUS ACRIS)

BIRD CHERRY (PRUNUS PADUS)

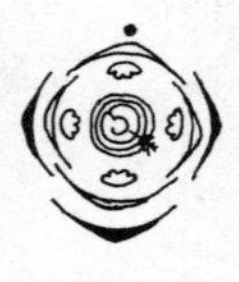

GREAT BURNET (SANGUISORBA OFFICINALIS)

MEADOW SWEET (FILIPENDULA ULMARIA)

A TYPICAL GRASS

A TYPICAL ORCHID CROSSES SHOW POSITIONS OF STERILE STAMEN-RUDIMENTS

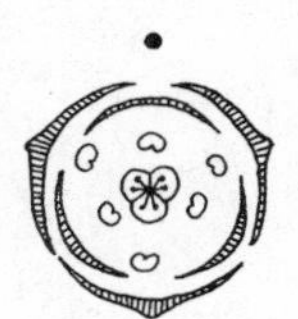
LILIUM

PEA (PAPILIONATAE)

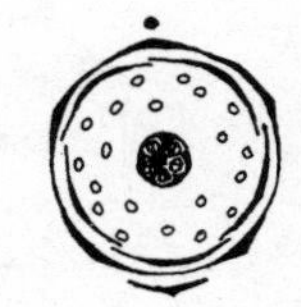
PEAR (PYRUS COMMUNIS)

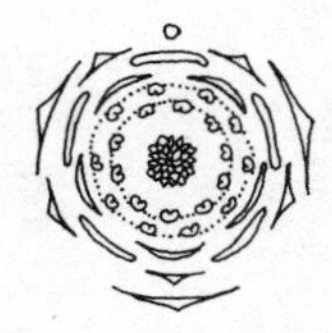
COMMON TORMENTIL (POTENTILLA ERECTA)

DOG ROSE (ROSA CANINA)

TULIP (TULIPA)

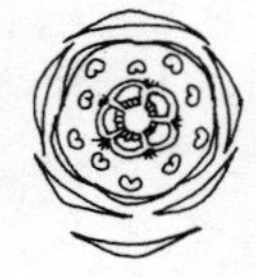
CRANBERRY (VACCINIUM)

VIOLA

WALLFLOWER (CHEIRANTHUS)

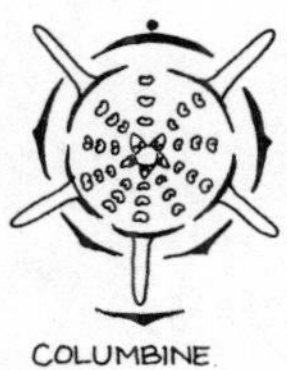
COLUMBINE (AQUILEGIA)

TRUE SERVICE (SORBUS DOMESTICA)

WHITE DEAD-NETTLE (LAMIUM ALBUM)

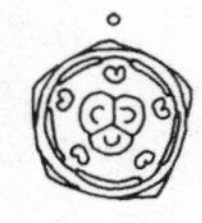
ELDER (SAMBUCUS NIGRA)

Types of Corolla

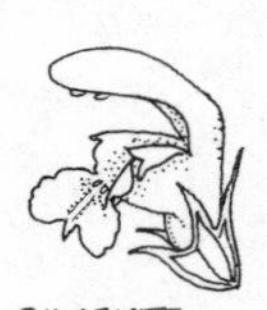
BILABIATE (LAMIUM)

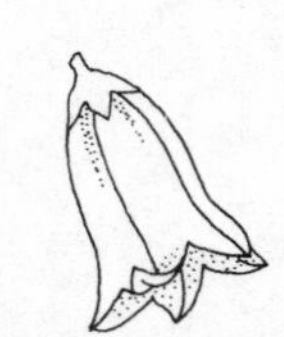
CAMPANULATE (CAMPANULA)

FUNNELFORM (CONVOLVULUS)

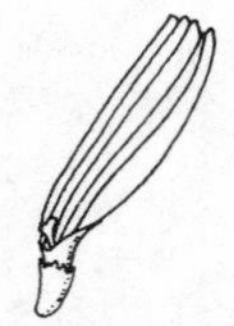
LIGULATE (HELIANTHUS)

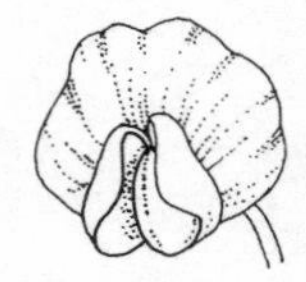
PAPILIONACEOUS (PEA FAMILY)

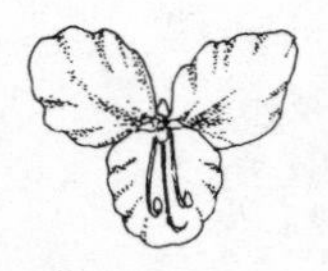
REGULAR (TRADESCANTIA)

IRREGULAR (ORCHIS)

ROTATE (SOLANUM)

URCEOLATE (ERICA)

SALVERFORM (PHLOX)

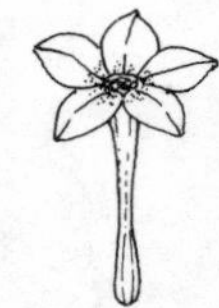
TUBULAR (NICOTIANA)

PETALS MANY (NYMPHAEA)

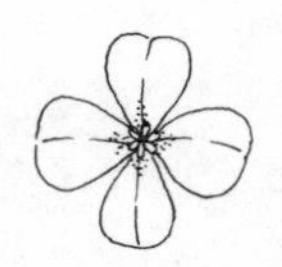
PETALS FOUR (BRASSICA)

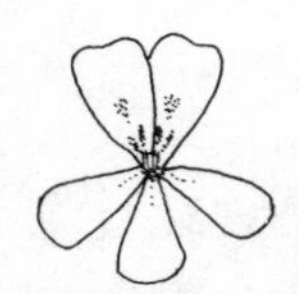
PETALS FIVE (PELARGONIUM)

Forms of Vegetative Propagation

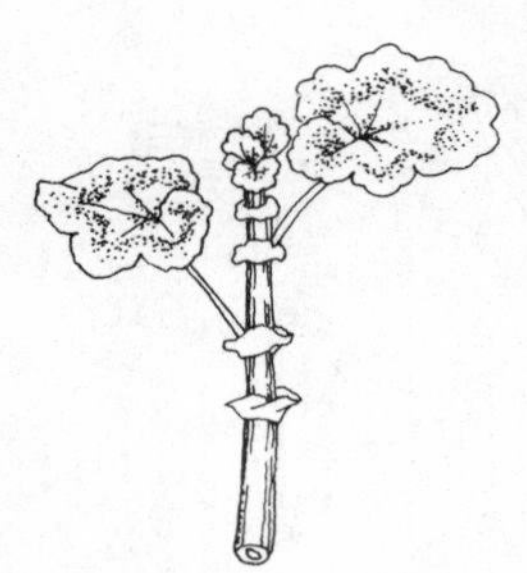

SOFTWOOD CUTTING OF PELARGONIUM

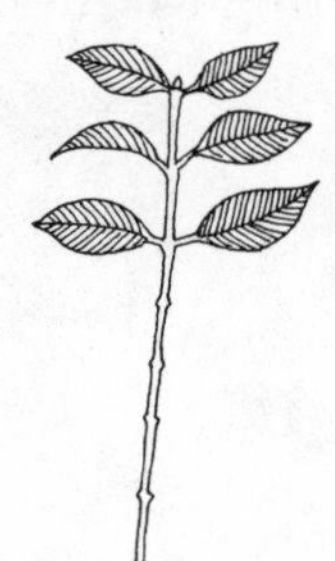

HARDWOOD CUTTING OF LIGUSTRUM

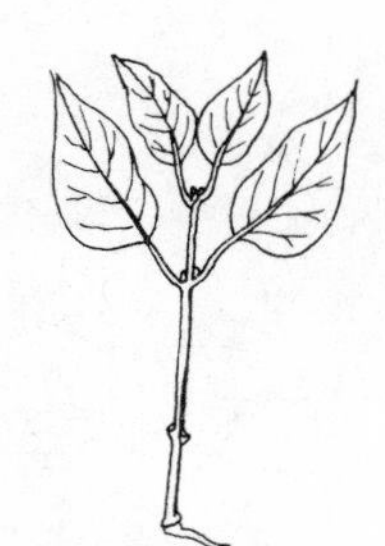

SEMI-HARDWOOD CUTTING WITH HEEL (HEEL IS TRIMMED OFF)

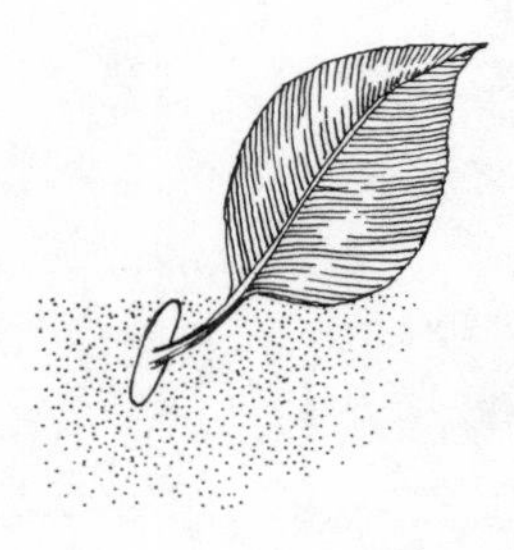

LEAF-BUD CUTTING OF CAMELLIA

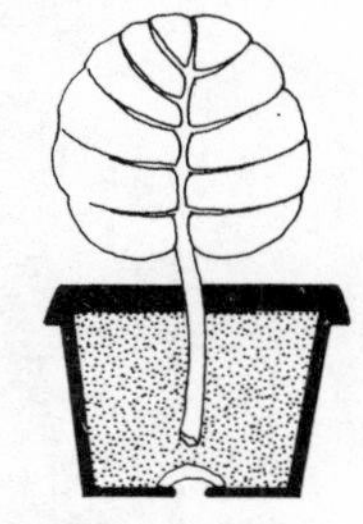

LEAF-CUTTING OF SAINTPAULIA

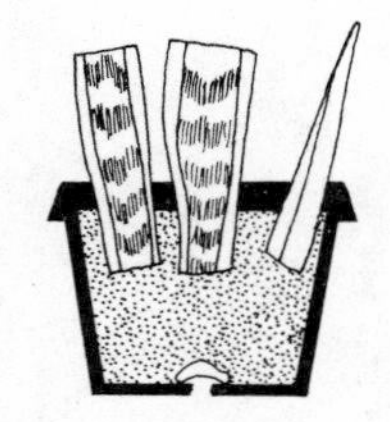

LEAF-CUTTINGS OF SANSEVIERIA

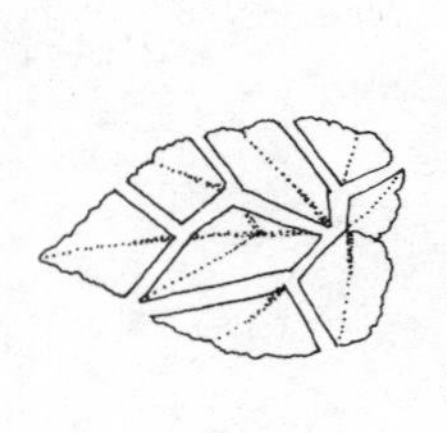

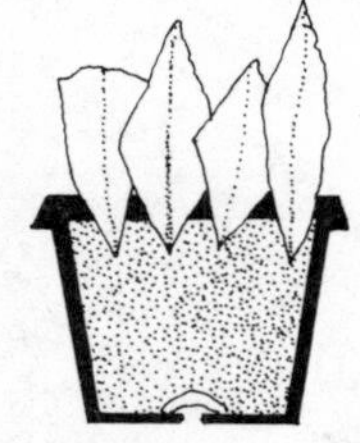

LEAF-CUTTINGS OF BEGONIA

Budding and Grafting Methods

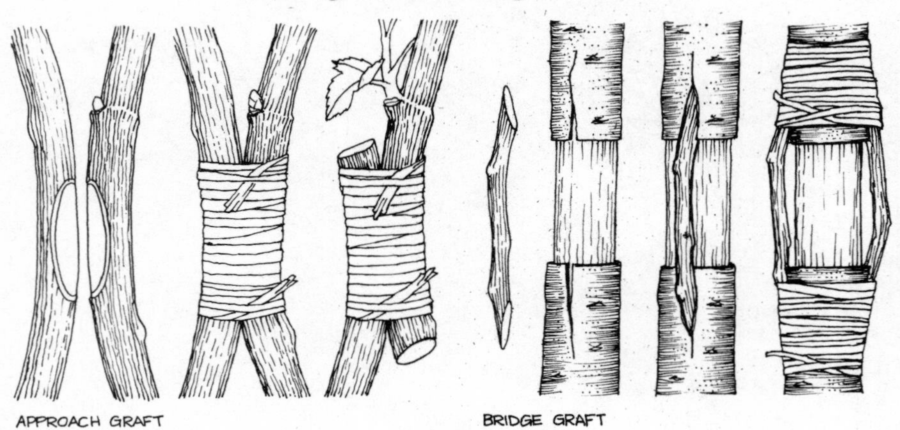

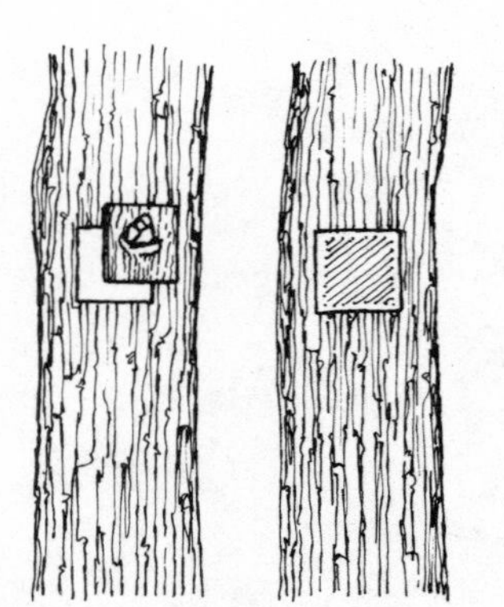

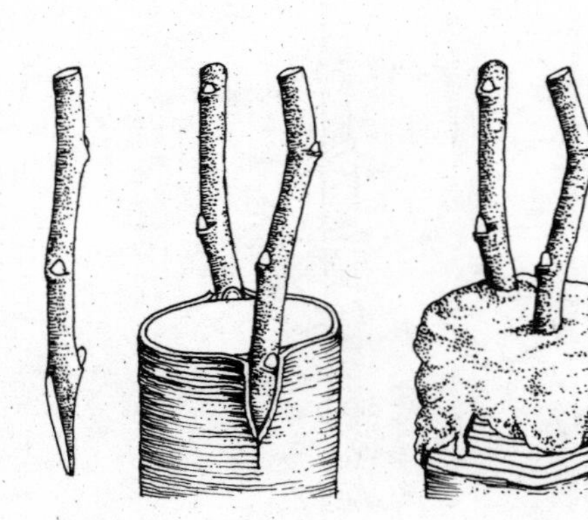

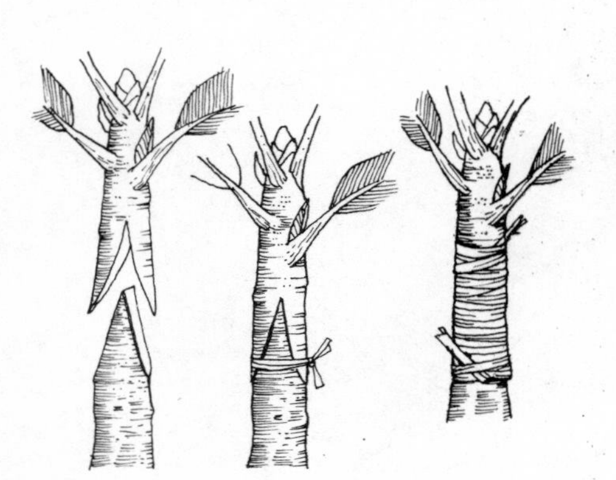

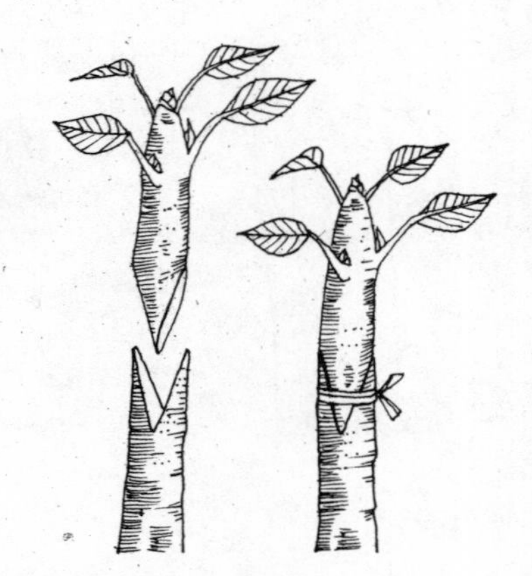